Neotropical Rainforest Mammals

A Field Guide

Neotropical Rainforest Mammals

A Field Guide

Text by
Louise H. Emmons
Color Plates and Illustrations by
François Feer

The University of Chicago Press
Chicago and London

Louise H. Emmons is research associate in the
Division of Mammals at the National Museum of
Natural History, Smithsonian Institution. She is also
research associate of the American Museum of Natural
History in New York and of the Primate Center and
Department of Zoology at Duke University. **François
Feer** earned his doctorate in zoology at the University
of Paris VI and the Museum of Natural History in
Paris, where he is currently affiliated with the
Laboratory of Ecology.

The University of Chicago Press, Chicago 60637
The University of Chicago Press, Ltd., London
© 1990 by the University of Chicago
All rights reserved. Published 1990
Printed in the United States of America
99 98 97 96 95 94 93 92 91 90   54321

Library of Congress Cataloging-in-Publication Data
Emmons, Louise H.
   Neotropical rainforest mammals : a field guide / text by Louise H.
Emmons ; color plates and illustrations by François Feer.
      p.   cm.
   Includes bibliographical references.
   ISBN 0-226-20716-1 (alk. paper). — ISBN 0-226-20718-8 (pbk. :
alk. paper)
   1. Mammals—Latin America—Identification.   2. Rain forest fauna—
Latin America—Identification.   I. Feer, François.   II. Title.
QL723.A1E44   1990
599.098—dc20   89-39353
                                                           CIP

This book is printed on acid-free paper.

Nuestra responsabilidad moral es grande frente a la extinción de las especies animales, sean éstas de valor económico o no, sea su desaparición por culpa directa de la actividad humana o no. Hay que meditar profundamente sobre el hecho de que el hombre no puede contemplar con indiferencia la desaparición de una especie a la que no podrá volver a crear, no importa cuán intensamente lo desee.

—Marc Dourojeanni

We must assume a profound moral responsibility for the preservation of animal species, whether they are of economic value or not, and whether their extinction results directly from our actions or not. One cannot contemplate with indifference the disappearance of a single species that can never be re-created, no matter how deeply we might wish to do so.

Contents

Illustrations

Maps

Only a sustained and serious commitment to research in tropical biology can provide a growing understanding and appreciation of the biological complexity of the tropics and contribute to the preservation and wise use of the diversity of life on earth. Thanks are due the Homeland Foundation for very generous support of the publication of the color plates in this book as an expression of that commitment.

Preface

This is the first broad regional field guide to Neotropical mammals. We created it from scratch because there are no reference books for the kinds of information we needed. The knowledge of Neotropical mammals has many glaring gaps, which will take decades to fill at current rates of research. This book reflects those gaps and has many others as well, because undoubtedly we failed to unearth some sources of information. We hope that, despite its many flaws, this book will fulfill its purpose of introducing the world's richest mammal fauna to its people, especially to those who live in the Neotropics. We describe the large array of mammals that live in New World lowland rainforests, what they look like, what they do, where they are, and which ones may be in danger and why. An appreciation of the mammals can come only from knowledge about them, and we hope that both knowledge and appreciation will help slow the forest destruction and overhunting that now jeopardize the future of many of the species described below.

Most of the information in this book is derived directly from specimens of mammals collected in museums. The descriptions of species, and most of the paintings in the plates, are based chiefly on material in three museums. First and foremost in its contribution to this book and in our thanks, was the Division of Mammals of the United States National Museum of Natural History (Smithsonian Institution). Its collections were the foundation of this volume. The Smithsonian also provided many other essential facilities, including the desk at which this book was written, a base during the years of fieldwork in the Neotropics when Emmons studied its mammals, and the unflagging support of its staff. The collections of Brazilian mammals in the Museu de Zoologia, São Paulo, were of primary importance in describing geographic variation of large species, and we owe special thanks to Paulo Vanzolini for his hospitality there. The American Museum of Natural History was the third major source of specimens described and painted. We also made important use of collections in the British Museum of Natural History, the Muséum National d'Histoire Naturelle in Paris, the Field Museum of Natural History in Chicago, and the Academy of Natural Sciences in Philadelphia. To all these museums and the many curators and staff who made us welcome and made this book possible, we express our deepest thanks. A grant from World Wildlife Fund–US supported essential museum work on the paintings.

A field guide is a product not only of its authors but of all the scientists and explorers who first discovered the species and described them, the many others who followed and collected the mass of information needed to draw each range map, and the ecologists, ethologists, and other biologists who studied their natural history. Many people freely shared their own superior knowledge of certain groups of mammals and donated their own unpublished observations or data to this book. J. L. Patton and P. Myers read the entire manuscript and provided much needed criticism and additional data; A. L. Gardner generously allowed use of unpublished manuscripts and range maps; and G. Dubost both contributed a large series of measurements and weights of large mammals and critically reviewed the sketches for the paintings. All of the following read and/or added their expertise to individual sections, provided information, or helped resolve taxonomic problems: J. Cadle, M. Carleton, G. Creighton, D. Decker, J. Dietz, H. Greene, C. Handley, Jr., T. McCarthy, J. Mead, K. Redford, R. Thorington, Jr., R. Voss, D. Wilson, and P. Wright.

What Is in This Book, Where It Comes from, and How to Use It

The information in the account of each animal has been arranged under a number of headings. If a heading is missing, either no information was found (for example, we do not know what calls some animals make) or none exists (most bats and mice have no specific local names).

Species included. This book includes the lowland rainforest mammals found in Central and South America at elevations below 1,000 meters (maps 1,2,3). These species for the most part form a discrete fauna: mammal communities within the rainforest region are very similar to each other in composition of monkeys, opossums, sloths, bats, deer, and rodents. Communities outside the region tend to be abruptly and radically different, although there can be zones of transition into dryer or higher habitat types. About five-hundred species are on the list of mammals for the region (Appendix F). The mammal fauna of all of the Americas south of the United States border is about 1,160 species; that of South America alone is about 800 species; the rainforest fauna thus includes just under half the mammals that occur south of the United States.

We treat the bats, murid rodents, and a few others to the level of genus only, because there are many similar species that are extremely difficult or impossible to identify with certainty in the hand and because their systematics is often still poorly understood.

A few species known only to occur above 1,000 meters have also been included. These are members of lowland rainforest genera; by including these few extra species, all members of a genus are described (e.g., by including one mon-

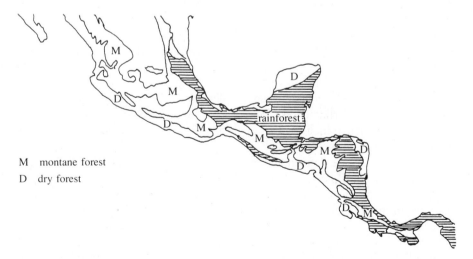

M   montane forest
D   dry forest

Map 1 Rainforest region of Central America (shaded), with general vegetation of bordering areas. Much of rainforest area shown has been deforested (after P. Wagner, 1964, *Handbook of American Indians,* vol. 1).

1

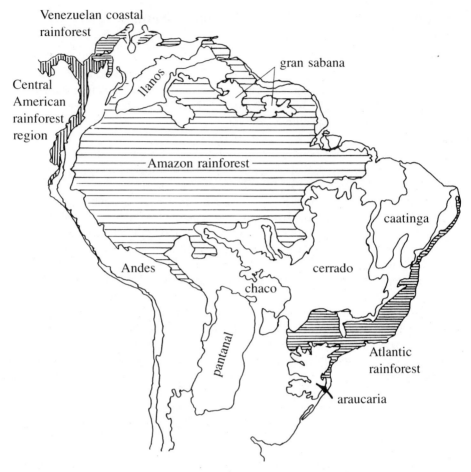

Map 2 The four rainforest regions of South America (shaded) and major bordering habitats (after morphoclimatic domains of Ab'Saber, 1977, Univ. São Paulo Inst. Geografia, no. 52).

tane species, all woolly monkeys and all New World monkeys are included). There are many borderline cases, and we have made some arbitrary decisions about which species to include. Our apologies if we have included our favorite borderline species but not yours.

The way species are defined and arranged in this book generally follows the most recent literature. A list of the more important sources used is found in Appendix E. The systematics of spiny rats and tree rats is based on Emmons's own research. In the frequent cases where there is currently no adequate systematic view of a taxon and current classifications seem wrong, we have followed the advice of colleagues or simply followed our own intuition in wending our way through the maze of conflicting views. This book is not intended to solve systematic problems or to be a species-level systematic reference for scientists. Identification for scientific purposes, where proof of an animal's identity is need-

Map 3 Major rivers of South America.

ed, requires far more detailed description than is possible in a field guide format, as well as examination of specimens with skulls, not just living animals. Many changes will be made in the taxonomy of Neotropical mammals as our understanding of them improves.

Common names. Many Neotropical mammals have no common names in English; some have none in any language. Because each account needed a common name, we invented many of those used ad hoc. We have tried to use descriptive or geographic names, or names derived from the scientific name. When accounts are to genus only, as in bats, it seemed most parsimonious to simply use the scientific names for individual species discussed rather than to invent common names which would have to be followed by the scientific name in every case. For help in understanding how animals are classified and how scientific names are composed and written, see the first section of Appendix C.

Arrangement of accounts. The accounts are arranged systematically to genus, subgenus, or group of closely related, similar species. Then species are arranged as follows: common species before rare or occasional species in a region; South American species before Central American species; and widespread species before species of limited distribution.

Measurements. Measurements and weights given in this book refer to adult, wild individuals. They give the usual or normal sizes of animals encountered; some individuals may fall outside these ranges, but they will be close to them. Most of the data were compiled directly from standard measurements taken in the field and recorded on museum specimen tags—the work of hundreds of individual collectors. In some cases published measurements from the scientific literature have been used. For a few species, the measurements are derived from few individuals, sometimes only one (single measurements in the text). When a species has a geographic range much larger than the region covered in this book, measurements refer only to animals from the rainforest region (e.g., Virginia opossum, northern raccoon). Figure 1 shows how the measurements are taken.

The following abbreviations are used to indicate measurements:
HB   Head and body length: tip of nose to inflection point of tail
T   Tail length: inflection point with body to tip of flesh
HF   Hindfoot length: heel to tip of claws, or if followed by (su) without claws
SH   Shoulder height: point of shoulder blade to tip of toe, stretched
FA   Forearm length (bats only): elbow to wrist of folded wing
E   Ear length: bottom of notch to tip, on inner side
WT   Whole body weight of nonpregnant adults

Metric units are used throughout; lengths are in millimeters.

Identification. The descriptions are intended to be useful both for animals seen from afar and for those in the hand. Much of the detail will not be visible from a distance. The most important distinguishing features are printed in boldface so that they can be rapidly scanned. The descriptions were taken directly from museum specimens and/or from life, with the exception of seven accounts for which no animals were available. For these the original scientific description of the species has been used, supplemented by other original published descriptions.

The number and arrangement of the mammae often differ between genera of small rodents, so the pattern of mammae can be helpful in identification (fig. 2). The mammae are easy to find only in adult females that have nursed young.

It is important to know whether an animal is adult. Signs of adulthood include the following: all molar teeth have erupted; body shape is that of an adult, with head and feet not disproportionately large; evidence of reproductive activity (in females the mammae of the young are small, flat, dotlike, and unpigmented, while the mammae of parous females that have nursed young are permanently elongated and often pigmented; adult males of most species have large, descended testes at least in the breeding season, although opossums have quite large descended testes as juveniles); fur of young is often woollier, grayer, or duller than that of adults; in living bats the long finger bones in the wings of adults have fused epiphyses, while those of subadults have a translucent wedge of growing cartilage.

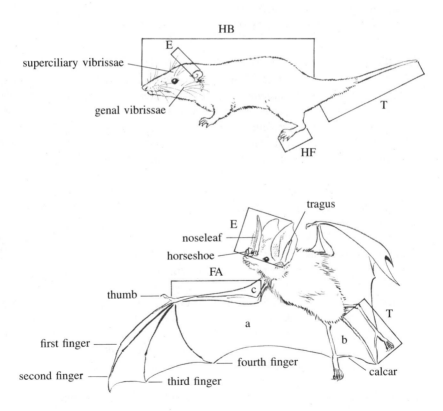

Figure 1. Measurements and parts of mammals: (*a*) wing membrane; (*b*) tail membrane; (*c*) antebrachial membrane or propatagium.

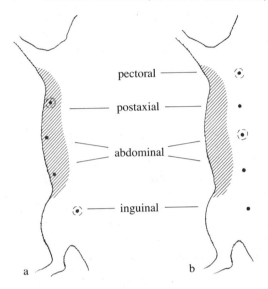

Figure 2. Mammae positions in ratlike rodents: (*a*) lateral and inguinal mammae
of spiny rats and tree rats (Echimyidae); (*b*) ventral mammae of murid rodents (Muridae).
Dotted lines surround mammae that are often absent; shading represents dorsal fur.

Variation. Our intent in this section is to give a feeling for the general range
and extremes of variation that can be encountered, as a help in identification.
Because the taxonomic problems concerning the definitions of species included
in this book mostly concern geographic variation in how species look, remarks
on variation in color or size and problems or controversies about what to call the
variants are treated together. The intricacies of disagreements between scientists
over how to classify certain populations of animals will be irrelevant to most
readers, but they become of the greatest importance when issues of conservation
and systematics are involved. We point out where major apparent discrepancies
might occur between this and other works. The second section of Appendix C
has more detailed discussion of variation.

Similar species. This section usually includes only species found in the same
geographic region as the animal in the account. Be sure to check range maps.

Natural history. We have tried to include only firsthand reliable information,
both from published sources and from direct observations made by Emmons and
scientific colleagues. Anecdotal information from hearsay or without a known
source has been excluded: much of such folklore is correct, but much is also
false, and it has an insidious way of converting itself into fact.

Geographic range. The entire known geographic range is described. The term
Central America as used here includes southern Mexico and the region more cor-
rectly defined as Mesoamerica. Our knowledge of species distributions is often

fragmentary, especially in the southern Amazon Basin, so some areas now blank on the maps will certainly be filled in the future. Conversely, some areas now shaded will prove to be blank. Range maps are compiled from single points where animals have been proven to occur: for many species of the Amazon Basin, known points of occurrence are often hundreds of kilometers apart, but nonetheless the areas between them are shaded in. The effects of recent deforestation or exploitation of animal species are usually not shown, so particularly for Central America, southeastern Brazil, and some sections of the Amazon Basin, the range maps are overly optimistic. Maps have been compiled from the literature and from museum specimens.

Status. Two levels of a species' status are included: (1) the 1988 official Convention on International Trade in Endangered Species (CITES), and the United States Endangered Species Act (US-ESA) listings; and (2) the general situation of a species or genus, apart from official classifications, and factors that contribute to its status.

CITES has three classifications of species: CITES Appendix I includes species currently threatened with extinction. Special permits from both the country of export and that of import are required for importing any part of any species so listed into a country that is signatory, regardless whether the country of export is signatory. Appendix II includes species that are not currently threatened but may become so unless trade is regulated. Appendix III includes species regulated for conservation purposes by particular countries. For species in Appendixes II and III, an export permit from the country of origin is needed for import to a signatory country. Abbreviations for countries with Appendix II and III restrictions are given below and are the same used to indicate country of use of local names.

US-ESA has two categories—endangered and threatened; these are defined in the same way as CITES Appendixes I and II and have similar import regulations, but they apply only to the United States. Of the species covered in this book, thirty-three are officially listed as endangered in at least part of their ranges and another thirty-seven as threatened, or a total of seventy species—14% of all the mammals.

The general status information is derived mainly from Emmons's observations and the frequency with which the species have been recorded by scientists. "Common" animals usually occur at high densities; they are found in many places and are the species most likely to be seen. "Rare" species usually occur at low densities and/or seem to be entirely absent from many areas (patchy); they are either unlikely to be seen anywhere or seen only in a few localities. "Uncommon" species generally occur at low densities in many places; they are likely to be seen, but much less often than "common" species. The "rareness" of a species of small mammal can also be a function of how easy it is to catch, since small mammals usually cannot be identified unless caught. Some "rare" species may simply live in inaccessible places, such as canopy bats, or tend to avoid traps. If a rare monkey cannot be found after a search for it, one can be fairly certain it is not present, but we do not yet know how, or where, to find rare small mammals in the rainforest. We thus cannot evaluate large and small mammals in the same way. We have nevertheless included statements about rarity of small mammals, both because people who find them should know to record any information they might get, and also because some of these species may be truly rare and at potential risk of extinction.

Local names. Only the more widely used common names are included. The countries of use are abbreviated as follows: Argentina (Ar), Belize (Be), Bolivia (Bo), Brazil (Br), Central America (Cent Am), Colombia (Co), Costa Rica (CR), Ecuador (Ec), El Salvador (ES), French Guiana (FG), Guatemala (Gu), Guyana (Gy), Honduras (Ho), Mexico (Me), Nicaragua (Ni), Peru (Pe), Panama (Pn), Paraguay (Pa), Surinam (Su), Trinidad (Tr), Uruguay (Ur), Venezuela (Ve). Languages are abbreviated as: Spanish (Span), Guarani (Gua); Saramaca (Sar), Tupi-Guarani (TGua), Mayan (May), Quichua (Qui); only native languages widely used in more than one country are included. For brevity, if a name is used in many Spanish-speaking countries, it will be listed as (Span) rather than listing individual countries, even if it happens originally to be a Guarani or Quichua name, such as *anta* or *danta* (tapir).

Local names are often nonspecific and should be interpreted with caution: the same name will be used for all members of a genus or for unrelated animals of vaguely similar habits. The same name can be used for different animals in different parts of a country; or the same animal can have different names in different regions of a country. Local people who are not hunters (and even some who are) often do not know the names or misapply them.

References. The references will help readers find more information, but they do not necessarily represent the sources of information used in the account, although they may do so. They sometimes represent an entirely different taxonomic view from that followed by the book. Many excellent, sometimes better, works have been omitted; often a recent, brief paper with an up-to-date bibliography has been included rather than a far more comprehensive older work, because the latter can be easily found by looking at the former, but not vice versa. We have tried to include a systematic reference and a behavior or ecology reference.

Species Accounts

Opossums (Marsupialia)

Didelphidae

Dental formula: I 5/4, C 1/1, P 3/3, M 4/4 = 50 teeth. All the teeth behind the canines are sharply pointed. Feet all with five toes; the first toe of the hindfoot is widely separated from the other digits, forming an opposable "thumb" used to grasp thin stems for climbing. Opossums are small (15–2,000 g) mammals, with pointed snouts, short legs, long tails, and usually soft, dense fur. The tail of most species is strongly prehensile, and even its extreme tip can tightly grip an object as thin as a wire, with many times the force needed to support the weight of the body. Some species will hang suspended only by the tail to reach a fruit, but this is an unusual posture. Most opossums have large, delicate ears that can be curled back in pleats, out of harm's way. At night they have bright eyeshine; the eyes appear small and far apart. Known diets are of insects and other invertebrates, small vertebrates, and some ripe fruit and nectar, but four species probably eat more fruit than animal matter, and one eats fish. All species give birth after a short gestation to tiny young that crawl up the mother's fur and attach by the mouth to a nipple, where they remain fastened for several weeks, until they are too large for the mother to carry them easily. The young of some species are protected within a pouch (marsupium) while they are attached to the nipples, but species in over half the genera have no pouch. Older young can be transported on the mother's back but she rarely carries them. For several weeks after the young detach from the nipples they continue to nurse, but the mother leaves them behind in a nest while she forages. Species for which nesting habits are known make nests of dead leaves in sheltered places. The leaves are carried to the nest clutched in a coil of the tail tip (known in three genera). Recent studies suggest that opossums rarely live for more than one breeding season after they become adult.

The mouse opossums, formerly all in the genus *Marmosa,* have recently been redivided into five genera, four of which occur in the rainforest region. The many species are similar externally, and caution should be used when accurate identifications are needed. Subadults tend to look like adults, and even experts sometimes mistake them in the hand for adults of a smaller, grayer species. In most South American lowland rainforests, one to two species of *Marmosa,* one each of *Micoureus* and *Marmosops* (and in Venezuela and SE Brazil, one *Gracilinanus*), occur together, along with one each of *Didelphis, Caluromys, Metachirus,* and *Philander,* and sometimes *Monodelphis.* The other genera are more restricted locally and geographically. In Central America one each of *Micoureus, Marmosa, Caluromys, Philander, Metachirus,* and *Didelphis* usually coexist. The family is restricted to the Americas, with one species in North America and about 70 species in Central and South America.

Western Woolly Opossum

Caluromys lanatus
Plate 1, map 4
Identification. Measurements: HB = 217–295; T = 330–435; HF = 39–50; E = 31–41; WT = 310–410 g.
Upperparts **red-brown** to pale brown, red brightest over shoulders, forearms, and hindlegs. Fur long, dense, and woolly. **Head gray, face with prominent dark stripe down center;** reddish brown eye rings extend in dark streak from corner of eye to nose; eye dark brown, eyeshine bright yellow-orange; ears naked, pinkish tan. **Tail thickly furred above for proximal half, and below for one-fifth; tail tip naked, whitish mottled with brown spots near base of naked area.** Feet red-brown or dark gray. Underparts yellowish white with more grayish midsection. Females seem to develop a pouch only when they are carrying young.
Variation. Animals from the Magdalena Valley of Colombia are paler, grayish brown, with gray sides, sometimes with completely gray underparts; those from Paraguay are pale brown, without mottling on tail. Rare individuals have a gray patch between the shoulders.
Similar species. Black-shouldered opossums (*Caluromysiops irrupta*) have black shoulders and tail furred dorsally to the tip;

bushy-tailed opossums (*Glironia venusta*) have tail furred dorsally to the tip and two black stripes on head. Woolly mouse opossums (*Micoureus spp.*) have no black stripe on face and tail furred only near base; bare-tailed woolly opossums (*Caluromys philander*) have tail fur ending near base.
Sounds. None usually heard in the field.
Natural history. Nocturnal; arboreal; solitary. Feeds on fruits, probably a few invertebrates, and in the dry season drinks flower nectar. Western woolly opossums favor dense, viny, midstory and canopy vegetation, but they are also seen in tall, open forest. Found in mature and secondary evergreen rainforest, disturbed forest, gallery forest, and gardens and plantations.
Geographic range. South America: east of the Cordillera Central of Colombia, to western Venezuela; east of the Andes in Ecuador, Peru, Bolivia, Paraguay, Argentina, and W Brazil. To at least 500 m elevation.
Status. Locally common in many areas; formerly hunted for its fur, which is no longer in demand.
Local names. Cuica lanosa (Ar); mucurachichica (Br); chucha lanosa (Co); mbicuré lanoso (Ec).

Bare-tailed Woolly Opossum
Caluromys philander
Plate 1, map 4
Identification. Measurements: HB = 160–240; T = 250–350; HF = 33–40; E = 30–36; WT = 140–270 g.
Upperparts uniform warm reddish brown. Head gray, face with dark brown stripe down center, dark brown eye rings and streak between eye and nose; ear brown, usually with pale yellow spot at upper and lower corners; eyeshine bright yellow, eyes appear small. Tail furred for first tenth of its length; **most of tail naked, usually heavily mottled cream and dark gray to tip, or completely gray,** rarely tip pale and without spots. Underparts completely orange, or midbody grayish. Feet gray or whitish. Fur short, dense, and soft in lowlands; longer at higher elevations. Females have poorly developed pouch only when they are carrying young.
Similar species. Woolly mouse opossums (*Micoureus cinereus*) have no central stripe on face and are gray-brown. Brown and

Map 4

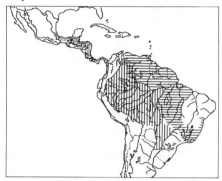

▨ Central American woolly opossum, *Caluromys derbianus*
▥ Western woolly opossum, *C. lanatus*
▤ Bare-tailed woolly opossum, *C. philander*

gray four-eyed opossums (*Metachirus nudicaudatus, Philander spp.*) have pale spots over eyes; western woolly opossums (*Caluromys lanatus*) have fur reaching halfway down tail; tree rats (*Echimys* spp., *Isothrix* spp.) have small ears and nonprehensile tails.
Sounds. None usually heard in field. Hisses when cornered, gives a loud distress call when taken by a predator.
Natural history. Nocturnal; arboreal; solitary. Feeds mainly on fruit and some nectar with about 20% of the diet invertebrates. Bare-tailed woolly opossums use the middle and upper levels of the forest, where they are sometimes seen running swiftly along a branch or vine or sitting motionless looking down. They seem to favor dense viny vegetation but also use the high open forest canopy. During the day they den in a leaf nest or tree hollow. Found in mature and secondary rainforest, gallery forest, and gardens and plantations.
Geographic range. South America: east of the Andes in E Venezuela, Trinidad, the Guianas, and northeastern and south-central Brazil. To at least 1,800 m elevation.
Status. Often common; thrives in disturbed vegetation.
Local names. Mucura-chichica (Br); awari (Su).
References. Charles-Dominique, P., M. Atramentowicz, M. Charles-Dominique, H. Gérard, A. Hladik, C. M. Hladik, and M. F. Prévost. 1981. Les mamifères frugi-

vores arboricoles nocturnes d'une forêt guyanaise: Inter-relations plantes-animaux. *Rev. Ecol.* 35:341–435.

Charles-Dominique, P. 1983. Ecology and social adaptations in didelphid marsupials: Comparison with eutherians of similar ecology. In J. F. Eisenberg and D. G. Kleiman, eds., *Advances in the study of mammalian behavior,* 395–422. Amer. Soc. Mammal. Spec. Pub., no. 7.

Central American Woolly Opossum
Caluromys derbianus
Plate 1, map 4
Identification. Measurements: HB = 225–300; T = 384–445; HF = 32–47; E = 35–40; WT = 245–370 g.
Upperparts rich reddish brown; back with pale gray patch between shoulders; hips and body behind shoulders pale gray; fur long, dense, soft, and woolly, slightly frosted. **Head pale gray with dark brown stripe down center of face,** brown eye rings merging to dark streak between eye and nose; ears naked and whitish or pink. **Forelegs and feet creamy white;** hindfeet brown. Underparts yellowish white. **Tail thickly furred above for basal 30–50% of length, below for 25%,** usually darker than back; **tail tip naked, pale, mottled with brown spots** especially near middle. Females have a pouch only when they are carrying young; the pouch apparently regresses to small folds at other times. Young slightly grayer than adults.
Variation. Races from E Nicaragua and mountainous areas of the Pacific slope of the Costa Rica–Panama border are mainly or entirely pale gray, with indistinct facial markings and a variably pronounced brownish tinge over shoulders and lower back. Occasional individuals lack the gray patch between the shoulders.
Similar species. No other mammals in range have tail half furred. Gray and brown four-eyed opossums (*Metachirus nudicaudatus, Philander opossum*) have pale spots above the eyes, nonmottled tails, and no pale patch between the shoulders. Woolly mouse opossums (*Micoureus cinereus*) are much smaller.
Sounds. None usually heard in the field.
Natural history. Nocturnal; arboreal; solitary. Feeds mostly on fruit, probably supplemented by nectar and invertebrates.

Map 5

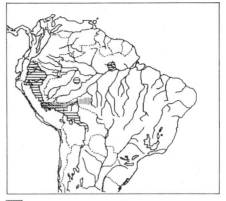

▦ Black-shouldered opossum, *Caluromysiops irrupta*
▤ Bushy-tailed opossum, *Glironia venusta*

Central American woolly opossums make nests of leaves and also den in tree holes. Found in mature and disturbed evergreen rainforest, dry forest, and gardens and plantations.
Geographic range. Central and South America: Veracruz, Mexico, south to the Cauca Valley of Colombia and western Ecuador. To 2,500 m elevation.
Status. Seems locally common; formerly hunted for the fur trade, but skins are no longer in demand.
Local names. Chucha lanosa (Co); zorra roja, comadreja roja (Pa); zorro de balsa (CR).
References. Bucher, J. E., and R. S. Hoffmann. 1980. *Caluromys derbianus.* Mammalian Species, no. 140.

Black-shouldered Opossum
Caluromysiops irrupta
Plate 1, map 5
Identification. Measurements: HB = 250–330; T = 310–340; HF = 47–52; E = 29–37. **Upperparts frosted grayish brown,** sides paling to buffy or whitish; **large black patches across shoulders extend backward in two narrow stripes beside the spine and fade to brown on rump, and forward as black inner forearm and top of forefoot;** fur long, dense, and soft. Head gray-brown, cheeks below ear pale buffy or dirty white, indistinct brown spots between eye and nose; **ears bright yellow on inside, round at**

tips; **nose pink;** eyeshine bright yellow. **Tail** tapered, **furred above to tip, below for basal 25%,** naked on distal 75%, frosted dark brown above for about 60%, dirty white at tip. Underparts buffy, especially bright on throat and inguinal region. Young like adults.

Similar species. Bushy-tailed opossums (*Glironia venusta*) are much smaller, with tan ears, no black on shoulders, and black stripes down the face. All other opossums in the region have naked tail tips. Tree rats (*Echimys, Isothrix* spp.) have small, inconspicuous ears and no black markings on shoulders. Squirrels have bushy tails.

Sounds. None heard in the field.

Natural history. Nocturnal; arboreal; solitary. In the dry season feeds on the nectar of flowers; presumably also eats fruit. Black-shouldered opossums use the upper levels of the forest and rarely seem to descend even to the middle levels. They move slowly and are most often seen sitting motionless on a branch. They will spend hours in the same flowering tree, feeding periodically on nectar, and will hang suspended only by the tail to reach flowers on the underside of a branch. Known only from mature rainforest.

Geographic range. South America: east of the Andes in SE Peru and adjacent W Brazil. Known from only three localities. Records from Leticia, Colombia, are questionable as animals from Peru are sent there for export. To 700 m elevation.

Status. Generally extremely rare and patchy in distribution, but in some years can be locally common. Known from fewer than 30 specimens.

References. Janson, C. H., J. Terborgh, and L. H. Emmons. 1981. Non-flying mammals as pollinating agents in the Amazonian forest. *Biotropica, Repro. Bot. Supp.,* 1–6.

Izor, R. J., and R. H. Pine. 1987. Notes on the black-shouldered opossum, *Caluromysiops irrupta. Fieldiana Zool.,* n.s., 39:117–24.

Bushy-tailed Opossum
Glironia venusta
Plate 1, map 5
Identification. Measurements: HB = 160–205; T = 195–225; HF = 27–31; E = 22–25.

Upperparts pale or cinnamon-brown; fur soft and dense, velvety or woolly. **Face with a broad black stripe on each side, from nose through eye to crown,** separated by a gray stripe down center of face from nose to ears; ears naked and brown or blackish. **Tail fully furred to tip** on top and sides, distal 60% below covered with short bristles, tail tip white or darker than base. Feet dirty white. Underparts gray or white tinged with orange. A female specimen has no pouch.

Similar species. Black-shouldered opossums (*Caluromysiops irrupta*) are larger, gray with black shoulders; all other opossums have naked tail tips; brush-tailed rats (*Isothrix* spp.) have small ears and blunt muzzles.

Natural history. Unknown; presumably nocturnal and arboreal. Emmons observed one for ten minutes in SE Peru, in Manu National Park. It was seen at night in dense, viny vegetation about 15 m above the ground. It ran around the vines quickly and with agility, often jumping from one to another in a manner unlike that of other arboreal opossums. It seemed to be hunting insects but at one point spent many seconds licking the surface of a branch. (Because no specimen has been collected from the locality, this cannot be considered an official record.) Known only from rainforest.

Geographic range. South America: the Amazon Basin of Brazil, Ecuador, Peru, and Bolivia. To elevations of 800 m. Known from only eight localities.

Status. Apparently extremely rare; known only from eight individuals, each from a different locality.

References. Marshall, L. G. 1978. *Glironia venusta.* Mammalian Species, no. 107.

Common Opossum
Didelphis marsupialis
Plate 1, map 6
Identification. Measurements: HB = 324–425; T = 336–420; HF = 51–70; E = 46–58; WT = 565–1,610 g; males larger than females.

Upperparts black or gray, fur in two layers, dense yellow or white underfur shows below long, coarse, black or gray outer guard hairs. Head dirty yellow, sometimes with indistinct black lines from

Map 6

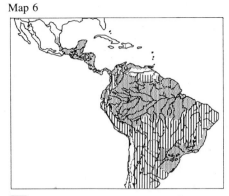

▨ Common opossum, *Didelphis marsupialis*
▥ White-eared opossum, *D. albiventris*

nose through eye nearly to ear, and down center of crown from between ears to eye; cheeks yellow, pale orange, or dirty white, not sharply contrasting with color of muzzle; nose pink; **ears** large, naked, **black;** feet black; **tail** usually longer than head and body, **naked, black with white tip,** black part usually shorter than white. Underparts usually like back, but paler, or orange. Females have a pouch. Typical posture a crouch with back humped high and long fur bristling, sometimes in a ridge along spine, and tail coiled. Small young blackish, with pale heads, sharp facial markings, and white-tipped ears.
Variation. Rainforest animals are usually black; gray animals seem more common in drier habitats, but black and gray phases occur together in the same populations. Some animals are brown from split ends on overhairs and fading with age. Animals from SE Brazil are sometimes considered a distinct species.
Similar species. Virginia opossums (*D. virginiana*) have white cheeks, sharply contrasting with the rest of the head, and the black part of the tail is usually equal to or longer than the white; white-eared opossums (*D. albiventris*) have white ear tips and sharp, black-and-white facial markings. Four-eyed opossums (*Philander* spp. and *Metachirus nudicaudatus*) have pale spots over eyes and short fur in only one obvious layer.
Sounds. Hisses with open mouth when cornered.
Natural history. Nocturnal; arboreal and terrestrial; solitary. Feeds chiefly on small animals: insects, worms, or small vertebrates, including snakes, with about a quarter of the diet consisting of fruit, and in the dry season sometimes nectar. Common opossums usually travel and feed on the ground but will climb to feed on fruits or nectar in the treetops. They also climb to escape danger and to rest in the middle of the night. They travel widely during a night's activity. They den in tree hollows, leaf nests in tree forks or dense vines, or sometimes in burrows (dug by other animals). Malodorous animals, they roll in fresh dung, and they squirt vile-smelling, burning urine and defecate when handled, distributing the effluent by twirling the tail. In some areas populations explode to high levels, followed by a crash and virtual absence of the species the following year. When handled, does not "play dead" but is generally aggressive and will threaten with open mouth, lunge, and bite when cornered or grasped. Found in humid forests and outlying gallery forests; thrives in secondary forests and around dwellings, where it feeds in garbage dumps.
Geographic range. Central and South America: Mexico to Bolivia, Paraguay, and NE Argentina, Trinidad, and the Lesser Antilles. To about 2,000 m elevation.
Status. Common, hunted for meat only where other game is scarce.
Local names. Comadreja grande (Ar) ; chucha común (Co); comedreja (Bo, Pa); mucura, zorro, muca (Br, Pe); raposa, yalu (Ec); zorra, zorro pelón, zarigüeya (Pn, CR); grote buidelrat, awari (Su); rabipelado (Ve); mbicuré (Gua); tacuazin (Cent Am); ux (May).
References. Gardner, A. L. 1973. The systematics of the genus *Didelphis* (Marsupialia: Didelphidae) in North and Middle America. Spec. Pub. Mus. Texas Tech., no. 4.
Sunquist, M. E., S. N. Austad, and F. Sunquist. 1987. Movement patterns and home range in the common opossum (*Didelphis marsupialis*). *J. Mammal.* 68:173–76.

White-eared Opossum
Didelphis albiventris
Plate 1, map 6
Identification. Measurements: HB = 305–437; T = 290–430; HF = 45–68; E = 41–60; WT = 500–2,000 g.
Similar to the common opossum except: head white with sharp black bars from ear through eye to nose, or black eye ring only; black stripe coming to a sharp point on crown of head at level of eyes; ears white, or black at base with broad white tips; throat white; tail usually slightly shorter than head and body. Overhairs often white. Underparts white, gray, or yellowish.
Similar species. See common opossum.
Natural history. Nocturnal. This opossum is not generally a rainforest species but occurs in more arid or temperate regions adjacent to lowland humid forests, where it may coexist with the common opossum. Lives in savannas, gallery forests, swamps, cultivated or deforested agricultural lands, and humid forests at high elevations and subtropical latitudes.
Geographic range. South America: in and east of the Andes around the periphery of the rainforest from W Venezuela and Colombia south to Ecuador, Peru, Bolivia, Paraguay, Uruguay, N Argentina, and SE Brazil. To at least 4,000 m elevation.
Status. Often common.
Local names. Saruê (Br); comadreja mora, comadreja overa (Ar, Ur); chucha de orejas blancas (Co); mbicuré (Gua). See names for common opossum.
References. Tyndale-Biscoe, C. H., and R. B. Mackenzie. 1976. Reproduction in *Didelphis marsupialis* and *D. albiventris* in Colombia. *J. Mammal.* 57:249–65.

Virginia Opossum
Didelphis virginiana
Plate 1, map 7
Identification. Measurements: HB = 370–501; T = 295–470; HF = 50–80; E = 34–60; WT = 500–2,300 g; males larger than females.
Externally similar to the common opossum in the rainforest region; **except: cheek below eye white, white extending posteriorly at least to group of whiskers below and forward of ear, and contrasting sharply with color of the rest of the**

Map 7

Virginia opossum, *Didelphis virginiana*

head; tail with black and white parts usually approximately equal, or black part longer than white part, or all black. Young black with white head prominently marked with a black bar from ear through eye to nose, and black of body extending forward as a sharp point above eyes.
Variation. Central American Virginia opossums are smaller and darker than those in North America, which are usually pale gray.
Similar species. See common opossum.
Sounds. Hisses with mouth open when cornered.
Natural history. Nocturnal; arboreal and terrestrial; solitary. Feeds on insects and other invertebrates, small vertebrates, carrion, fruit and other plant matter. Virginia opossums usually travel and feed on the ground but will climb readily. By day they den in tree holes, hollow logs, brush piles, and rock outcrops. When disturbed they may "play dead" by lying inert on the side with mouth open and salivating.
Found in many types of forested and open habitats, both humid and arid, and often in towns and around garbage dumps.
Geographic range. North and Central America: southern Canada to northwestern Costa Rica. In southern part of its range to 3,000 m elevation.
Status. Common.
Local names. Zorro, zorra (Span); tlacuache (Me); ux (May).
References. Gardner, A. L. 1973. The systematics of the genus *Didelphis* (Marsu-

pialia: Didelphidae) in North and Middle America. Spec. Pub. Mus. Texas Tech., no. 4.

————. 1982. Virginia opossum. In J. A. Chapman and G. A. Feldhamer, eds., *Wild mammals of North America*, 3–36. Baltimore: Johns Hopkins University Press.

Common Gray Four-eyed Opossum
Philander opossum
Plate 1, map 8

Identification. Measurements: HB = 250–302; T = 253–315; HF = 35–50; E = 25–40; WT = 200–660 g.

Upperparts uniform sooty gray; fur short, dense, and soft, finely grizzled at close range by white bands on hairs. **Face with black mask around eyes and across crown, large white spots over each eye and at base of ear;** chin and lower cheeks creamy white; ears large, naked, rims black, centers pale, nose, lips above chin, and toes pink; lips at corner of mouth black. **Tail dark gray, naked, tip white sharply demarcated in a line, thickly furred with continuation of body fur for first 5–8 cm.** Underparts yellowish white, orange, or chest and belly pale gray. The female has a pouch, stained orange if she has had young. Scrotum of males black. Young like adults.

Variation. Upperparts vary from pale to dark gray; animals from south of the Amazon in Pará, Brazil, are brownish gray; those from Paraguay are pale gray; those from west of the Andes in N Ecuador are dark gray, with gray bellies and completely dark gray tails. There is dispute over the generic name of this taxon, which is sometimes given as *Metachirops*.

Similar species. Anderson's gray four-eyed opossums (*P. andersoni*) are black or have a black middorsal streak contrasting with paler sides, and ears always > 35 mm; brown four-eyed opossums (*Metachirus nudicaudatus*) are brown with buff grizzling in fur and tail without sharply demarcated white tip; water opossums (*Chironectes minimus*) have broad black bands across the back.

Sounds. No sounds usually heard in the field; hisses with open mouth when cornered; young emit sharp clicks.

Natural history. Nocturnal; arboreal and terrestrial; solitary. Feeds chiefly on inver-

Map 8

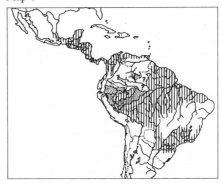

▨ Anderson's gray four-eyed opossum, *Philander andersoni*

▥ Common gray four-eyed opossum, *P. opossum*

tebrates and small vertebrates, supplemented by fruit. Four-eyed opossums use the ground to middle vegetation levels of the forest; they are most common around treefalls and especially in dense undergrowth near water, where they forage like small carnivores, winding in and out of fallen brush, along fallen logs, and among low vegetation. They hunt frogs by homing in on their calls. When disturbed they often climb up into a tree, and they rest during the night on elevated perches. They swim well and voluntarily. Their nests may be 8–10 m above the forest floor, in hollow trees or in tree forks, or occasionally in hollow logs or treefalls on the ground. Found in mature and secondary rainforest, gardens and plantations, and gallery forests.

Geographic range. Central and South America: S Mexico to Panama, west of the Andes to N Peru, east of the Andes to Paraguay and NE Argentina. To 1,500 m elevation.

Status. Often common.

Local names. Guaiki (Ar); cucha gris de cuatro ojos (Co); mucura-de-cuatro-olhos (Br); cuica común, zorro (Ec); pericote (Pe); quica opossum, fo-ai awari (Su); zorro de cuatro ojos (Pa, CR); mbicuré (Gua).

References. Atramentowicz, M. 1986. Dynamique de population chez trois marsupiaux didelphidés de Guyane. *Biotropica* 18:136–49.

Anderson's Gray Four-eyed Opossum
Philander andersoni
Plate 1, map 8
Identification. Measurements: HB = 223–307; T = 255–332; HF = 34–51; E = 35–42; WT = 225–425 g.

Like the common gray four-eyed opossum except: **back along midline** from the neck to the furred portion of the tail, **black or blackish in a narrow to broad strip, with long glossy black guard hairs, distinctly darker than sides.** Ears usually larger. **Underparts pale buff,** grayish buff, **gray, gray-brown,** or blackish with an orange tinge, or gray mottled with cream. **Feet black.**

Variation. Color darkens geographically, north to south, from pale gray or gray-brown on sides, and grayish or buff belly in Venezuela and Ecuador, to dark gray or blackish in Ucayali, Peru, and Acre, Brazil, to black with a blackish belly on the Alto Purús in Peru.

Similar species. See common gray four-eyed opossum (*P. opossum*). Anderson's may be the only gray four-eyed opossum in parts of its range; in central Peru it can occur with the common gray four-eyed opossum.

Natural history. Nocturnal; terrestrial and arboreal; solitary. Anderson's opossums favor dense viny vegetation near water. They are often confused with common gray four-eyed opossums, with which they probably share many characteristics. Mature and disturbed lowland rainforest.

Geographic range. South America: east of the Andes in the upper Amazon Basin of Venezuela, probably Colombia, Ecuador, Peru to Ucayali Department, and Acre, Brazil.

Status. Locally common, except where it occurs with the common gray four-eyed opossum, when it seems to be rare.

Local names. The same as common gray four-eyed opossum.

References. Gardner, A. L., and J. L. Patton. 1972. New species of *Philander* (Marsupialia: Didelphidae) and *Mimon* (Chiroptera: Phyllostomidae) from Peru. Occ. Pap. Mus. Zool. Louisiana State Univ., no. 43.

Map 9

Water opossum, *Chironectes minimus*

Water Opossum
Chironectes minimus
Plate 1, map 9
Identification. Measurements: HB = 260–298; T = 327–420; HF = 60–74; E = 25–32; WT = 590–700 g.

Upperparts frosted silver-gray with four broad black or dark brown bars across back joined by a narrow black stripe down spine, **looks marbled;** fur dense and soft. **Head black, with U-shaped pale band from ear to ear above eyebrows,** cheeks white; ears short, eyes black; chin and area around mouth almost naked; whiskers long and stiff. Tail furred for first tenth, the rest naked with large scales, black with a short white tip. **Hindfeet large, completely webbed between all toes,** black; forefeet not webbed, toes long, thin, with tips expanded into large pads; a (pisiform) bone of the hand is enlarged to form an accessory sixth "finger"; surface of palms rough like sandpaper, fingers and palms pale. Underparts creamy white or pale yellow. Both sexes have a pouch. Young similar, but blacker than adults.

Variation. Some animals from drier areas have dark brown markings; the black parts of preserved skins of this species usually fade to brown.

Similar species. These are the only opossums with webbed feet. Gray four-eyed and common opossums (*Philander* spp., *Didelphis* spp.) have no bars across back and have large ears; all other genera of large opossums are brown. Small carnivores have fully furred tails.

Sounds. When threatened gives sharp, screechlike spitting barks.

Natural history. Nocturnal; terrestrial and semiaquatic; solitary. Feeds on fish, crustaceans, and invertebrates that it catches in the water. The long fingers and sandpaper-like palms of water opossums are used to probe for prey and grasp slippery fish. These opossums are found in and near rivers and streams. The female's pouch has a watertight seal to keep the young dry when she swims, and the male's scrotum is said to be pulled into the pouch when the animal is in water. They build nests of leaves and grasses, and by day they den in burrows in stream banks or sometimes in nests on the ground surface. Found in tropical forests and cleared areas in tropical forest regions. Most records are from clear, fast-flowing rivers and streams in hilly areas; may be rare or absent from sluggish, silt-laden lowland watercourses.

Geographic range. Central and South America: S Mexico and Belize to W Venezuela, Colombia, Ecuador, Peru, and W Brazil; the Guianas and mouth of the Amazon; and SE Brazil, Paraguay, and N Argentina. To elevations of 1,800 m.

Status. Seems rare in most regions, but is common in some Central American rivers.

Local names. Cuica de agua (Ar); mucura do fundo, cuíca-d'agua (Br); chucha de agua (Co); zorro de agua (Pa, CR); watra-stonawari, watra-alata (Su); yapó (Gua).

References. Marshall, L. G. 1978. *Chironectes minimus*. Mammalian Species, no. 109.

Brown Four-eyed Opossum
Metachirus nudicaudatus
Plate 1, map 10

Identification. Measurements: HB = 250–280; T = 280–325; HF = 38–45; E = 28–36; WT = 300–450 g; males slightly larger than females.

Upperparts brown; center of back sometimes blackish; fur short, dense and soft, finely grizzled with buff. **Head with a large pale yellow spot above each eye** and at the upper base of the ear; **crown black,** sometimes extending as a black stripe on neck; **corner of eye to nose black;** ears naked, brown; nose gray; eyes dark brown, eyeshine bright orange-yellow. **Tail naked from near its base, entirely brown or**

Map 10

Brown four-eyed opossum, *Metachirus nudicaudatus*

gray or gradually paling toward tip. Underparts and lower cheeks uniform clear pale yellow or orange, scrotum pinkish. The female has no pouch. A large, long-legged, cursorial opossum.

Variation. Some animals from Pará, Brazil, and Paraguay are grayish brown, with sides tinged pale orange-brown.

Similar species. Water opossums (*Chironectes minimus*) have black bars across back; gray four-eyed opossums (*Philander* spp.) are gray with white grizzling on fur, and abruptly demarcated white tail tips; no other opossums have pale spots above the eyes.

Sounds. Chatters teeth rapidly when cornered, but normally silent.

Natural history. Nocturnal; terrestrial; solitary. Appears to feed chiefly on termites, supplemented by other insects and invertebrates and fruit. Brown four-eyed opossums seem to favor mature forest with little undergrowth, but they are also present in dense habitats. Unlike other opossums, these are extremely nervous and wary. They travel on the ground and along the tops of fallen logs, trotting or walking quickly and silently. When alarmed they run rapidly and soundlessly away on the ground. They are difficult to observe because they run at the slightest disturbance, but seem most active and easy to see immediately following rainstorms. Their nests are in well-hidden hollows in or near the ground. Found in mature and disturbed rainforest, secondary forest, and gallery forest.

Geographic range. Central and South America: Costa Rica south to Paraguay and NE Argentina.
Status. Widespread and often common; lives at low population densities.
Local names. Cuica común (Ar); mucura de cuatro-olhos (Br); chucha mantequera (Co); zorricí (CR); pericote (Pe); froktu-awari, bruine opossum (Su); zorra morena (Pa); yupatí (Gua).
References. Miles, M. A., A. A. De Souza, and M. M. Póvoa. 1981. Mammal tracking and nest location in Brazilian forest with an improved spool-and-line device. *J. Zool., Lond.* 195:331–47.

Woolly Mouse Opossum
Micoureus cinereus
Plate 2, map 11
Identification. Measurements: HB = 158–210; T = 225–270; HF = 25–30; E = 29–34; WT = 80–152 g; males larger than females.
Upperparts smoky gray-brown; fur long, soft, and woolly. Eye rings large, black and prominent; cheeks to below ears pale orange; ear naked, tan; eyeshine bright yellow. **Tail thickly haired for first 3–4 cm,** rest naked, **gray-brown, usually with long whitish tip, mottled** with dark spots around dark/white junction, or entirely dark, sometimes with white spots near tip. Feet yellowish. **Underparts orange,** yellowish or gray with orange wash. Females have no pouch. Scrotum of males blue. Young grayer than adults, with shorter fur and base of tail less furry. A large mouse opossum.
Variation. South American members of this genus have recently been redivided into three species—*M. regina, constantiae,* and *cinereus*—on the basis of color and fur patterns. However, there is variation both among populations and among individuals within populations, and the distinctions between the three forms are not always clear. They are here considered subspecies. Animals from Central America are considered by some to be a separate species (*M. alstoni*). Some animals from the Andes of Peru and Ecuador, and Bolivia, Paraguay, and SW Brazil have whitish underparts only faintly tinged orange.
Similar species. Other lowland rainforest mouse opossums are all smaller, have

Map 11

Woolly mouse opossum, *Micoureus cinereus*

short, velvety fur, or are tiny, with no white tail tip; bare-tailed woolly opossums (*Caluromys philander*) are similar, but larger, with a black stripe down the center of the face and short body fur in lowlands.
Sounds. None usually heard; hisses when cornered.
Natural history. Nocturnal; arboreal; solitary. Feeds on insects and small animals, fruit and nectar. Woolly mouse opossums are usually seen in the middle to upper levels of the forest, but they descend to the ground when food is scarce in the dry season. They favor dense, viny vegetation with many palm trees but are also found in open, high forest. They build a nest of dead leaves in the crown of a palm tree or in a vine tangle. This is an aggressive species that will bite fiercely and hang on tenaciously if grasped. Found in mature and secondary evergreen rainforest, gallery forest, and gardens and plantations.
Geographic range. Central and South America: Belize south to N Argentina, Paraguay, and SE Brazil. To 2,200 m elevation in Andean foothills.
Status. Widespread and often common.
Local names. Zorra, zorricí (CR); rapposa (Ec, Pe); muca (Pe); moismoisi-awari (Su); anguyá-guaik (Gua).
References. Charles-Dominique, P., M. Atramentowicz, M. Charles-Dominique, H. Gérard, A. Hladik, C. M. Hladik, and M. F. Prévost. Les mamifères frugivores arboricoles nocturnes d'une forêt guyanaise: Inter-relations plantes-animaux. *Rev. Ecol.* 35:341–435.

Miles, M. A., A. A. de Souza, and

M. M. Póvoa. 1981. Mammal tracking and nest location in Brazilian forest with an improved spool-and-line device. *J. Zool., Lond.* 195:331–47.

White-bellied Slender Mouse Opossum
Marmosops noctivagus
Plate 2, map 12
Identification. Measurements: HB = 125–160; T = 160–190; HF = 19–22; E = 18–25; WT = 55–85 g; males larger than females.
Upperparts slate brown, sides of neck and body slightly paler and with more orange tones; fur short, soft, and velvety. **Eye rings black, often indistinct;** eyes black; ears brown. Tail thinly haired for first 1 cm, the rest naked, uniform dark brown. **Underparts and cheeks pure creamy white** to hair base except nipple area orange on parous females. Feet whitish, hands brown above, toes pale. Females have no pouch. A relatively large mouse opossum.
Similar species. See woolly mouse opossum (*Micoureus cinereus*). Other mouse opossums in region are smaller if they have white underparts, or are warm brown, or have gray or partly gray underparts; short-tailed opossums *(Monodelphis* spp.) have short tails. There are two other similar species in the genus that are not described in this book: a rare, smaller, paler species, *M. dorothea*, is found only in montane humid forests of N and E Bolivia (map 12); it is probably not distinguishable in the field. *M. impavidus* is found at higher elevations in the Andes and possibly sometimes below 1,000 m; it is dark slate gray, with gray patches on the belly.
Sounds. None heard in field. Hisses when cornered.
Natural history. Nocturnal; arboreal and terrestrial; solitary. Feeds on insects and fruit. These mouse opossums use the lower vegetation levels and ground of the forest, where they run around rapidly when active, especially on fallen trees, in dense undergrowth thickets, and up and down vines. They seem to favor dense *platanillos (Heliconia* spp.) in swamps and watersides. They build a nest of dead leaves. Found in mature, disturbed, and secondary forests.
Geographic range. South America: Ama-

Map 12

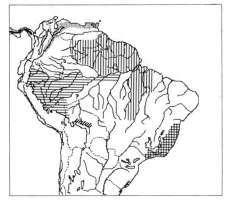

▦ Gray-bellied slender mouse opossum, *Marmosops fuscatus*
▤ White-bellied slender mouse opossum, *M. noctivagus*
▥ Delicate slender mouse opossum, *M. parvidens*
▨ Dorothy's slender mouse opossum, *M. dorothea*
▦ Gray slender mouse opossum, *M. incanus*
■ Slaty slender mouse opossum, *M. invictus*

zon Basin of Ecuador, Peru, W Brazil south of the Amazon, and possibly Bolivia.
Status. Common.
Local names. Rapposa (Pe); achocaya (Qui).
References. Creighton, G. K. 1984. Systematics and taxonomy of the marsupial family Didelphidae. Ph.D. diss., University of Michigan, Ann Arbor.

Gray-bellied Slender Mouse Opossum
Marmosops fuscatus
Map 12
Identification. Measurements: HB = 120–155; T = 148–182; HF = 18–26; E = 22–29; WT = 41–85 g.
Upperparts gray-brown; fur short and slightly stiff. Eye rings black, prominent or indistinct, face between eyes pale, **cheeks whitish.** Tail robust, gray, becoming paler toward tip, slightly bicolored pale below. Feet and lower legs whitish. **Underparts gray frosted with white or yellowish;** tail base dark brown below. Females have no pouch. Young gray. Large, robust mouse opossums.
Similar species. These are the only large, grayish, short-furred, gray-bellied mouse opossums in their range: Robinson's (*Mar-*

mosa robinsoni) and murine (*Marmosa murina*) mouse opossums have pale pink, orange, or yellow bellies with fur pale at base; northern gracile mouse opossums (*Gracilinanus marica*) are smaller and have long soft fur; all are warm brown.
Natural history. Nocturnal; arboreal and terrestrial. Probably feeds on insects and fruit. Gray-bellied slender mouse opossums seem to use the ground and lower vegetation levels and are often found near water. They mainly inhabit montane forest and cloud forest above 1,000 m but can occur lower in some localities. Found in evergreen forests and clearings within it.
Geographic range. South America: northern mountains of Venezuela, Colombia, and Trinidad. To 2,400 m elevation.
Status. Locally common.
Local names. Chucha ratón (Co).

Delicate Slender Mouse Opossum
Marmosops parvidens
Plate 2, map 12
Identification. Measurements: HB = 95–105; T = 130–160; HF = 14–17; E = 19–23; WT = 15–27 g.
Upperparts deep velvety smoky brown. Dark eye rings variable, but **often inconspicuous;** muzzle and face between eyes pale, sometimes with a narrow dark central stripe above nose; cheeks white or gray; ears dark gray. Tail completely naked, uniform gray-brown. **Underparts usually a mixture of gray and white:** from gray with a white stripe down center of belly to mostly white, usually with gray spots on chest. Feet white. Females have no pouch. A **tiny opossum with a very long tail, long, pointed muzzle, matchstick-thin legs, and tiny hands and feet** with thin digits. Resembles a juvenile of other species.
Variation. A closely related and similar form, *M. cracens,* is known only from three specimens from the type locality in Falcón, Venezuela. Similar animals from the mountains of Peru may belong to another species, *M. juninensis.*
Similar species. See woolly mouse opossum (*Micoureus cinereus*). *M. cracens* may not be distinguishable in the field. Other slender mouse opossums in the region are larger.
Sounds. None usually heard.

Natural history. Nocturnal; arboreal and terrestrial; solitary. Feeds on insects and fruit. Delicate mouse opossums use the forest understory near the ground, usually in tall, open-understory, terra firme forest. They are more slow-moving than most other mouse opossums, and they are often found sitting still perched on a low shrub or a branch of a treefall. When disturbed they run a few feet up into a sapling, where they stop and can be caught by hand. Found in mature, closed canopy, evergreen forest, not often in disturbed or secondary forest.
Geographic range. South America: east of the Andes in the Amazon Basin of Venezuela, Colombia, the Guianas, and Brazil (possibly Peru above 1,000 m).
Status. Usually uncommon, occasionally locally common.
References. Pine, R. H. 1981. Reviews of the mouse opossums *Marmosa parvidens* Tate and *Marmosa invicta* Goldman (Mammalia: Marsupialia: Didelphidae) with a description of a new species. *Mammalia* 45:55–70.

Gray Slender Mouse Opossum
Marmosops incanus
Plate 2, map 12
Identification. Measurements: HB = 130–158; T = 155–185; HF = 20–27; E = 19–27; WT = 54 g.
Upperparts dark gray-brown. Eye rings dark, prominent; face between eye rings pale; cheeks below eyes cream. **Underparts creamy white.** Scrotum well furred. Females have no pouch. A large mouse opossum.
Variation. This species has distinct winter and summer coats: in the austral winter (May–Aug.), the fur is long, soft, and brown. In the summer (Dec.–Apr.), the coat is short, stiff, and more grayish. In September–December the molt pattern may show a "collar" of new hair. This molt stage was once described as a distinct species.
Similar species. Agile gracile mouse opossums (*Gracilinanus agilis*) are smaller, with orange bellies; Brazilian gracile mouse opossums (*G. microtarsus*) are smaller, redder, with grayish white bellies.
Natural history. Found in humid Atlantic coastal forests.

Geographic range. South America: Brazil, Bahía to São Paulo.

Slaty Slender Mouse Opossum
Marmosops invictus
Map 12

Identification. Measurements: HB = 104–113; T = 124–147; HF = 17–19; E = 18–22. **Upperparts dark slate gray** sometimes tinged with brown, especially on forequarters; fur short. **Face dark, eye rings inconspicuous against dark background; cheeks gray.** Tail dark gray. Lower legs dusky, feet paler. **Underparts slate gray, frosted white.** Females have no pouch. Small, shrewlike mouse opossums.
Similar species. These are the only dark gray, gray-bellied mouse opossums in their range; sepia short-tailed opossums (*Monodelphis adusta*) are dark brown, with short ears and tail much shorter than head and body.
Natural history. Nocturnal; terrestrial and arboreal. Feeds on insects and plant material. Slaty slender mouse opossums live in the forest understory, especially around fallen logs. They are chiefly montane, cloud forest dwellers, but can be found to below 500 m elevation.
Geographic range. Central America: montane areas of Panama, likely to be found in Colombia and Costa Rica. About 500–1,300 m elevation.
Status. Rare, geographic range small.
References. See delicate slender mouse opossum (*M. parvidens*).

Murine Mouse Opossum
Marmosa murina
Plate 2, map 13

Identification. Measurements: HB = 125–150; T = 170–198; HF = 20–24; E = 20–25; WT = 43–60 g; males larger than females.
Upperparts uniform warm brown; fur short, velvety; Eye rings large, prominent, black; center of face between eye rings pale; cheeks pale yellow or orange. **Underparts delicate salmon or seashell pink, yellowish or cream, not pure white.** Tail with short hair for about 1.5 cm at base, rest naked, tan. Feet whitish. Females have no pouch. Scrotum of males blue.
Variation. Animals from Matto Grosso,

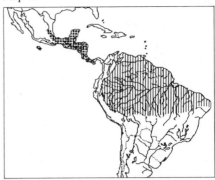

Map 13

⊞ Mexican mouse opossum, *Marmosa mexicana*
▥ Murine mouse opossum, *M. murina*

Brazil, are pale fawn. The delicate, luminous pink of the underparts of live animals fades and eventually disappears from skins.
Similar species. See woolly mouse opossum (*Micoureus cinereus*); red mouse opossums (*Marmosa rubra*) are bright chestnut, with a bicolored tail; northern gracile mouse opossums (*Gracilinanus marica*) have long fur and gray-based fur on belly; other small mouse opossums in the region have white or gray bellies, except young woolly mouse opossums, which have woolly fur.
Sounds. None usually heard in field; makes explosive clicks in threat.
Natural history. Nocturnal; arboreal; solitary. Diet consists of about two-thirds insects and other small animals, one-third fruit. Murine mouse opossums use the ground and lower levels of the forest, almost always below 5 m. They cruise the short undergrowth or walk slowly on the forest floor, exploring the undersurfaces of living and dead leaves for invertebrate prey. Especially common in disturbed zones of swampy *platanillos*, extremely dense vines, thickets, river edges, and secondary forest, and usually rare in tall, open-understory, terra firme forest. Inhabits mature and secondary rainforest, plantations, and camps, gardens, and houses.
Geographic range. South America: east of the Andes in Venezuela, the Guianas, the Amazon Basin of Ecuador, Peru, and Brazil, and Tobago. To at least 1,365 m elevation.
Status. Widespread and often common.

Local names. Gambasinha (Br); muca, rapposa (Pe); boesi-mois-moisi (Su).
References. Charles-Dominique, P., M. Atramentowicz, M. Charles-Dominique, H. Gérard, A. Hladik, C. M. Hladik, and M. F. Prévost. 1981. Les mamifères frugivores arboricoles nocturnes d'une forêt guyanaise: Inter-relations plantes-animaux. *Rev. Ecol.* 35:341–435.

Red Mouse Opossum
Marmosa rubra
Plate 2, map 14
Identification. Measurements: HB = 121–165; T = 180–220; HF = 22–26; E = 18–23; WT = 59–67 g.
Upperparts deep chestnut red; fur velvety. Eye rings black, prominent, especially pronounced between nose and eye; face with dark stripe on center above nose, snout appears black, cheeks yellow-orange; ears brown, relatively short for a mouse opossum. **Tail** with short furred base darker than back above, rufous below, rest naked, **bicolored dark above, pale below. Underparts rich orange;** chin, throat, inguinal region, and stripe down midventer paler orange-yellow. Outer hind- and forelegs and backs of hands blackish, toes pale. The female has no pouch.
Similar species. Murine mouse opossums (*M. murina*) are paler brown, less red, with face distinctly pale from nose to between eyes. See little rufous mouse opossum (*M. lepida*).
Natural history. Has been trapped on the ground and in the thatched roof of a house. Found in lowland rainforest and secondary forest.
Geographic range. South America: the Amazon Basin of Ecuador and Peru.
Status. Can be locally common, may have a patchy distribution.

Little Rufous Mouse Opossum
Marmosa lepida
Plate 2, map 14
Identification. Measurements: HB = 97–120; T = 140–150; HF = 16–19; E = 15–18; WT = 10 g.
Upperparts uniform bright chestnut or orange-red; fur long and dense. **Eye rings pitch black, prominent;** ear short, red-brown, reaches middle of eye when laid forward. **Tail very long, slender, pale**

Map 14

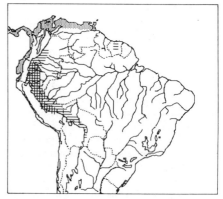

Robinson's mouse opossum, *Marmosa robinsoni*
Red mouse opossum, *M. rubra*
Little rufous mouse opossum, *M. lepida*

brown, lightly coated with white hairs beneath. Underparts creamy white or tinged with pinkish; scrotum thickly furred with cream hair. Feet cream. Females have no pouch. Young like adults. A **tiny** opossum.
Similar species. These are the only tiny red mouse opossums in their range. Red mouse opossums (*M. rubra*) are larger, with an orange belly and bicolored tail; murine mouse opossums (*M. murina*) are larger and browner; Emilia's mouse opossums (*Gracilinanus emiliae*) are smaller, with shorter fur, smaller hindfeet and browner, narrower eye rings, and are probably indistinguishable in the field, but the ranges are not known to overlap.
Natural history. Found in rainforest.
Geographic range. South America: the Amazon Basin of Colombia, Ecuador, Peru, Bolivia, and Brazil; and from Surinam.
Status. Rare; known from fewer than 20 specimens.

Robinson's Mouse Opossum
Marmosa robinsoni
Plate 2, map 14
Identification. Measurements: HB = 128–200; T = 170–239; HF = 20–30; E = 23–30; WT = 36–132 g; males larger than females.
Upperparts cinnamon-brown; fur short. Eye rings black, large and prominent; face between eye rings pale, cheeks to be-

hind ear pale orange-yellow; ears tan, large. Tail naked, tan, relatively thick and robust. Feet large, cream or whitish above. **Underparts pale orange-yellow,** nipple area of parous females bright orange; scrotum of males blue. Females have no pouch. Young like adults.
Variation. Animals from coastal Venezuela and Ecuador are pale cinnamon-gray, with yellow-white underparts.
Similar species. Murine mouse opossums (*M. murina*) are smaller, with pinker underparts; see that species and keys for other comparisons.
Sounds. None usually heard in field, but makes explosive clicks in threat and an insectlike twitter when excited.
Natural history. Nocturnal; arboreal and terrestrial; solitary. Feeds mainly on insects, with some fruit. Robinson's mouse opossums use the low undergrowth and ground. Found in a wide variety of habitats from evergreen or cloud forest to deciduous and thorn forest and secondary vegetation.
Geographic range. Central and South America: Belize and Honduras south, west of the Andes from Colombia to N Peru, and east of the Andes in a narrow coastal strip from N Colombia east to the mouth of the Orinoco in N Venezuela; Trinidad and Tobago. To 2,000 m elevation.
Status. Locally common.
Local names. Marmota (Pn).
References. O'Connell, M. A. 1983. *Marmosa robinsoni*. Mammalian Species, no. 203.

Mexican Mouse Opossum
Marmosa mexicana
Map 13
Identification. Measurements: HB = 85–144; T = 145–190; HF = 18–25; E = 19–25; WT = 40–100 g; males larger than females.
Upperparts cinnamon to warm reddish brown, **sides paler;** fur short. **Eye rings prominent,** dark, face between eye rings pale, cheeks yellow or pale orange. **Tail densely furred for a short section of base,** rest naked, solid gray-brown, relatively thick and robust. Feet whitish. **Underparts pale yellow or orange, usually mixed with gray around center of body** between fore- and hindlegs, nipple area of parous

females stained orange. Females have no pouch.
Variation. Animals from Nicaragua are rich cinnamon; those from Mexico are browner.
Similar species. Probably cannot be distinguished from Robinson's mouse opossums (*M. robinsoni*) in the field, but the two do not overlap geographically. See other mouse opossums.
Sounds. None usually heard in field.
Natural history. Nocturnal; arboreal and terrestrial; solitary. Feeds on insects and probably fruit. Mexican mouse opossums are found on or near the ground. A leaf nest was in a burrow in the ground. Most common in moist evergreen forests, but also found in plantations and even arid grasslands.
Geographic range. Central America: Taumalipas, Mexico, south to W Panama. To 1,675 m elevation.
Status. Locally common.
Local names. Ratón tlacuache (Me); zorra (Pa, CR).
References. Hall, E. R., and W. W. Dalquest. 1963. The mammals of Veracruz. *Univ. Kansas Pub. Mus. Nat. Hist. 14:165–362.*

Northern Gracile Mouse Opossum
Gracilinanus marica
Map 15
Identification. Measurements: HB = 96–116; T = 131–151; HF = 15–18; E = 15–22.
Upperparts pale warm brown; fur long and soft. Eye rings dark, prominent; face pale brown between eyes; cheeks pale buff. Tail slender, brown, fur of back not extending onto base. Feet brownish, lower legs usually furry almost to feet. **Throat and often chest pale buff with hairs pale to base; rest of underparts pale buff with fur dark gray at base. Tiny, delicate, fluffy** mouse opossums.
Variation. A similar and closely related form, *G. dryas,* is known only from the Merida Andes of Venezuela; it may be a subspecies.
Similar species. These are the only mouse opossums in their range with long, pale warm brown fur and belly hair dark gray at base; only the delicate mouse opossums (*Marmosops parvidens*) are this tiny, and

Map 15

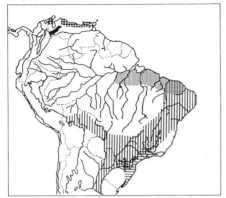

▨	Emilia's gracile mouse opossum, *Gracilinanus emiliae*
▤	Brazilian gracile mouse opossum, *G. microtarsus*
▥	Agile gracile mouse opossum, *G. agilis*
▦	Northern gracile mouse opossum, *G. marica*
■	Wood-sprite gracile mouse opossum, *G. dryas*

they are dark gray-brown; *G. dryas* are probably indistinguishable in the field, but the ranges apparently do not overlap.
Natural history. Arboreal. Found in lowland to montane rainforest, deciduous forest, and savanna.
Geographic range. South America: northern coastal Venezuela and Colombia. To 2,100 m elevation.
Status. Rare.
References. Gardner, A. L., and G. K. Creighton. 1989. A new generic name for Tate's (1933) *Microtarsus* group of South American mouse opossums (Marsupialia, Didelphidae). *Proc. Biol. Soc. Washington* 102:3–7.

Emilia's Gracile Mouse Opossum
Gracilinanus emiliae
Map 15
Identification. Measurements: HB = 72–75; T = 128–142; HF = 14–15; E = 16. **Upperparts bright red-orange; fur relatively short.** Eye rings narrow, black, sharply contrasting. **Tail exceptionally long,** brown. Underparts cream to orange. **Hindfeet tiny, narrow. A tiny opossum.**
Variation. The type specimen was described as brown; possibly the bright orange of the specimens described above is partly due to bleaching.
Similar species. Little rufous mouse opossums (*Marmosa lepida*) have a relatively

shorter tail, larger hindfoot, longer fur, and seem to be larger; delicate mouse opossums (*Marmosops parvidens*) are slaty, with white and/or gray underparts; other mouse opossums in range are larger.
Natural history. Found in lowland rainforest and perhaps other habitats.
Geographic range. South America: Brazil south of the Amazon and east of the Tapajós to Ceará.
Status. Rare, known only from a few individuals and localities.

Brazilian Gracile Mouse Opossum
Gracilinanus microtarsus
Plate 2, map 15
Identification. Measurements: HB = 105–110; T = 145–153; HF = 16–21; E = 18–19; WT = 19–29 g.
Upperparts reddish brown; fur long and soft. Eye rings large, black, prominent. Forelegs and neck under ear tinged orange. Tail brown. Feet whitish. **Underparts cream, neck and chest tinged orange, hairs gray at base.** Scrotum well furred. **A small, red-brown mouse opossum.**
Similar species. Gray slender mouse opossums (*Marmosops incanus*) are larger and gray; agile gracile mouse opossums (*G. agilis*) are indistinguishable in the field; they have fur of underparts white at base and apparently occupy dryer habitats; Emilia's gracile mouse opossums (*G. emiliae*) are smaller and bright red.
Natural history. Found in wet evergreen Atlantic coastal forest.
Geographic range. South America: Brazil, Minas Gerais to Rio Grande do Sul.

Agile Gracile Mouse Opossum
Gracilinanus agilis
Map 15
Identification. Measurements: HB = 90–106; T = 109–132; HF = 15–17; E = 15–20; WT = 20–30 g.
Upperparts drab gray-brown to red-brown. Eye rings black, prominent; face above nose sometimes with faint dark line down center, cheeks with pale orange not reaching to below ear; ears light tan. Tail relatively short, undersurface with fine silky coat of white hairs. **Underparts pure pale orange** to cream, or lower belly gray-orange, base of hairs not gray on most of underparts. Hands and feet whitish. Females have no pouch.

Variation. Animals from dry areas of Argentina and Paraguay are smaller than ones from farther north.
Similar species. See Brazilian gracile mouse opossum (*G. microtarsus*); this species is grayer and smaller than murine mouse opossums (*Marmosa murina*); delicate mouse opossums (*Marmosops parvidens*) have gray and white underparts.
Natural history. Nocturnal; arboreal. Agile mouse opossums frequent the forest understory, where they use slender branches and vines. Found in evergreen and gallery forests along the S fringe of the Amazon forest and in Caatinga, Cerrado, and Chaco.
Geographic range. South America: east of the Andes in Bolivia, SE Brazil, Paraguay, Uruguay, and Argentina.
Status. Locally common.

Red-legged Short-tailed Opossum
Monodelphis brevicaudata
Plate 3, map 16
Identification. Measurements: HB = 134–183; T = 76–105; HF = 20–26; E = 13–20; WT = 46–150 g.
Upperparts grizzled dark gray from nose to tail base; sides from nose to rump red or rusty. Ear naked, tan. Tail densely furred for first 2–3 cm, fur tapering to fine covering on dark tail. Feet dark gray. Underparts glowing, violaceous pink, cream, or pale orange, grayish orange, or gray. Females have no pouch. Young like adults; tail less thickly furred at base.
Variation. Highly variable in color; ranges from almost black back with deep red sides and legs, in the Guianas and parts of Venezuela, to pale gray with pale rusty sides and legs, south of the Amazon. Some animals from N Venezuela and Pará are almost entirely red, sometimes with a cream patch on the face. Probably several species will eventually be defined from within this variable group. The glowing violaceous color of the underparts of living animals fades and eventually disappears from skins.
Similar species. Emilia's (*M. emiliae*) and shrewish (*M. sorex*) short-tailed opossums have red tops of rumps; Emilia's also has a red head and is smaller and darker in area of geographic overlap; mouse opossums have long tails, large ears, and black eye rings; Macconnell's rice rat (*Orzyzomys macconnelli*) has a long tail and large ears.
Sounds. None usually heard in field.

Map 16

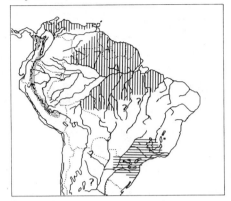

	Sepia short-tailed opossum, *Monodelphis adusta*
	Red-legged short-tailed opossum, *M. brevicaudata*
	Shrewish short-tailed opossum, *M. sorex*

Natural history. Diurnal; terrestrial; solitary. Apparently insectivorous. Red-legged opossums are found on the ground, especially near fallen logs or occasionally on them. In mature, disturbed, and secondary rainforest, gardens and plantations; rarely in deciduous or dry forests.
Geographic range. South America: east of the Andes in Colombia, Venezuela, the Guianas, and Amazonian Brazil and N Argentina. To at least 1,160 m elevation.
Status. Locally common.
Local names. Colicorto (Ar); catita (Br); koort-staart opossum, moismoisi-awari (Su).
References. Charles-Dominique, P., et al. 1981. Les mamifères frugivores arboricoles nocturnes d'une forêt guyanaise. *Rev. Ecol.* 35:341–435.

Emilia's Short-tailed Opossum
Monodelphis emiliae
Plate 3, map 17
Identification. Measurements: HB = 120–158; T = 50–70; HF = 17–24; E = 14–18; WT = 60 g.
Back grizzled gray; entire rump, hindlegs, and head forward of ears rufous. Snout short and broad. Tail with rump hair extending far down length, gradually decreasing to near tip. Forearms and hands gray or gray-buff. Underparts pale orange, with violet- or rose-colored wash. Females have no pouch.
Variation. The violet wash on the belly is

Map 17

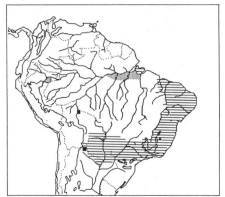

▦   Emilia's short-tailed opossum, *Monodelphis emiliae*

▤   Gray short-tailed opossum, *M. domestica*

▪   Pygmy short-tailed opossum, *M. kunsi*

a "fugitive" coloration that vanishes from skins after a year or two.

Similar species. See red-legged short-tailed opossum (*M. brevicaudata*).

Natural history. Lowland evergreen rainforest.

Geographic range. South America: Brazil south of the Amazon and east of the Madeira, and Peru south of the Amazon near Iquitos.

Status. Apparently rare; known from only a few individuals and localities.

Local names. Catita (Br).

References. Pine, R. H., and C. O. Handley, Jr. 1984. A review of the Amazonian short-tailed opossum *Monodelphis emiliae* (Thomas). *Mammalia* 48:239–45.

Sepia Short-tailed Opossum
Monodelphis adusta
Plate 3, map 16

Identification. Measurements: HB = 84–114; T = 51–67; HF = 11–17; E = 10–14; WT = 15–35 g.

Upperparts completely dark brown, darker on rump than foreparts, **fur very short** (~ 3 mm). Ears naked, dark brown-gray. Tail short, dark brown. Underparts gray or orange-gray, sometimes with a cream streak on midline. Females have no pouch.

Variation. Another species of short-tailed opossum, *M. peruanus*, also occurs at higher elevations in the Andes: it is gray-brown with medium-length fur and whitish underparts.

Similar species. See other short-tailed opossums. Grass mice (*Akodon* spp.) and long-nosed mice (*Oxymycterus* spp.) have large ears and long fur.

Natural history. Terrestrial. Feeds on invertebrates. Sepia short-tailed opossums have been captured in rocky terrain and in a house. In captivity they are in constant motion during activity, always investigating their surroundings. This is chiefly a montane species of wet forests at midelevations (1,400–2,200 m), but in some places is found as low as 200 m.

Geographic range. Central and South America: the northern Andes and their eastern slopes from E Panama to Colombia, Ecuador, Peru, and Bolivia.

Status. Apparently rare.

Local names. Zorra de cola corta (Pa).

References. Handley, C. O. 1966. Checklist of the mammals of Panama. In R. L. Wenzel and V. J. Tipton, eds., *Ectoparasites of Panama*, 753–95. Chicago: Field Museum of Natural History.

Pygmy Short-tailed Opossum
Monodelphis kunsi
Map 17

Identification. Measurements: HB = 71; T = 42; HF = 12; E = 12.

Upperparts uniform warm brown. Tail bicolored dark above, buff below. Underparts with whitish areas. **Very small.**

Similar species. Sepia short-tailed opossums (*M. adusta*) are larger, with dark, very short, velvety brown fur. This is the smallest short-tailed opossum. Grass mice (*Akodon* spp.) have long tails, large ears, and long, somewhat grizzled fur.

Natural history. One individual was trapped in second-growth brush.

Geographic range. South America: Bolivia, 200–640 m elevation. Likely to be found in N Argentina and Brazil.

Status. Apparently rare, known from only two specimens.

References. Anderson, S. 1982. *Monodelphis kunsi*. Mammalian Species, no. 190.

Long-nosed Short-tailed Opossum
Monodelphis scalops
Map 18

Identification. Measurements: HB = 133; T = 71; E = 8.

Head, rump, and tail bright rufous; forward part of back and shoulders grizzled

Map 18

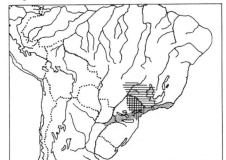

Long-nosed short-tailed opossum, *Monodelphis scalops*

One-striped short-tailed opossum, *M. unistriata*

Chestnut striped short-tailed opossum, *M. rubida*

olive-gray. Snout long and narrow. Forearm and hand red. Rump hair extending only onto base of tail. Underparts grizzled olive-gray. Females have no pouch.
Similar species. Shrewish short-tailed opossums (*M. sorex*) have a gray crown.
Natural history. Found in lowland Atlantic coastal forest.
Geographic range. South America: SE Brazil in Rio de Janeiro and NE Argentina.
Status. Apparently rare.
Local names. Catita.

Shrewish Short-tailed Opossum
Monodelphis sorex
Map 16
Identification. Measurements: HB = 110–130; T = 65–85; HB = 13–16; E = 10–13; WT = 48 g.
Top of head, neck, and forequarters gray finely grizzled with tawny yellowish; cheeks, sides of neck and body, and rump dull rusty; fur short. Tail thinly covered with fine rusty hairs. Feet reddish. Underparts pale orange, demarcated from sides, hairs gray at base on midsection only, scrotum black furred with orange. Females have no pouch. Small, dull-colored opossums.
Similar species. Often confused with southern short-tailed opossums (*M. dimidiata*), a grassland species, known with certainty only from Argentina and Uruguay, that has long fur and whitish or buffy feet; gray short-tailed opossums (*M. domestica*) are gray, with large ears; red-legged short-tailed opossums (*M. brevicau-*

data) have gray top of rump and tail base and are larger; long-nosed short-tailed opossums (*M. scalops*) have a red head.
Natural history. Found in Atlantic coastal rainforests.
Geographic range. South America: SE Brazil, Minas Gerais to Rio Grande do Sul, E Paraguay, and NE Argentina.
Status. Apparently rare.
Local names. Catita.

Gray Short-tailed Opossum
Monodelphis domestica
Plate 3, map 17
Identification. Measurements: HB = 130–191; T = 70–106; HF = 17–22; E = 20–28; WT = 36–98 g.
Upperparts entirely pale gray. Face sometimes with faint dark stripe on center; ears quite large, naked, pale gray-tan. Tail well furred for first 1–2 cm, the rest thinly haired, either completely blackish (E Brazil) or blackish above, paler below (Bolivia, W Brazil). Feet whitish. Underparts dirty white to pale gray with an orange tinge, paler than back. Females have no pouch.
Variation. An animal from Pernambuco, Brazil, has large black patches on neck and rump and black streaks from nose to eyes. A distinctive population on Ilha Marajó, Brazil, is gray to gray-brown, with orange tinge on flanks; it is sometimes considered a separate species, *M. maraxina*.
Similar species. All other short-tailed opossums in range either are striped or have reddish color on body; mouse opossums have long tails, black eye rings, and large ears; mice and rats have four front toes, nonprehensile tails, two large incisors, and no canine teeth.
Natural history. Nocturnal and diurnal; terrestrial; solitary. Feeds mostly on insects and small animals. Gray short-tailed opossums are found in deciduous forest, and around and in houses and agricultural land. Probably not a rainforest species, but a scrub species that has followed man into cleared areas of rainforest regions.
Geographic range. South America: the southern rim of the Amazon Basin, from Bolivia and Paraguay to SE Brazil and north to Ilha Marajó.
Status. Locally common.
Local names. Colicorto gris (Ar); catita (Br); mbicuré-í (Pa).

Map 19

▦ Three-striped short-tailed opossum, *Monodelphis americana*
⬚ Ihering's short-tailed opossum, *M. iheringi*
⬚ Theresa's short-tailed opossum, *M. theresa*

Three-striped Short-tailed Opossum
Monodelphis americana
Plate 3, map 19
Identification. Measurements: HB = 101–105; T = 45–55; HF = 16–18; E = 15; WT = 23–35 g.
Upperparts tawny brown faintly grizzled, rump slightly reddish; back with three prominent black stripes, a central stripe from nose to tail bordered by two lateral stripes from shoulder to tail, central stripe indistinct on face. Ears naked, brown. Tail bicolored, pale below, blackish above. Underparts pale orange-gray. Females have no pouch. Young like adults. **A very small opossum.**
Similar species. Two other three-striped species are present in SE Brazil; they are much smaller but otherwise almost indistinguishable: *M. theresa*, from Rio de Janeiro and Minas Gerais, is tiny (HB = 92), with more whitish underparts; *M. iheringi*, from São Paulo to Rio Grande do Sul (map 19) is miniscule (HB = 77) and has the central black stripe ending between ears. Young of *M. americana* are probably indistinguishable from these in the field. See other short-tailed opossums.
Natural history. Terrestrial; probably diurnal. Feeds on insects; in captivity will voraciously kill and eat small vertebrates. Found in Atlantic coastal forests, usually near water.

Geographic range. South America: Brazil, coastal forests from near Belém south to at least São Paulo.
Status. Apparently rare.
Local names. Catita.

Chestnut Striped Short-tailed Opossum
Monodelphis rubida
Map 18
Identification. Measurements: HB = 125–137; T = 56–60; HF = 15–18; E = 12; WT = 45–46 g.
Upperparts entirely warm red-brown, near chestnut; neck behind ear brighter, almost orange; **back with three faint, darker brown stripes;** fur short. Tail looks naked, body hair does not extend far onto base. Forefeet dusky, hindfeet yellow on inner sides, outer like back. Underparts gray, frosted yellow-white throughout, scrotum black with yellow-white hairs. Females have no pouch.
Variation. This form is sometimes considered a subspecies of the three-striped opossum *M. americana*.
Similar species. Other three-striped short-tailed opossums have distinct black stripes and are tawny brown.
Geographic range. South America: Brazil, Goiáz, Minas Gerais, and São Paulo.
Status. Apparently uncommon or rare.
Local names. Catita.

One-striped Short-tailed Opossum
Monodelphis unistriata
Map 18
Identification. Measurements: HB = 140; T ~ 60.
Upperparts rusty gray; midback with a single chestnut stripe. Sides, feet and underparts yellowish orange. Tail bicolored, brown above yellowish below. Females have no pouch.
Similar species. These are the only short-tailed opossums with a single stripe on back; Atlantic forest mice (*Delomys* spp.) have long tails and large ears.
Geographic range. South America: Brazil, São Paulo.
Status. Apparently rare.
Local names. Catita.

Anteaters, Sloths, and Armadillos (Xenarthra)

This order (sometimes called Edentata) includes four families of mammals so dissimilar externally that they hardly appear related. The characteristics that unite them include simple peglike teeth (when teeth are present), with no development of milk teeth and no true incisors, canines, or premolars; extra joint surfaces between the vertebrae, and other characters of the skeleton; and features of the reproductive tract and circulatory system that are unique among mammals. They are the last living remnants of a large group of species that evolved in South America when it was an isolated island continent (Appendix C). Most living members of this order are specialized feeders that eat mainly ants and termites or rainforest canopy leaves. A general reference is G. G. Montgomery, ed., *The Evolution and Ecology of Armadillos, Sloths, and Vermilinguas* (Washington, D.C.: Smithsonian Institution Press, 1985).

Anteaters (Myrmecophagidae)

The true anteaters have no teeth; they have long, tubelike snouts and fairly small eyes and ears. The tongue can be greatly extended to reach into inaccessible crevices and is covered with sticky saliva which traps insects. The powerful curved front claws are so large that anteaters cannot walk on the soles of their forefeet; the claws are folded inward, and they walk on the outside of the hand, which has a thickened pad. The three arboreal species have strongly prehensile tails. These animals feed on social insects (ants, termites, bees), and their anatomy is specialized for opening the nests of their prey and licking up the occupants. The prehensile tail acts as a fifth paw to anchor the anteater firmly while it uses both forepaws to rip open an arboreal nest. Anteaters give birth to a single young, which rides on the mother's back, clinging tightly to her fur when she travels. Because she has no teeth or fingers, the mother cannot pick up her young; it must crawl up onto her back by itself. There are three genera and four species of true anteaters, all restricted to the New World tropics and subtropics and all found in rainforest.

Giant Anteater
Myrmecophaga tridactyla
Plate 4, map 20
Identification. Measurements: HB = 1,000–1,900; T = 640–900; HF = 150–180; E = 35–50; WT = 22–39 kg.
Upperparts grizzled gray-brown; hairs banded black, brown, and white; **shoulder to chest and neck with wide black stripe bordered with white;** midback with crest of hair increasing in length from crown to tail tip; fur thick, coarse, and dull. **Head greatly elongated, narrow,** and convex; grizzled gray, covered with short, thin, stiff hairs; nose black; ears small, rounded, gray. **Tail long, nonprehensile, very bushy with long, coarse, drooping hair.** Chest grizzled; belly and hindlegs black. **Forelegs white, wrist crossed by black band;** elbows fringed with long hair; **forefeet with three greatly enlarged** and two smaller **claws,** hindfeet with five short

Map 20

Giant anteater, *Myrmecophaga tridactyla*

claws. Young like adults. A large shaggy animal that walks on its knuckles with an ambling gait.
Similar species. Collared anteaters (*Tamandua* spp.) are much smaller and not

31

shaggy; no other large terrestrial mammal has a long, shaggy tail.

Sounds. Usually silent, but adults can roar when disturbed; isolated young give a shrill whistle.

Natural history. Nocturnal and diurnal; terrestrial; solitary. Feeds mainly on ants that are licked out of nests or tunnels opened with the claws. Giant anteaters travel widely and feed from many ant colonies in a day, taking a few ants from each. Normally they walk or amble along, but they are able to gallop with surprising speed. They have a keen sense of smell, which is used to find prey. At rest they bed down on the surface of the ground in an open or sheltered spot. Giant anteaters are normally harmless, but if attacked they may rear up on the hindlegs and slash and grasp with the formidably muscled and armed forelegs; they can kill large enemies. Found in a wide range of habitats from wet or dry grasslands to rainforest.

Geographic range. Central and South America: S Belize and Guatemala to N Ecuador west of the Andes and east of the Andes to N Argentina and Uruguay.

Status. CITES Appendix II. Extinct in parts of its range; widespread, but rare and wantonly killed or captured. Records from Central America are historic or anecdotal, and it is not known if any remain there. Always rare in rainforest; more common and easy to see in grasslands with many ant mounds.

Local names. Oso hormiguero, tamandua de bandera (Ec, Pe); tamandua bandeira, tamanduá-açu, papa-formigas (Br); ant bear (Be); oso caballo (CR, Ho, Pn); yurumí, tamanduá (Gua); zam hool (May); tamanúa (Sar).

References. Shaw, J. H., J. Machado-Neto, and T. S. Carter. 1987. Behavior of free-living giant anteaters (*Myrmecophaga tridactyla*). *Biotropica* 19:255–59.

Collared Anteater or Southern Tamandua

Tamandua tetradactyla
Plate 4, figure 3, map 21

Identification. Measurements: HB = 535–880; T = 400–590; HF = 87–105; E = 47–60; WT = 3.6–8.4 kg.

Head, legs, and hindparts typically pale golden yellow with a black "vest" over the belly, lower back, and forward in a band over the shoulders and around the chest (but see variation below); fur stiff, glossy bristles. **Head long, narrow, and convexly curved;** muzzle to level of eyes naked, blackish; eyeshine almost none, reddish. **Tail long,** thick, **prehensile, base furred, tip naked, with mottled skin. Forefeet with four long, massive, claws;** hindfeet with five smaller claws. Young colored like adults.

Variation. The "vested" coloration described above is typical of animals from the eastern half of Brazil. Those from N Brazil and the Guianas **may be pure blond,** with no vest; animals from central and western Amazonia may be vested, **partially vested,** blond, **or completely black** (fig. 3). The dark parts of the body may be either black, sooty gray, or brown; blond parts may be silvery white to red-gold. Some native peoples believe that blond and black forms are different species; they occur together in parts of Peru.

Similar species. Monkeys have short muzzles; giant anteaters (*Myrmecophaga tridactyla*) have bushy tails and are much larger.

Sounds. No calls are commonly heard, but feeding tamanduas noisily rip and tear apart insect nests and rotten wood, which rains to the forest floor if they are in a tree. Sounds of tearing wood at night almost always lead to a tamandua; by day, to a tamandua or to brown capuchin monkeys.

Natural history. Diurnal and nocturnal; arboreal and terrestrial; solitary. Feeds mainly on ants, termites, and bees extracted after ripping apart their nests with the foreclaws. Tamanduas can be seen foraging on the ground or in the canopy anywhere in the forest, but seem most common beside watercourses and in viny, epiphyte-laden habitats, where their prey may be concentrated. They move slowly and awkwardly on the ground and appear to see poorly. When alarmed they stand upright on their hindlegs, raise the nose to sniff, then climb a tree or amble away. When attacked or cornered they rear up and slash with the foreclaws and can inflict serious wounds. By day in rainforest they are accompanied by a dense cloud of flies and mosquitoes and often brush their eyes with

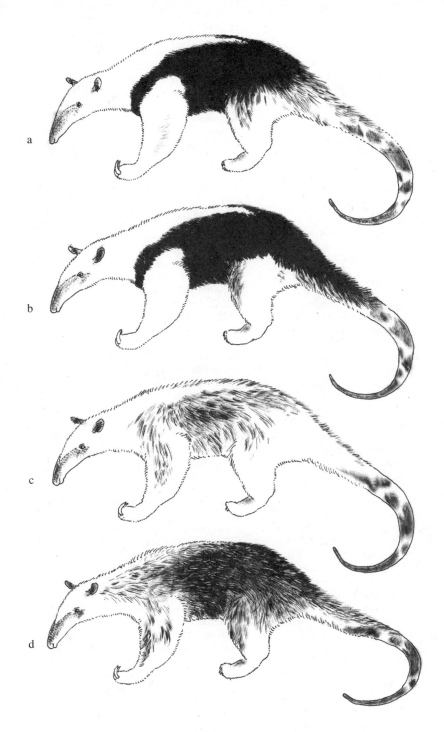

Figure 3. Color patterns of tamanduas: (*a*) **northern tamandua,** *Tamandua mexicana;* (*b–d*) **southern tamandua,** *Tamandua tetradactyla:* (*b*) Brazil, (*c*) Venezuela, (*d*) Colombia. Patterns are neither limited to those areas nor the only ones found there.

Map 21

Northern tamandua, *Tamandua mexicana*
Collared or southern tamandua, *T. tetradactyla*

a forepaw. When not active, tamanduas rest in hollow trees, burrows of other animals, or other natural shelters. Found in many habitats from mature and secondary rainforest and plantations to gallery forest and arid savannas.
Geographic range. South America: east of the Andes, Venezuela south to N Argentina and Uruguay. To about 2,000 m elevation.
Status. CITES Appendix II (SW Br). Widespread, always present in rainforest, but usually uncommon. Hunters often kill tamanduas, claiming that they kill dogs. They are rarely used for food, but their tail tendons are sometimes used to make ropes.
Local names. Oso colmenero, tamanduá (Pe, Ec, Co); tamanduá-colete, mambira, tamanduá-mirim (Br); kaguaré (Gua); termieteneter, mierenfluiter, mirafroiti (Su).
References. Montgomery, G. G. 1985. Movements, foraging and food habits of the four extant species of Neotropical vermilinguas (Mammalia; Myrmecophagidae). In G. G. Montgomery, ed. *The evolution and ecology of armadillos, sloths, and vermilinguas,* 365–77. Washington, D.C: Smithsonian Institution Press.
   Wetzel, R. M. 1982. Systematics, distribution, ecology, and conservation of South American edentates. *Pymatuning Symp. Ecol.* 6: 345–75.

Northern Tamandua
Tamandua mexicana
Figure 3, map 21
Identification. **Externally identical to southern tamandua,** northern tamandua is

distinguished from it by characters of the skull. **All individuals are black vested.**
Similar species. See southern tamandua.
Natural history. Similar to southern tamandua.
Geographic range. Central and South America: SE Mexico south through all of Central America, and South America west of the Andes from N Venezuela to N Peru.
Status. CITES Appendix III (Gu). Threatened by habitat destruction in much of its range.
Local names. Oso hormiguero común (Span); ant bear (Be); tamanduá (Co): brazo fuerte (Me); oso mielero (CR); chab (May).
References. Montgomery, G. G., and Y. D. Lubin. 1977. Prey influences on movements of Neotropical anteaters. In R. L. Phillips and C. Jonkel, eds., *Proceedings of the 1975 Predator Symposium,* 103–31. Missoula, Mont.

Silky or Pygmy Anteater
Cyclopes didactylus
Plate 4, map 22
Identification. Measurements: HB = 154–205; T = 163–225; HF = 25–38; E = 6–15; WT = 155–275 g.
Upperparts smoky gray-gold with silver iridescence, midback usually with dark brown stripe from shoulder to rump; **fur long, soft, silky,** and slightly wavy. **Head usually gold; ears tiny, buried in fur; button eyes round and black; nose elongated but blunt, lower jaw protruding slightly,** chin prominent; nose, lips, palms, and soles pink. **Tail long, tapered, prehensile, fully furred to tip dorsally, naked for last third of underside,** tip sometimes whitish. Legs gold or gray; feet with large naked pads, hindfoot with four long claws, **forefoot with one huge and one smaller claw.** Underparts gray or buff, midline usually with dark brown stripe; chest sometimes dark brown or with brown spot. Young like adults; but center of back often much darker and fur longer. **A beautiful tiny golden anteater;** several of its local names translate to "little angel."
Variation. Animals from Mexico are entirely orange-gold, with shorter fur; those from the western Amazon Basin may lack the middorsal stripe; those from N Venezuela and Trinidad have brown chests.
Similar species. No other small, long-

Map 22

 Silky or pygmy anteater, *Cyclopes didactylus*

furred animals have long claws on the fore-feet. Woolly oposums (*Caluromys* spp.) have large ears and naked tail tips; black-shouldered opossums (*Caluromysiops ir-rupta*) have large ears and black shoulders; small monkeys have flat faces and nonpre-hensile tails.

Sounds. Said to make a soft whistling sound.

Natural history. Nocturnal; arboreal; soli-tary. Feeds mainly on ants, but also on oth-er insects. Silky anteaters travel and feed above the ground on small stems and lian-as, which they grasp in a tight fold of the hind sole. They climb slowly, with two feet always in contact with the branch. To feed, they slit open an ant-filled hollow stem with the large foreclaw and lick out the ants from within. By day they rest curled into a tight ball, usually in a tangle of lianas. Young are left in a tree at night while the mother forages. They nurse on their mother's milk until they are old enough to forage for ants on their own. Adult females have large territories; a male territory includes those of several females. They defend themselves by pinching part of the attacker with the foreclaw. Found in lowland rainforests; the distribution in dif-ferent forest types is not documented.

Geographic range. Central and South America: SE Mexico south to SE Bolivia and E Brazil.

Status. Unknown, widespread. Not hunt-ed. These animals are rarely seen but may not be uncommon.

Local names. Tamandua-í (Br); ceibita (CR); angelito (Co); serafín (Ec, Pe); flor de balsa (Ec); tapacara, gato balsa (Pn).

References. Best, R. C., and Y. Harada. 1985. Food habits of the silky anteater (*Cyclopes didactylus*) in the central Ama-zon. *J. Mammal.* 66:780–81.

Montgomery, G. G. 1983, *Cyclopes di-dactylus.* In D. H. Janzen, ed., *Costa Ri-can natural history,* 461–63. Chicago: University of Chicago Press.

Sloths (Bradypodidae, Megalonychidae)

Dental formula: 5 simple, peglike teeth on each side in the upper jaw and 4 in the lower = 18; no true canines or incisors: the large, bladelike "canines" of two-toed sloths are actually derived from other teeth. Feet with no free toes, but two or three long, curved claws that form a hook by which sloths can hang passively from a branch or clasp objects against the palm with a pincerlike grip. Sloths have long limbs, short bodies, and stumpy tails. To accommodate their upside-down lifestyle, the fur grows from the belly toward the back, and the head can rotate over 90 degrees. They feed chiefly on forest canopy leaves, which they digest by bacterial fermentation in a many-chambered stomach. Their hair has microscopic grooves and notches which provide a home for greenish algae that camouflage the fur among the canopy leaves. These algae in turn provide food for partic-ular species of moths. Sloths move slowly and little, and they spend much time resting. The single young spends its first 6 to 9 months clinging to its mother, usually to her chest. Their silence, immobility, and camouflaged fur make sloths extremely difficult to see from the ground, and they are usually much more common than they seem. Eagles, however, are able to find them, and they prey on them extensively, as do jaguars in some localities. Recent studies suggest that the two genera of sloths are only distantly related: the two-toed sloths belong to the family of the giant ground sloths (Megalonychidae), while the three-toed sloths belong to the Bradypodidae (where the two-toed sloths were formerly placed). The two genera and five species of sloths are restricted to New World tropical rainforests.

Brown-throated Three-toed Sloth
Bradypus variegatus
Plate 4, map 23

Identification. Measurements: HB = 400–750; T = 38–90; HF = 90–180; E = 8–22; WT = 2.3–5.5 kg.

Upperparts usually **pale brown with large patches of dirty white** concentrated on **lower back** and hindlegs; **fur long, coarse and wavy, shaggy** except on face, tinged greenish. **Head small and round, ears not visible, face whitish or brownish, often with darker band across brow; eye surrounded by a black stripe** that extends back toward ear; mouth slightly smiling; nose black. **Throat and chest brown. Tail short** and stumpy. **Limbs long; feet each with three, long, curved, hooklike claws.** Back of males between shoulders has large patch of short, orange fur with brown stripe down middle (the speculum). Young like adults. Usually hanging upside down by its claws or sitting in a tree fork with head between forelegs.

Variation. Color varies from very pale, almost whitish in some Central American animals to dark red-brown.

Similar species. Pale-throated three-toed sloths (*B. tridactylus*) have pale throat; two-toed sloths (*Choloepus* spp.) have no pale patches on back and no tail; all other large arboreal mammals have longer tails.

Sounds. None usually heard, but can emit a shrill whistle.

Natural history. Nocturnal and diurnal; arboreal; solitary. Feeds on the leaves of many species of trees. Three-toed sloths usually feed high in the forest canopy, where they are very difficult to see. They prefer trees with crowns exposed to the sun and may sometimes be spotted on exposed branches as they warm themselves in the early morning sunlight. On the ground they are helpless and virtually unable to walk, but they can swim and are sometimes found crossing rivers. They have the astonishing behavior for an arboreal animal of descending to the ground to defecate: while clinging to a tree trunk they dig a hole with the tail, defecate in it, and cover it over. The necessity arises only about once a week. Found in both evergreen and dry forests, and even in isolated trees in pastures. Sloths occur throughout primary and secondary forests, but they are most easily

Map 23

▦ Brown-throated three-toed sloth, *Bradypus variegatus*

☰ Pale-throated three-toed sloth, *B. tridactylus*

■ Maned three-toed sloth, *B. torquatus*

seen in low-canopy second-growth habitats. People often pick them up from roadsides and later release them in city gardens and parks, where they can thrive.

Geographic range. Central and South America: Honduras south, west of the Andes to S Ecuador, east of the Andes from N Venezuela and Colombia to Bolivia and N Argentina, except area east of the Rio Negro and north of the Amazon. To at least 1,100 m elevation.

Status. CITES Appendix II. Widespread and common; unlikely to be threatened except where habitat is destroyed. Sloths are hunted widely for meat, but they are difficult to see, which both protects them from extermination and makes it difficult to establish their status in any forest. When forests are flooded by dams or cut, sloths are often found in enormous numbers.

Local names. Perezoso de tres dedos (Span); Preguiça-de-bentinho (Br); aí aí (Qui, Gua); q'oral (May).

References. Montgomery, G. G., and M. E. Sunquist. 1978. Habitat selection and use by two-toed and three-toed sloths. In G. G. Montgomery, ed., *The ecology of arboreal folivores*, 329–59. Washington, D.C.: Smithsonian Institution Press.

Pale-throated Three-toed Sloth
Bradypus tridactylus
Plate 4, map 23
Identification. Measurements: HB = 445–554; T = 31–75; HF = 90–140 (cu); E = 10–15; WT = 3.3–6 kg.
Upperparts grizzled, smoky gray-beige or chocolate with large cream- or buff-tinged splotches usually concentrated on lower back and rump; top of head and foreparts more uniform, darker tan or brown than hindparts; fur long, shaggy, wavy, tinged green in wild. Head conspicuously round; face flat, with short hair, ears not visible; forehead pale orange, yellow, or whitish in a band from brow to behind ear, continuous with pale cheeks and throat; eyes in the center of dark stripes; mouth slightly smiling; nose black. Tail short and stumpy. Limbs very long, fore and hindfeet with three large curved claws. Back of males with large patch of short, orange fur with dark brown stripe down center. Young grizzled, shaggy, gray-tan like adults. Body shape and posture like brown-throated three-toed sloth.
Variation. Color variable; dark parts from pale tan to chocolate, pale parts from dirty white to orange.
Similar species. See brown-throated three-toed sloth (*B. variegatus*).
Sounds. None usually heard, but can emit a shrill whistle.
Natural history. Probably diurnal and nocturnal; arboreal; solitary. Feeds on leaves, usually high in the forest canopy. Most knowledge of the behavior of three-toed sloths is derived from the brown-throated sloth; the pale-throated sloth probably has similar habits. These sloths seem most numerous in dense forests of second-growth trees or viny disturbed habitats, where there is a good food supply of young leaves and the sloth's main enemy, the harpy eagle, is often absent. Found in mature and secondary lowland rainforest.
Geographic range. South America: E Venezuela, the Guianas, and NE Brazil north of the Amazon and east of the Rio Negro, and in a narrow band south of the Amazon from the Rio Negro to Belém.
Status. Widespread and sometimes common; hunted for meat; see brown-throated three-toed sloth.
Local names. Perezoso de tres dedos (Span); Preguiça-de-bentinho (Br); driteen-luiaard, sonlori (Su); aí (Gua, Br).
References. Wetzel, R. M., and F. de Avila-Pires. 1980. Identification and distribution of the recent sloths of Brazil (Edentata). *Rev. Brasil. Biol.* 40:831–36.

Maned Three-toed Sloth
Bradypus torquatus
Plate 4, map 23
Identification. Measurements: HB = 450–500; T = 48–50; HF = 100–115; WT = 3.6–4.2 kg.
Head and body uniform grizzled tan (a mixture of tan and cream hairs); back of neck from nape to shoulder with pitch black "mane" of hairs to 15 cm long, projecting as plumes over shoulders; fur long, coarse, wavy, and shaggy. Face and chin brown. Shape like other three-toed sloths. No middorsal speculum on male.
Similar species. Other sloths lack black mane; all other large arboreal mammals have long tails.
Natural history. From Atlantic coastal forests.
Geographic range. South America: Brazil, in the remaining fragments of coastal forest in Bahía, Espírito Santo, and Rio de Janeiro.
Status. US-ESA endangered. Extinct on much of its former range because of deforestation.
Local names. Preguiça-preta, aí-pixuna (Br); aí-igapó (TGua).
References. See pale-throated three-toed sloth.

Hoffmann's Two-toed Sloth
Choloepus hoffmanni
Plate 4, map 24
Identification. Measurements: HB = 520–700; T = 14–30; HF = 100–150; E = 20–37; WT = 4.5–8.1 kg.
Upperparts completely tan, legs or forequarters sometimes brown; fur long, coarse, and wavy, slightly greenish in wild. Throat pale, contrasting with chest. Head round, often paler than body, muzzle protruding, brown; face pale, with short hair; ears not visible; anteriormost teeth long and sharp, caninelike. Limbs long; hindfeet with three, forefeet with two long, curved claws. Tail not visible. Young unlike adult: dark brown, with

Map 24

Hoffmann's two-toed sloth, *Choloepus hoffmanni*
Southern two-toed sloth, *C. didactylus*

short, woolly fur. Usually hanging upside down by its claws.

Variation. Color varies from brown to very pale, almost whitish. Central American animals have dark brown eye rings.

Similar species. Difficult to distinguish in the field from southern two-toed sloths (*C. didactylus*), which have a brown throat; three-toed sloths (*Bradypus* spp.) have large pale splotches on back, visible tail, whitish brow, and three claws on front foot. All other large arboreal animals have longer tails.

Sounds. None usually heard, but hisses in defense; low bleats in distress.

Natural history. Nocturnal; arboreal; solitary. Feeds on leaves and fruit. Two-toed sloths favor trees covered with many vines and with crowns exposed to sun. They are often high in the forest canopy and very difficult to see. At night they sometimes descend to hang in lianas of the understory, where they are more visible. Like three-toed sloths, they also descend to the ground level to defecate, but they do not dig a hole. These sloths do not attain such high numbers as three-toed sloths in secondary and disturbed habitats, but they are seen as often in primary Amazon Basin forest. More aggressive than three-toed sloths, they will defend themselves by slashing with the foreclaws and biting. Found in mature and secondary rainforests and deciduous forests.

Geographic range. Central and South America: Nicaragua south, west of the Andes to NW Ecuador; west of the Cordillera Oriental in Venezuela; and a disjunct population in S Peru, W Brazil, Bolivia, and likely in adjacent Brazil. To 1,800 m elevation.

Status. CITES Appendix III (CR). Status unknown, widespread, hunted for meat.

Local names. Perezoso de dos dedos (Span); preguiça real, unau (Br); perico ligero (Co); cucala (Ho); aí (TGua); pelejo (Pe); intillama (Qui).

References. Wetzel, R. M. 1985. The identification and distribution of recent Xenarthra (= Edentata). In G. G. Montgomery, ed., *The evolution and ecology of armadillos, sloths, and vemilinguas*, 5–21. Washington, D.C.: Smithsonian Institution Press.

Montgomery, G. G., and M. E. Sunquist. 1978. Habitat selection and use by two-toed and three-toed sloths. In G. G. Montgomery, ed., *The ecology of arboreal folivores*, 329–59. Washington, D.C.: Smithsonian Institution Press.

Southern Two-toed Sloth
Choloepus didactylus
Plate 4, map 24

Identification. Measurements: HB = 462–860; T = 14–33; HF = 110–170; E = 20–35; WT = 4.1–8.5 kg.

Generally **the same as Hoffmann's two-toed sloth, except throat the same color as upper chest;** fur often brown with long, cream tips, looks variegated. Face often same color as body. Legs often darker brown than body.

Variation. An animal from Colombia has a white throat.

Similar species. See Hoffmann's two-toed sloth (*C. hoffmanni*).

Sounds. None usually heard in field.

Natural history. Nocturnal; arboreal; solitary. Habits probably similar to Hoffmann's two-toed sloth. Found in mature, disturbed, and secondary rainforest.

Geographic range. South America: east of the Andes, E Venezuela, and the Guianas south to Ecuador and Peru and the Amazon Basin of Brazil, reaching south of, but near, the main river.

Local names. Same as Hoffmann's two-toed sloth; tweeteenluiaard, skapoloiri (Su).

References. See Hoffmann's two-toed sloth.

Armadillos (Dasypodidae)

Dental formula: no incisors or canines and seven or more small, peglike teeth on each side of each jaw. There are three to five toes on the forefeet and five toes on the hindfeet. Armadillos have the dorsal surface of the body covered with bony armor plates that shield the head, back and sides, and sometimes the legs and tail. Around the center of the body the armor is arranged into rows or "bands" of plates separated by soft skin; this allows the animal to bend its body. The number of these bands is used to distinguish between some species. The back is smoothly rounded and the legs are short and strong, with stout claws on the toes. The belly is soft and naked. Most species have little or no hair, but one montane species has dense fur covering the armor. "Hairy" armadillos have a thin sprinkling of long, stiff bristles. Armadillos are mainly insectivorous; they feed chiefly on ants and termites, but they also eat many other kinds of small animals, carrion, and sometimes plant matter. With their broad, inflexible backs and short legs, armadillos trot with a rolling or scuttling gait, some like windup toys, snuffling and grubbing with their noses and forepaws and seemingly unaware of anything more than a foot or two away. They have a good sense of smell but poor eyesight and will often run right into a person standing still in their path. All species seem to sleep and raise their young in burrows that they dig themselves; the burrow of each species has a characteristic size and shape. An armadillo burrow can be recognized by its smooth, dome-shaped roof, polished by and fitting its owner's carapace. The litter size is 1–12 young; in some long-nosed armadillos each litter is a set of four identical quadruplets formed from a single egg. There are 8 genera and 20 species, all in the New World, and 4 genera and 8 species in the lowland rainforest.

Yellow Armadillo
Euphractus sexcinctus
Plate A, map 25
Identification. Measurements: HB = 401–495; T = 119–241; HF = 78–92; E = 32–47; WT = 3.2–6.5 kg.
Upperparts pale yellow, tan, or reddish tan; **head and body covered with armorplates; sparsely but prominently covered with long, stiff, whitish hairs;** carapace at center of body with six or seven movable bands. **Head triangular from in front, broad between eyes and ears; ears short, not protruding above crown, widely separated by more than length of ear on top of head.** Tail moderately long, cylindrical, with two or three rings of scales near base. Feet with five toes; claws not greatly enlarged.
Similar species. Long-nosed armadillos (*Dasypus* spp.) have ears set close together and long narrow muzzle; naked-tailed armadillos (*Cabassous* spp.) have large ears and greatly enlarged foreclaws.
Natural history. Mostly diurnal; terrestrial; solitary. Feeds on a wide range of foods including much plant material, ants, other insects, small vertebrates, and carrion. Active yellow armadillos trot rapidly about, often stopping to dig up a food item. They den in burrows with a single entrance

Map 25

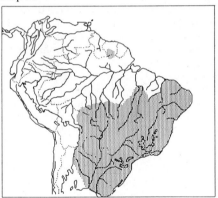

Yellow armadillo, *Euphractus sexcinctus*

about 19 cm high and 21 cm wide. Unlike most other armadillos, these will bite when handled. This is mainly a species of savannas, brushy grasslands, cerrados, chaco, gallery forests, and forest edges, but there are rare reports of its occurrence deep in the lowland rainforest.
Geographic range. South America: savannas of S Surinam, and Pará, Brazil, south through SE Brazil and most of the Rio Paraguay Basin in Matto Grosso, Brazil, Paraguay, NE Argentina, and Uruguay. Distri-

bution in rainforest poorly known.
Status. Locally common, hunted for meat.
Local names. Gualacate (Ar); tatu peba, tatu peludo (Br); peludo (Ur); tatú-podyú (Gua).
References. Redford, K. H., and R. M. Wetzel. 1985. *Euphractus sexcinctus.* Mammalian Species, no. 252.

Southern Naked-tailed Armadillo
Cabassous unicinctus
Plate A, map 26
Identification. Measurements: north of Amazon: HB = 347–445; T = 165–200; HF = 70–84; E = 30–40; WT = 2.5–3.6 kg; south of Amazon: HB = 290–345; T = 87–140; HF = 65–76; E = 25–30; WT = 1.6–4.8 kg; females larger than males.
Upperparts naked, **covered with bony armor; dark gray with sharply demarcated yellow border on lower edge of carapace;** midbody with 10–13 **movable bands (not sharply demarcated);** scales on body and movable bands alike, squarish and fingernail-sized. Head broad, blunt-nosed; **ears large, rounded, funnellike,** can fold back to cover meatus, often frayed at edges, **set widely apart on crown, armor shield extending between ears** back to cover nape. **Tail long and narrow, appears naked** of armor (some small inconspicuous scales), gray usually with a paler tip. Feet with five pale claws, **central claw of forefoot greatly enlarged.** Underparts completely naked, tan.
Variation. The race found south of the Amazon in Brazil (*C. u. squamicaudis*) is much smaller than that north of the river; the two hybridize south of the river mouth.
Similar species. Giant armadillos (*Priodontes maximus*) are much larger, with huge hindfeet; long-nosed armadillos (*Dasypus* spp.) have ears almost touching each other at base; yellow armadillos (*Euphractus sexcinctus*) have short ears, white hairs protruding between scales of back and face, and small front claws.
Sounds. Piglike grunts when handled.
Natural history. Nocturnal; terrestrial; solitary. Feeds on ants and termites. Built for heavy-duty digging, naked-tailed armadillos move slowly. Burrows have a single small round entrance about 16 cm diameter, usually dug into flat ground in an old termite

Map 26

░ Southern naked-tailed armadillo, *Cabassous unicinctus*
≡ Northern naked-tailed armadillo, *C. centralis*

mound or into a bank. A burrow is usually used only for one night; it has a strong musky odor when the animal is in residence. Southern naked-tailed armadillos are extremely rare in rainforest, where their habits are unknown. Found in a wide range of habitats from rainforest to grassland.
Geographic range. South America: east of the Andes, N Colombia and Venezuela south to Matto Grosso and Minas Gerais in Brazil.
Status. Unknown; apparently uncommon or rare everywhere.
Local names. Tatu-rabo-de-couro, cabassú, tatu-de-rabo-mole (Br); lugubre, tatu iba (Ec); cabassou (FG); tatuaí (Gua).
References. Carter, T. S., and C. D. Encarnaçao. 1983. Characteristics and use of burrows by four species of armadillos in Brazil. *J. Mammal.* 64:103–8.
Wetzel, R. M. 1980. Revision of the naked-tailed armadillos, genus *Cabassous* McMurtrie. *Ann. Carnegie Mus.* 49:323–57.

Northern Naked-tailed Armadillo
Cabassous centralis
Map 26
Identification. Measurements: HB = 305–400; T = 130–183; HF = 60–74; E = 31–37; WT = 2–3.5 kg.
Similar to southern naked-tailed armadillo, but smaller. At night, naked tail appears pale. Walks on the tips of its foreclaws, with hindfeet pigeon-toed.
Similar species. Nine-banded armadillos

(*Dasypus novemcinctus*) have ears set close together on head and tail ringed with armor.
Sounds. Growls or squeals when restrained.
Natural history. Nocturnal; terrestrial; solitary. Feeds mainly on ants and termites. These armadillos move slowly; a characteristic view is the distinctive pale, naked tail disappearing under a log. When burrowing, they rotate the body from side to side, so that the forward-projecting edge of the carapace acts as an auger. Naked-tailed armadillos are strong smelling, and if picked up they urinate copiously and sometimes defecate, twirling the tail. Found both in rainforest and drier habitats, often in rocky areas.
Geographic range. Central and South America: Honduras south to NE Colombia on both sides of the Andes, and NW Venezuela west of Cordillera Oriental.
Status. CITES Appendix III (CR). Appears rare everywhere.
Local names. Armado de zopilote (CR); tumbo armado (Ho); armadillo rabo de puerco, morrocoy (Pa).
References. Merritt, D. A., Jr. 1985. Naked-tailed armadillos, *Cabassous* sp. In G. G. Montgomery, ed., *The evolution and ecology of armadillos, sloths and vermilinguas,* 389–91. Washington, D.C.: Smithsonian Institution Press.

Giant Armadillo
Priodontes maximus
Plate A, map 27
Identification. Measurements: HB ~ 750–1,000; T ~ 500–550; HF ~ 180–190 ; E ~ 47–59; WT ~ 30 kg.
Upperparts naked and **covered with bony armor; carapace looks** several sizes **too small, does not cover lower sides or legs;** carapace gray with pale yellow border on lower edge, but color often obscured by clay from digging. **Head small, ears set very wide apart,** with armor plates between; snout conical. Tail long and tapered, covered with small pentagonal scales. Underparts naked, pinkish tan. **Legs and feet,** especially hind, **enormous; forefoot with greatly enlarged central claw.**
Similar species. No other armadillo approaches this size: naked-tailed armadillos (*Cabassous* spp.) are similar but much

Map 27

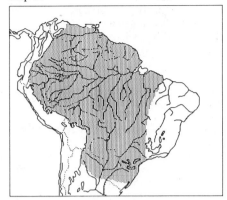

Giant armadillo, *Priodontes maximus*

smaller, have carapace covering sides, and normal-sized legs; great long-nosed armadillos (*Dasyspus kappleri,* often locally called "giant") have ears set close together, carapace covering sides, rings of armor on tail, and normal-sized legs.
Natural history. Nocturnal; terrestrial; solitary. Feeds mainly on ants and termites, which it obtains by digging into their nests, and other insects. Giant armadillos use their great weight to dig in firm ground: they rock back onto their outsize hindfeet and throw the head and forequarters violently forward, claws extended as if diving, and ram them with all their weight into the hole. They then rock backward, raking the loosened dirt with the forefeet to the hindfeet, which simultaneously kick the dirt behind the body. When disturbed, they raise their forequarters off the ground, sit balanced on the huge hindfeet and tail, and lift the snout to sniff the air. The burrows have one entrance, 45 cm wide by 30 cm high, usually dug in a flat termite mound. Found in many habitats from rainforest to grassland.
Geographic range. South America: east of the Andes from N Venezuela and Colombia south to N Argentina, Paraguay, and S Amazon Basin of Brazil.
Status. US-ESA endangered, CITES Appendix I. Rare and endangered, extinct from overhunting for meat in parts of its range, including most areas near human settlements and savannas where they are unable to hide.

Local names. Tatu carrera (Ar); tatu canastra, tatu-açu (Br); yungunturu, carachupa maman (Pe); tatu gigante (Ec); tatu carreta (Pa); reuzengordeldier, granmankapasi (Su); tatú-guazú (Gua); tatou géant (FG).
References. Clark, T. S., and C. D. Encarnaçao. 1983. Characteristics and use of burrows by four species of armadillos in Brazil. *J. Mammal.* 64:103–8.

Nine-banded Long-nosed Armadillo
Dasypus novemcinctus
Plate A, map 28
Identification. Measurements: HB = 356–573; T = 245–450; HF = 75–110; E = 35–57; WT = 2.7–6.3 kg.
Upperparts hairless, **covered with bony armor; usually nine (8–10), movable bands around midbody;** scales on movable bands narrow triangles, scales on rest of armor small and roundish; upperparts usually gray, paling gradually to yellowish sides, but color often hidden by clay from burrowing. **Muzzle long, narrow; ears large, set close together almost touching at their bases,** with no armor between. **Tail long,** tapered, **armored with distinct rings for first 60%,** rings often with contrasting pale border. Hindfeet with five claws; forefeet with four; central claws somewhat enlarged. Underparts almost hairless, pinkish yellow.
Similar species. Great long-nosed armadillos (*D. kappleri*) are larger, with projecting scales on knees; seven-banded long-nosed armadillos (*D. septemcinctus*) have six or seven bands around midbody; other armadillos have ears wide apart at base, with armor between.
Sounds. No calls usually heard, snuffles while foraging; one of the noisiest travelers in the forest: treads heavily on dead leaves and shoves through the undergrowth, a fast-moving nocturnal crashing sound is generally a long-nosed armadillo.
Natural history. Chiefly nocturnal, sometimes diurnal; terrestrial; solitary. Feeds mostly on ants, termites, and other insects, but will eat many kinds of small animal prey, carrion, and some fruit, fungus, and other plant material. Nine-banded armadillos are found throughout the forest but most often seen in thickets and dense vegetation on sloping, well-drained terra firme; rarely on flat plains subject to flooding. Small diggings in the soil show where an

Map 28

Nine-banded long-nosed armadillo, *Dasypus novemcinctus*

armadillo has been foraging. They walk or trot rapidly, often using well-worn pathways. They see poorly and will run into a stationary person at night. When alarmed they explode crashing away through the undergrowth for a short distance or, if surprised at close range, may make a high vertical leap into the air. They dig burrows with several entrances of about 20 cm diameter, usually in a bank or slope. Well-packed, narrow runways radiate from a burrow throughout the surrounding area, especially along steep hillsides. This is the most commonly seen armadillo. Found in a wide range of mature and secondary habitats from deep rainforest to grassland and dry scrub.
Geographic range. North, Central, and South America: S United States through Central America west of the Andes to N Peru, east of the Andes to Uruguay and Argentina, including the islands of Grenada, Margarita, Trinidad, and Tobago. To at least 1,500 m elevation.
Status. Common and widespread; hunted extensively for its excellent meat, which is often an important food source; sometimes scarce in areas populated with subsistence hunters, but seems to withstand heavy hunting.
Local names. Armadillo (Span); dilly (Be); tatu-galinha (Br); cachicame (Co); mulita (Ec); tatou à neuf bandes (FG); cusuco, pitero (Ho); tochi (Me): carachupa (Pe); negengordelig gordeldier, gewone kapasi, diki-diki (Su); tatueté (Gua).
References. Wetzel, R. M., and E. Mon-

dolfi. 1979. The subgenera and species of long-nosed armadillos, genus *Dasypus* L. In J. F. Eisenberg, ed., *Vertebrate ecology in the northern Neotropics*, 43–63. Washington, D.C.: Smithsonian Institution Press.

Redford, K. H. 1985. Food habits of armadillos (Xenarthra: Dasypodidae). In G.G. Montgomery, ed., *Evolution and ecology of sloths, armadillos and vermilinguas*, 429–37. Washington, D.C.: Smithsonian Institution Press.

Great Long-nosed Armadillo
Dasypus kappleri
Plate A, map 29

Identification. Measurements: HB = 510–575; T = 406–456; HF = 110–135; E = 48–55; WT = 8.5–11.8 kg.
Similar to nine-banded armadillo, but much larger; seven or eight movable bands at midbody. Hind knees with two rows of long, downward-projecting, spurlike scutes, bordered by other scale rows that do not protrude. Often covered with clay from burrowing.
Similar species. Giant armadillos (*Priodontes maximus*) have small, widely spaced ears, thick hindlegs, huge claws on forefeet, and undersized carapace; all other armadillos are smaller; this is the only species with projecting scutes on knees.
Sounds. Travels noisily, like nine-banded armadillo.
Natural history. Nocturnal; terrestrial; solitary. Feeds on ants, termites, and a wide variety of other insects; there is a record of one eating a caecilian. These armadillos seem to favor the vicinity of water. They den in burrows with more than one entrance, dug into steep banks on stream edges; the floor of the entrance may be below water level. Networks of well-worn paths branch out from the burrow. Their behavior when seen is much like that of nine-banded armadillos. The spurs on the hindlegs may allow these armadillos to crawl on their knees into narrow tunnels or to wedge themselves in. Confusing local names often translate to "giant" armadillo. Found only in lowland rainforest.
Geographic range. South America: east of the Andes, Colombia, and S Venezuela to N Bolivia and N Pará, Brazil.
Status. Populations seem patchy, quite common in some areas, rare or absent in

Map 29

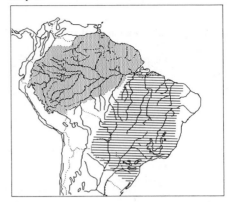

░░ Great long-nosed armadillo, *Dasypus kappleri*
▤ Seven-banded long-nosed armadillo, *D. septemcinctus*

others; hunted for meat.
Local names. Tatu, tatu quinze quilos, tatu canastra (Br); carachupa (Pe); maka kapasi (Sar).
References. Barreto, M., P. Barreto, and A. D'Alessandro. 1985. Colombian armadillos: Stomach contents and infection with *Trypanosoma cruzi*. *J. Mammal.* 66:188–93.

Seven-banded Long-nosed Armadillo
Dasypus septemcinctus
Plate A, map 29

Identification. Measurements: HB = 240–305; T = 125–170; HF = 45–75; E = 30–38; WT = 1.5 kg.
Externally similar to nine-banded armadillo, except: much smaller, six or seven movable bands around midbody, more dark than pale areas on sides, and tail relatively shorter.
Similar species. See nine-banded armadillo (*D. novemcinctus*).
Sounds. Clicks, growls, grunts, and squeals have been noted in captivity.
Natural history. Most records are from dry habitats, but a few seem to be from rainforest; most of range is outside of rainforest regions.
Geographic range. South America: eastern Amazon Basin from the mouth of the Amazon south to N Argentina, and west to Matto Grosso, Brazil.
Status. Apparently rare.
Local names. Tatuí (Br).
References. See nine-banded armadillo.

Bats (Chiroptera)

Bats are the only mammals that can fly. Their forelimbs are true wings that are flapped in powered flight. Their wing membranes are made of remarkable skin that is strong, soft, flexible, and can heal if punctured. The skin is amazingly elastic: when a bat closes its wings, the membrane does not fold but contracts like a sheet of rubber, yet the wing opens with almost no force required to stretch it. All New World bats belong to the suborder Microchiroptera. These bats all use a kind of sonar, or echolocation. High-frequency sounds are emitted through the mouth or nose, and the returning echoes inform the bat about the characteristics of nearby objects. This enables bats not only to navigate in the complete darkness of caves and low light of forest understory, but also to home in on flying insects. Most echolocation sounds are ultrasonic (above the human hearing range), but people with good hearing can hear a faint ticking sound as a bat flies close by. A few species that feed on large insects echolocate with audible sounds. Members of different families are often specialized for different styles of echolocation, with distinctive configurations of the ears to gather the sounds and structures around the mouth and nose to emit them. Most echolocating bats also see quite well, especially in dim light. Despite these senses, bats flying down a familiar path may not be paying attention or from curiosity may fly close to people and may bump them or brush them with their wings; a bat will not get tangled in human hair (an old superstition), and it will not bite spontaneously. But like many animals, a bat will bite in its own defense if it is restrained in the hand. Relatively few species of bats in Neotropical rainforest have ever been found to carry rabies, but there is a risk, especially in populated regions with many cattle and dogs. The best policy is either not to handle live bats or to use heavy gloves when doing so. Bats are beneficial; they eat enormous quantities of insects, pollinate many important plants such as balsa, chicle, and calabash, and help regenerate the forest by dispersing seeds. In numbers of species, bats are the most important order of mammals in Neotropical rainforests: in a given forest there are generally as many species of bats as there are of all other species of mammals combined; 39% of all mammal species in the region are bats. Worldwide, bats are the second largest order of mammals, with about 950 species.

Sheath-tailed Bats (Emballonuridae)

Dental formula: I1/3, C1/1, P2/2, M3/3 = 32. The New World members of this family are all small (3–23 g). They have plain faces with no elaborate structures or warts. The nose is simple, the muzzle a little elongated and upturned, with nostrils slightly tubelike and pointing forward, the snout extending well beyond the lower lip. The ear is a simple, broad triangle, slightly pointed or rounded at the tip, often deeply ridged with parallel folds on its inner, rear half and edge; the tragus is a short, simple stub. The eye is quite large and almond-shaped, coming to a gradual point in the inner corner. The tail is sheathed in the tail membrane for most of its length, but the tip sticks out freely above from the middle of the membrane. The extended tail membrane is as long as or longer than the legs. These bats all have long, fine lax fur. In many species, either just males or both sexes have a moist, glandular pocket in the wing membrane forward of the elbow (propatagium). Bats of this family are all aerial insectivores: New World species all have a slow, fluttering, butterflylike flight and tend to forage in "beats," repeating the same flight path again and again, night after night. They do not have as strong an aversion to daylight as do most bats, and many species roost in open, well-lit places. They often fly when there is some daylight at dusk or dawn. The roosting posture of bats of this family is distinctive: the bat hangs head downward from a vertical or slanted surface, with the feet spread apart and the front of the body supported against the surface by the thumbs on the tips of the folded wings that are held out at an angle to the body (plate 5); the head is raised up and facing straight outward; the tail membrane is contracted and the tail protrudes; the roosting bat often rocks from side to side if disturbed. The family is worldwide, tropical and subtropical, and includes about 48 species, 18 of them in the New World.

Long-nosed Bat
Rhynchonycteris naso
Plates B, 5; map 30
Identification. Measurements: HB = 37–53; T = 9–19; HF = 7–9; E = 10–14; FA = 35–41; WT = 3–7 g.

Upperparts brown heavily frosted or grizzled with dirty white, brown base of fur showing through gives slightly variegated look; **lower back with two indistinct wavy white lines; forearm with little white tufts of hair evenly spaced along wingbone; wings and tail membrane brown. Nose elongated, proboscislike, extending far beyond mouth.** Tail membrane longer than legs, calcar about three times as long as foot; membrane thinly haired for about length of tail on upper side. Underparts whitish, hair brown at base.

Variation. Lines on back may be prominent or almost invisible; color varies from tawny brown to whitish frosted.

Similar species. These are the only little bats with an elongated nose: see white-lined sac-winged bats (*Saccopteryx* spp.).

Natural history. Feeds on tiny insects (about 2 mm). Long-nosed bats forage over water, such as rivers, streams, lakes, ponds, and marshes. They are one of the most commonly seen and locally well known bats. They roost in the open over or beside water, on the lower side of a large branch, fallen tree, leaning tree trunk, steep face of a bank, or below a bridge. They roost in groups of 3–45, usually about 12, and hang in a characteristic straight line, with each bat spaced 2–4 cm from its neighbors. The frosted fur blends perfectly with tree bark, but the profile of evenly spaced little bumps is a giveaway. They sometimes start to rock back and forth, with all group members swaying like leaves in the wind. If approached too closely, the entire group flies off like a little flock of moths. They can be seen on almost any river trip in the rainforest region. At dusk they are the first bats to be seen out foraging over the water surface. Found in rainforest, dry forest, gallery forest, secondary vegetation, gardens, or pastures.

Map 30

Long-nosed bat, *Rhynchonycteris naso*

Geographic range. Central and South America: Veracruz, Mexico, south to N Bolivia, and Matto Grosso and Minas Gerais, Brazil.

Status. Often common; widespread.

References. Bradbury, J., and S. L. Vehrencamp. 1976. Social organization and foraging in emballonurid bats. *Behav. Ecol. Sociobiol.* 1:337–81.

White-lined Sac-winged Bats
Saccopteryx spp.
Plates B, 5, map 31
Identification. Measurements: HB = 40–67; T = 11–20; HF = 6–13; FA = 35–50; WT = 3–12 g.

Upperparts dark, black-brown to tawny brown, usually with two wavy white lines from shoulder to end of rump. Wing sac well developed in males, close to forearm bone beyond elbow. Tail membrane longer than legs, calcar about two times as long as foot. Underparts usually pale brown, fur darker brown at base.

Species. There are four species: *S. bilineata* is the largest (7–12 g), almost black, with blackish wings and tail and prominent white lines; *S. leptura* (4–7 g) is chocolate-brown with chestnut tints, distinct yellowish lines on back, brown wings and tail membrane; *S. canescens* (3–4 g) is tawny brown, frosted with grayish or yellowish especially on head, indistinct white lines on back, brown wings and tail membranes;

Map 31

White-lined sac-winged bats, *Saccopteryx*

S. gymnura (3–4 g) is dark chocolate-brown, with no stripes and underparts only slightly paler than back, wings and tail membranes brown.
Similar species. *S. canescens* is like the long-nosed bat (*Rhynchonycteris naso*) but has short muzzle and no tufts of hair on forearm; the position of the wing sac distinguishes *S. gymnura* from doglike (*Peropteryx* spp.) or chestnut (*Cormura brevirostris*) sac-winged bats.
Sounds. *S. bilineata* often and *S. leptura* rarely "sing" at their roosts with a cricket-like twittering and chirping, and both species chirp loudly during encounters with conspecifics while foraging.
Natural history. White-lined bats feed on tiny insects (2–4 mm). They come out to forage during the last daylight and fly in beats, repeating the same path again and again. They forage at dusk inside the forest (*S. leptura*), in small openings in the forest, the forest edges, around camps (where they often fly through buildings), river edges and pastures (*S. bilineata, S. canescens*). Later in the evening they may move higher and higher, to forage above the forest canopy. They roost in small groups, each bat spaced a few centimeters from its neighbors, inside hollow trees, or on the outside of a tree under an overhanging branch or knot (*S. leptura*), or in cavelike cavities between giant buttresses of trees such as figs (*S. bilineata*), under bridges or the outer eves of houses in forest. *S. bilineata* has a complex social system in which a male defends a tiny territory at the roost site, around his harem of one to eight

females. Several harems may roost side by side on a tree. The male displays by waving his folded wing, with the gland open, at another bat, or sometimes he hovers in flight, singing to his mates. It is easy to see these behaviors by watching discreetly, especially at dawn or from four to six in the afternoon. The harem of females together defends a territory where they feed. *S. leptura* has a simpler social system; that of the other two species is unknown. Found in lowland mature and secondary rainforest, dry forest, gallery forest, plantations and gardens, and pastures.
Geographic range. Central and South America: *S. bilineata* ranges from S Mexico to SE Brazil; *S. leptura* from S Mexico to Bolivia and to Bahía, Brazil; *S. canescens* from the Amazon Basin and Colombia, Venezuela, and the Guianas; *S. gymnura* is known only from Brazil near the Amazon River from the Rio Tapajós to Belém.
Status. All are widespread and sometimes common except *S. gymnura,* which appears to be restricted and rare.
References. Bradbury, J. W., and L. H. Emmons. 1974. Social organization of some Trinidad bats. I. Emballonuridae. *Z. Tierpsychol.* 36:137–83.
   Bradbury, J., and S. L. Vehrencamp. 1976. Social organization and foraging in emballonurid bats. *Behav. Ecol. Sociobiol.* 1:337–81.

Shaggy Bat
Centronycteris maximiliani
Plate B, map 32
Identification. Measurements: HB = 46–57; T = 20–32; HF = 7–9; E = 17–19; FA = 43–49; WT = 5–6 g.
Upperparts smoky brown with yellow-orange tint; hair very long and woolly. Head woolly with hair extending to nose and chin; ear with prominent parallel ridges on inner rear edge. **Tail long, emerging about halfway down tail membrane; tail membrane much longer than legs when extended; strongly patterned with parallel rows of dots crossed by prominent branching veins;** reddish body hair extends onto the base of the membrane. Underparts slightly more yellowish than back. No wing sac. **Bat like a small ball of fluff.**

Map 32

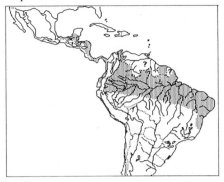

Shaggy bat, *Centronycteris maximiliani*

Map 33

Doglike sac-winged bats, *Peropteryx*

Similar species. Long-nosed bats have a long nose and variegated fur.

Natural history. Shaggy bats forage in beats, with a slow fluttering flight; they have been noted flying in late afternoon. They roost in tree holes. Found in mature rainforest and secondary forest.

Geographic range. Central and South America: Veracruz, Mexico, south to Ecuador and N Peru, on both sides of the Andes, and all of N South America north of the Amazon and south to coastal Bahía. Poorly known.

Status. Apparently rare everywhere in large geographic range.

Doglike Sac-winged Bats
Peropteryx spp.
Plate B, map 33

Identification. Measurements: HB = 41–64; T = 8–17; HF = 6–12; E = 10–20; FA = 38.5–53.6; WT = 3–11 g.

Upperparts chocolate-brown, grayish smoky brown, or yellowish smoky brown; fur long and lax. **Head with a prominent tuft on crown, this tuft slightly wavy, ending abruptly over naked face; mouth fringed with a narrow mustache of stiff hairs; snout short and broad; ear pointed, with fold from inner rim extending partly or completely across forehead.** Wings and tail membrane brown, or wings white. **Tail membrane long, calcar twice as long as foot; membrane with rows of fine dots below, undersurface covered with short, fine, stiff hairs, feels rough; above slightly hairy with body hair extending onto membrane for length of tail,** about one-third of length. **Glandular sac on forward edge of propatagium.** Underparts like back or paler.

Species. There are three species: *P. kappleri* is large (6–11 g) and chocolate-brown; *P. macrotis* is small (3–8 g)and chocolate or smoky to tawny brown; *P. leucoptera* is intermediate, with white wings and a membrane or fold completely across forehead between ears above naked face; it is sometimes placed in a separate genus (*Peronymus*).

Similar species. Chestnut sac-winged bats (*Cormura brevirostris*) have no hair on top of tail membrane and a line of hair running down muzzle; least sac-winged bats (*Balantiopteryx* spp.) have the wing sac in the center of the propatagium, not at its forward edge, and often have a white rear edge on wing and a long snout.

Natural history. *P. macrotis* feeds mainly on tiny beetles and flies. *P. kappleri* roosts in small groups of one to six near the ground in hollow logs, moist hollow buttressed trees, or caves, where the bats hang in contact in a line, each touching with its belly the back of the one in front (they separate if disturbed). *P. macrotis* roosts in rock piles, rock crevices, caves, hollow logs, houses, or tunnels, and *P. kappleri* also sometimes uses these roosts. *P. kappleri* forages in the rainforest subcanopy, flying in long, slow beats 20–30 m long; it more rarely uses croplands. *P. macrotis* usually forages in rainforest but also feeds in savannas, plantations, and dry forests. *P. leucoptera* is poorly known.

Geographic range. Central and South

Map 34

Chestnut sac-winged bat, *Cormura brevirostris*

America: *P. kappleri* and *P. macrotis* from
S Mexico to Paraguay and SE Brazil;
P. kappleri not in the Amazon Basin.
P. leucoptera from the Guianas, the Ama-
zon Basin, and coastal Brazil to Bahía.
Status. Widespread but may be restricted
by roosting sites; generally uncommon or
rare.
References. Bradbury, J., and S. L. Veh-
rencamp. 1976. Social organization and
foraging in emballonurid bats. *Behav.
Ecol. Sociobiol.* 1:337–81.

Chestnut Sac-winged Bat
Cormura brevirostris
Plate B, map 34
Identification. Measurements: HB = 50–
65; T = 11–16; HF = 7–10; E = 13–
16; FA = 41–50; WT = 8–11 g; females
slightly larger than males.
**Upperparts rich, chestnut red-brown, or
dark, black-brown;** fur medium length.
**Wings and tail membrane black; hair of
body does not extend onto base of tail
membrane. Ear tip rounded, inside of
ear with deep, parallel ridges; fur of
crown gradually decreasing in length
down muzzle;** lower face naked. Wing sac
long; a slit in the center of the propatagium
is present in both sexes. Tail short, pierces
the tail membrane at one-fourth to one-
third its length. Calcar about two times as
long as foot; tail membrane extends that
much beyond toes when stretched. Underparts slightly paler or more chestnut than
upperparts.
Variation. Blackish and chestnut colors
occur in the same populations.

Similar species. Least sac-winged bats
(*Balantiopteryx* spp.) and doglike sac-
winged bats (*Peropteryx* spp.) have a tuft
of fur on crown that ends abruptly at naked
face and brown wings; they have gray
tones or are chocolate-brown, respectively.
Natural history. Feeds aerially on tiny insects. These bats live in the forest and forage in the forest undercanopy. They roost
under logs, especially logs over streams, or
inside hollow logs or trees. Found in lowland rainforest.
Geographic range. Central and South
America: central Nicaragua south through
Panama; South America east of the Andes
in the Amazon Basin from Peru to Pará,
and Columbia, Venezuela, and the Guianas. To about 1,000 m elevation.
Status. Sometimes common; widespread.

Least Sac-winged Bats
Balantiopteryx spp.
Plate B, map 35
Identification. Measurements: HB = 45–
53; T = 12–21; HF = 6–11; E = 11–
16; FA = 35–46; WT = 5–8 g; females
larger than males.
**Upperparts dark brown, smoky brown,
or gray.** Crown with tuft of fur that ends
abruptly on naked face. Muzzle broad and
triangular; ear broadly triangular, round at
tip. **Wing and tail membranes brown; tail
membrane reaching only slightly beyond
tip of foot, calcar only slightly longer
than foot; edge of wing may be white** between leg and fingers. **Sac in center of
propatagium, opening facing toward
body, well developed in males, rudimentary in females.** Underparts the same as
back, or slightly paler.
Species. There are three species: *B. plicata*
has a white rear-wing margin and is smoky
to gray; *B. io* is dark brown; and
B. infusca is dark chestnut-brown.
Similar species. See doglike (*Peropteryx*
spp.) and chestnut (*Cormura brevirostris*)
sac-winged bats.
Natural history. Feeds aerially on tiny insects. Natural history well known only for
B. plicata. Least sac-winged bats roost in
small to large colonies in well-lighted parts
of caves, rock crevices, culverts, or tree
hollows. The roosting bats space themselves about 20 cm apart. *B. plicata* occupies dry forest and arid habitats, where it

Map 35

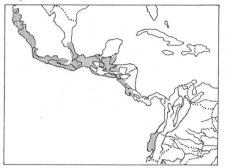

Least sac-winged bats, *Balantiopteryx*

Map 36

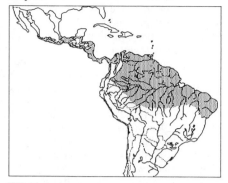

Ghost bats, *Diclidurus*

forages with slow flight in long, unbroken
sallies at 15–25 m above the surface of the
ground or forest canopy. It may forage sin-
gly, or several may use overlapping beats.
B. io and *B. infusca* are found in rainfo-
rest; their habits are undescribed.
Geographic range. Central and South
America: *B. plicata* from the tip of Baja,
and S Sonora, Mexico, to Costa Rica;
B. io from Veracruz, Mexico, to Guatemala and Belize; and *B. infusca* only from the
NW coast of Ecuador west of the Andes.
To at least 1,500 m elevation.
Status. *B. plicata* is widespread and sometimes common; *B. infusca* may be rare.
References. Arroyo-Cabrales, J., and
J. K. Jones, Jr. 1988. *Balantiopteryx plica-
ta.* Mammalian Species, no. 301.
   Bradbury, J., and S. L. Vehrencamp.
1976. Social organization and foraging in
emballonurid bats. *Behav. Ecol. Sociobiol.*
1:337–81.

Ghost Bats
Diclidurus spp.
Plates B, 5, map 36
Identification. Measurements: HB = 55–
79; T = 17–26; HF = 10–14; E = 13–
23; FA ~ 53–71; WT = 12–23 g; females larger than males.
There are two color patterns: (1) **Upperparts pure white, or white with gray hair
base showing through;** underparts white
with gray hair base showing through;
**wings and tail membrane translucent,
without pigment, yellowish,** bones pale
brown; **or** (2) **forequarters white, lightly
frosted with pale brown, hindquarters**

pale brown, hair white at base; underparts white tinged with gray-brown, hairs
brown at base; **wings and tail membrane
pale brown, semitransparent.** All have the
following: **ear short and rounded,** lower
edge extending forward in a flange to below the eye, upper edge extending as a
membrane partially across the forehead;
**thumb enclosed in the wing membrane,
with only the claw emerging;** no wing
sac, **tail membrane with prominent,
brown sacs around tail near tip,** these
most highly developed in males; fur long
and silky.
Species. There are four species:
D. scutatus, D. ingens, and
D. albus have color pattern (1); they differ
in size. *D. isabellus* has color pattern (2)
and is sometimes placed in a separate genus (*Depanycteris*).
Similar species. Honduran white bats (*Ectophylla alba*) are the only other white bats
in the region; they are much smaller (5–6
g) and have a noseleaf.
Sounds. Emits a musical twittering while
in flight.
Natural history. Ghost bats fly high in
open spaces, such as above rivers, streams,
and lagoons, and probably above the forest
canopy. They are attracted to concentrations of insects around high floodlights and
are among the many species swooping
around such lights, even in towns. Their
roosts are not well known; they have been
found in abandoned towers. *D. albus*
roosts between the leaves of palms. *D. isabellus* is known only from lowland rainforest habitat (over water); *D. scutatus* oc-

curs only in the rainforest region. The
other species are found in rainforest, dry
forest, plantations, and towns.
Geographic range. Central and South
America: only *D. albus* in Central America, from Nayarit, Mexico, south down the
coast to N Peru, and the Amazon Basin of
Brazil, to Bahía; *D. scutatus* and *D. ingens*
in the Amazon basins of Brazil, Venezuela,
Columbia, NE Peru, and the Guianas; *D.
isabellus* is known only from the Amazon
Basin of Venezuela and a small part of adjacent Brazil. All geographic ranges poorly
known, from few, scattered records.
Status. The high roosting and flying behavior of these bats makes detecting them
extremely difficult, and it is not possible to
say if they are common or rare.
Local names. Murciélago blanco (or albi-
no) (Span); jumbie bat (Tr).
References. Ceballos, G., and R. A. Medellín. 1988. *Diclidurus albus*. Mammalian
Species, no. 316.

Smoky Bat
Cyttarops alecto
Map 37
Identification. Measurements: HB = 50–
55; T = 20–25; HF = 8; E = 10–13;
FA = 45.8–47.
**Upperparts dull, dark, smoky gray, al-
most black,** palest on upper back and
shoulders; **wings and tail membranes
black; fur long and silky. Ear short and
rounded, with margins as in ghost bats;
tragus with a distinct projection on lower, outer margin;** nostrils tubelike and
slightly diverging; face thickly furred down
center to above nose, almost naked on side

Map 37

● Smoky bat, *Cyttarops alecto*

from ear to nose. No glandular sacs on
wings or tail. Underparts like back.
Similar species. White-lined sac-winged
bats (*Saccopteryx* spp.) have white lines on
back; doglike (*Peropteryx* spp.) and least
(*Balantiopteryx*) sac-winged bats have
longer, more pointed ears, brown wings,
and a naked face with a tuft on crown.
Natural history. Smoky sac-winged bats
have been found roosting in groups of one
to four under fronds of coconut palms near
buildings, and flying in an oil-palm grove
and in a garden.
Geographic range. Central and South
America: known only from Costa Rica,
Nicaragua, Guyana, and near Belém,
Brazil.
Status. This is one of the rarest bats,
known from fewer than 20 individuals
from five lowland rainforest localities.
References: Starrett, A. 1972. *Cyttarops
alecto*. Mammalian Species, no. 13.

Bulldog Bats (Noctilionidae)

Dental formula: I1/1, C1/1, P1/2, M3/3 = 28. This small family derives its name from
the distinctive, bulldoglike face, with downcurved, drooping lips and jowls that expose
large canine teeth. The characteristics of the family are described below. The family includes one genus and two species restricted to the Neotropics and subtropics.

Bulldog or Fishing Bats
Noctilio spp.
Plates B, 5, map 38
Identification. Measurements: HB 64–
109; T = 14–37; HF = 16–39; E = 22–
31; FA = 60–88; WT = 19–88 g; males
larger than females.

**Upperparts bright rust-orange, rust-red,
chestnut-brown, dull tawny brown, or
silvery gray; midback with a single pale
stripe** from neck or shoulder to rump; sides
of neck often paler, more yellowish; **fur
extremely short, sides of lower back naked** (wings attach high on sides). Wings

Map 38

Bulldog or fishing bats, *Noctilio*

and tail membrane brown, semitransparent. **Head with large, forward-pointing nosepad, no noseleaf; lips split near nose and forming downward-drooping, bulldoglike folds** that form cheek pouches and expose large canine and pointed upper incisor teeth; **ears long, sharp, narrow, forward-slanting.** Tail membrane longer than hindlegs; tail shorter than membrane, enclosed within it except for short stub emerging from upper surface. **Feet large, robust, toes long, claws large, strongly curved and sharp.** Underparts paler than back, rich bright orange to pale orange or pale, tawny brown in brown animals.
Species. The lesser bulldog bat (*N. albiventris*) is smaller (FA = 60–65; WT = 20–40 g), with relatively smaller feet; it has an extreme color form of brown frosted with silver or white on the neck and shoulders. The greater bulldog or fishing bat (*N. leporinus*) is one of the largest New World bats (FA = 77–88; WT = 60–90 g). It has enormous feet, with scimitar-shaped, gaff-like claws. The two species have similar, highly variable coloration; some populations include both brown and orange individuals, with males often more orange and females browner, while other populations are more homogeneous.
Similar species. All other bats as large as

bulldog bats have large, prominent, spear-shaped noseleaves, or long tails extending well beyond the edge of the tail membrane.
Natural history. Feed on insects (both species) and fish (*N. leporinus*). Bulldog bats usually forage over water, where they can be recognized from a distance by their large size, red color, agile flight, pointed wings, and behavior of dipping their feet and skimming the water as they fly (many species of bats come out at nightfall and drink on the wing, dipping their lower jaw into the water). Both species also forage over dry land. Greater bulldog bats are specialized for fishing: with echolocation they detect fish disturbing the water surface, gaff them with the feet, and transfer them to the mouth. Lesser bulldog bats are insectivorous; they eat many water beetles. They come out just at nightfall and often forage in small flocks of several bats that circle back and forth close to the water surface for a few minutes in one area, then move on to another. Both species roost in large groups in hollow trees; greater bulldog bats also roost in caves and lesser bulldog bats in houses. Usually found around water with a still surface in rainforest, dry forest, coastal marine habitats, marshes, and disturbed habitats such as pastures and plantations.
Geographic range. Central and South America: Pacific coastal Mexico (greater species) or Guatemala (lesser species) south to N Argentina and Uruguay, and Greater and Lesser Antilles. Lowlands only.
Status. Both species widespread and often common.
Local names. Murciélago pescador (Span); hazemond vleermuis (Su); mbopí pytá (Gua).
References. Hood, C. S., and J. Pitocchelli. 1983. *Noctilio albiventris.* Mammalian Species, no. 197.
  Hood, C. S., and J. K. Jones, Jr. 1984. *Noctilio leporinus.* Mammalian Species, no. 216.

Leaf-chinned, Mustached, and Naked-backed Bats (Mormoopidae)

Dental formula: I2/2, C1/1, P2/3, M3/3 = 34. These are small- to medium-sized bats with lips expanded into flaring plates and mouth flanked on its upper sides by a mustache of stiff, forward-pointing hairs. Two "naked-backed" species have wing membranes that are attached along the midline of the back, covering the fur beneath. The lower rim of the ear is prolonged forward as a low membrane to the corner of the mouth. The tail is enclosed in the tail membrane except for its tip, which emerges free about halfway down the membrane. The calcar is about twice as long as the foot, and the tail membrane is longer than the legs. These bats emit their echolocation calls through the mouth, with the enlarged lips acting as a megaphone. They are strong, fast flyers, with long narrow wings. All species seem to roost in dark, warm, humid caves, and their local distributions may be limited by the availability of appropriate roost sites. There are two genera and 8–10 species in the New World tropics and subtropics.

Mustached and Naked-backed Bats
Pteronotus spp.
Plates B, 5, map 39

Map 39

Mustached or naked-backed bats, *Pteronotus*

Identification. Measurements: HB = 38–80; T = 16–28; HF = 8–17; E = 12–20; FA = 35–56; WT = 3–28 g.
Upperparts rust-orange, gold-orange, gold-brown, smoky brown, tawny brown, or gray-brown. Back either naked, covered by wing membranes to behind shoulders, or with sides of lower back "naked" (wings attach high over furry sides); fur on lower back short, velvety, and fur on neck, throat, and shoulders longer, often paler, forming a paler collar. Face with lips flared out forward in meaty plates, lower lip with small round tubercules; **muzzle with stiff, forward-pointing mustaches above; ear narrow, pointed, flared into funnel at base, pointing forward;** eye small, sometimes almost lost in fur, surrounded by ear funnel. Tail membrane longer than legs, **tail tip emerging freely from middle of top surface of membrane.** Underparts usually paler than back, silver, yellowish, beige, orange, or gray, fur on belly dark at base.
Species. There are about six species; two have naked backs: *P. davyi* (FA = 41–49; WT = 6–11 g) and *P. gymnonotus* (FA = 49–56; WT = 12–16 g). The two furry-backed mainland species can also be separated by size: *P. parnellii* (FA = 49–65; WT = 11–28 g) and *P. personatus* (FA = 40–47; WT = 6–9 g). Two small species, *P. macleayii* and *P. quadridens,* are found only in Cuba and Jamaica, and one of them also in Haiti and Puerto Rico. In all species there is much color variation both within and sometimes between populations. Some variation may be due to length of time since the last molt, the fur bleaching with age.
Similar species. Lesser bulldog bats (*Noctilio albiventris*) have the upper lip split from below the nose and a pale stripe on back.
Natural history. All mustached bats are aerial insectivores. They often forage over water, up and down forest trails, or in clearings and plantations. They roost in warm, humid caves in colonies of up to thousands. At roost they hang free by their hind feet, back rounded, in tight clusters of individuals in direct contact. Rarely they are found in other types of roosts such as old buildings, culverts, or tree hollows. Their scarcity in the central Amazon Basin may be due to the lack of caves and rocks. The naked-backed species are said to be agile at running and hopping quadrupedally on the ground, on feet and folded wings. In rainforest and deciduous forest, clearings and plantations.
Geographic range. Central and South America: Mexico south to N Bolivia, the base of the Andes and N coast of South America through the Guianas, generally absent from the central Amazon Basin. Large Caribbean islands. To 3,000 m elevation.
Status. Often common, especially *P. parnellii.*

References. Herd, R. M. 1983. *Pteronotus parnellii.* Mammalian Species, no. 209.
Silva, G. 1979. *Los murciélagos de Cuba.* Havana: Editorial Academia.

Leaf-chinned Bat
Mormoops megalophylla
Plate B, map 40

Identification. Measurements: HB = 61–76; T = 19–31; HF = 10–13; E = 13–17; FA = 51–59; WT = 12–20 g.
Upperparts chestnut, chocolate-brown, tawny yellow-brown, pink-beige, or mouse gray; fur long; base of neck between shoulders with stiff, straight, back-slanting hair often contrasting pale yellow, pink, or pale brown; area behind ears this same color; fur on neck sometimes sparse, emphasizing shoulder crest. **Face with complex skin folds: chin with two large concave plates connected to a series of folds below and beside chin; nostrils set wide apart in short tubes, side of nose with folds; ears short, rounded, connected by a large fold across forehead that is deeply notched on midline, inside of ear with complex, deep folds; lower edge of ear extending forward in a fold to corner of mouth; eye medium-sized, surrounded by ear funnel.** Tail membrane longer than legs, tail enclosed within it, with only tip emerging. Underparts paler than back, buffy yellow, pink-brown, or reddish, fur long. Body long and slender.
Variation. Color varies greatly between populations. Another species, *M. blainvillei*, is found only in the Greater Antilles; it is smaller (FA = 43–50; WT = 6–11 g).
Similar species. Wrinkle-faced bats (*Centurio senex*) and vampires (*Desmodus, Di-*

Map 40

Leaf-chinned bat, *Mormoops megalophylla*

phylla spp.) have simple, upstanding ears without complex folds.
Natural history. Feeds on insects. Leaf-chinned bats often forage over water. They roost in deep caves, where they hang singly, separated from their neighbors, in colonies of up to several thousand, but usually of only a few individuals. They seem to be nomadic. Chiefly found in arid habitats, rarely in rainforest.
Geographic range. North, Central, and South America: S Texas and Arizona south through Honduras; absent from the Isthmus of Panama and present in NE Colombia, N Venezuela, Trinidad, and W coastal Ecuador and Peru.
Status. Generally uncommon, rarely in large numbers.
References. Barbour, R. W., and W. H. Davis. 1969. *Bats of America.* Lexington: University Press of Kentucky.
Smith, J. D. 1972. *Systematics of the chiropteran family Mormoopidae.* Univ. Kan. Mus. Nat. Hist. Misc. Publ., no. 56.

Leaf-nosed Bats (Phyllostomidae)

Dental formula highly variable. In most genera, the nostrils are surrounded by a fleshy fold that has a dishlike horseshoe below and/or beside the nostrils and a freestanding, spear-shaped leaf above and behind the nostrils. These structures act as a horn or megaphone to focus the echolocation sounds, which are emitted through the nostrils (although the bats can emit them through the mouth also). The point of the chin usually has a naked, V-shaped groove, often bordered by an array of warts or tubercules. The wings are broad in outline, so that the bats are able to maintain flight at slow speeds, and they can maneuver among thick vegetation. Leaf-nosed bats represent one of the largest living radiations of mammals. They have evolved a great diversity of specialized feeding habits and associated morphological adaptations and social behaviors. They give birth to single young. The family is restricted to the New World and includes about 50 genera and 140 species, mostly in the tropics, with a few species reaching temperate North America. The family is treated by subfamilies below. Members of a subfamily generally have similar food habits.

Spear-nosed Bats (Subfamily Phyllostominae)

The spear-nosed bats range from the largest New World bats to some of the smallest. They all have large, spear- or sword-shaped noseleaves with a free horseshoe at least partially around the nostrils; narrow muzzles; long tail membranes; tail enclosed within the membrane except for the extreme tip, which may protrude on the top of the membrane; and usually large to very large ears. The colors of many species are variable, with brown, chestnut, and rust-orange individuals found in the same roosts or populations. Among greater spear-nosed bats in Trinidad, some individuals are orange when young and change to dark brown when they are several years old, while other young in the same social group and presumably with the same father are brown. The spear-nosed bats include all genera of leaf-nosed bats that are true carnivores and kill vertebrate prey; they also include the most strongly insectivorous genera, as well as mixed feeders that eat both insects and fruit. The subfamily includes 10 genera and about 33 species in the New World tropics and subtropics; 9 genera and about 31 species are found in the lowland rainforest region; one genus is restricted to arid habitats. Recent evidence suggests that the genera in this subfamily are not all closely related and should probably be split into a least two groups.

Little Big-eared Bats
Micronycteris spp.
Plate C, map 41
Identification. Measurements: HB = 46–88; T = 5–16; HF = 8–20; E = 13–31; FA = 32–58; WT = 4–30 g.
Generally, **ears very large, usually rounded,** tragus short; noseleaf medium-sized, spear-shaped; **face and muzzle narrow; naked chin tip with single, long, smooth pads on each side angled to a V- or Y-**shape; eyes small; **fur long** and slightly wavy; face hairy; **thumb with first joint equal to or longer than second joint;** tail short, enclosed in tail membrane except for tip, which protrudes slightly on top surface; **tail membrane the same length as or shorter than legs;** calcar about one half to one times the length of hindfoot; **size small,** with one exception. There are four pelage patterns: (1) Upperparts plain brown to chestnut or orange, **long hair on shoulders with long white base that shows prominently** through parts in fur; **ears often with a low membrane partly or entirely joining bases** across forehead; **ear hairy, with long hairs on back or both surfaces;** underparts like back or contrasting white, silvery, beige, yellowish, or or-

Map 41

Little big-eared bats, *Micronycteris*

ange; hair on throat pale to base. **Size small** (4–16 g). (2) Upperparts chestnut to pale brown or gray-brown, **hairs look unicolor,** lower back sometimes with faint gray stripe; underparts slightly paler whitish or yellowish. Ears not hairy on back except on base. (3) Upperparts dark brown, **hairs on shoulder prominently tricolor,** dark brown at base and tip, with broad, paler tawny or silvery midsection; underparts paler tawny yellow or gray-brown, hairs dark at base. Ears not hairy.

(4) Upperparts uniform **gray-brown**, underparts paler; **fur long, dense and woolly**; ears slightly hairy; **size large** (FA > 50; WT = 19–30 g).
Species. There are about 10 species, 5–6 of them have color pattern (1); *M. nicefori* has pattern (2); *M. sylvestris* has pattern (3); *M. (Barticonycteris) davesi* (plate C), the only large member of the genus, has pattern (4).
Similar species. Short-tailed fruit bats (*Carollia* spp.) have shorter, triangular ears and chin-tip **V** bordered by row of tubercles; Mexican big-eared bats from arid habitats (*Macrotis,* not treated in this book), with color pattern (1), have long tails that extend all the way to the end of the tail membrane and beyond it; round-eared bats (*Tonatia* spp.) and all other similar-looking bats have rows of round tubercles on chin.
Natural history. These bats feed on large insects and occasionally fruit, but there is little information about most species. Some species seem to forage by gleaning large insects such as cockroaches, dragonflies, and katydids from the vegetation, and they carry their prey back to a feeding roost before eating. Several species are deep rainforest inhabitants; they tend to fly along streams, gullies, and paths well shaded by the forest canopy. Big-eared bats roost in small groups, often a half-dozen or fewer, in damp hollows near the ground. These may be hollow logs or trees, cavities under roots, fallen logs, or overhanging dirt banks, and culverts. The roosts are often well enough lit so that the bats can be seen without a flashlight. They hang freely suspended and spaced apart and always seem alert. They twist and turn and vibrate their large ears and will readily take flight if disturbed. Found in many habitats including mature and disturbed rainforest, gallery forest, dry forest, cloud forest, and gardens and plantations.
Geographic range. Central and South America: S Mexico south to N Bolivia and SE Brazil. All but two species are found in both Central and South America. To 2,000 m elevation.
Status. Three species are apparently rare: *M. davesi, M. pusilla,* and *M. behni,* the latter two with restricted distribution.
References. Wilson, D. E. 1971. Food

Map 42

Sword-nosed bats, *Lonchorhina*

habits of *Micronycteris hirsuta* (Chiroptera: Phyllostomidae). *Mammalia* 35:107–10.
Medellín, R. A., D. E. Wilson, and D. Navarro L. 1985. *Micronycteris brachyotis.* Mammalian Species, no. 251.

Sword-nosed Bats
Lonchorhina spp.
Plate C, map 42
Identification. Measurements: HB = 49–74; T = 46–63; HF = 9–20; E = 19–38; FA = 41–59; WT = 8–32 g.
Upperparts dark brown, chestnut, rust-red, or pale beige; neck and shoulders often paler, with hair pale at base; fur deep and plushy. **Noseleaf extremely long (18–22 mm) and narrow, about as long as ears,** with smooth edges; folds around nostrils at base complex; **ears very large, pointed,** hairy inside and out on upper edge, **wide at base, with lower rim extending as a fold to below eye; tragus very long and narrow, pointed, at least half the length of ear. Legs very long, tail membrane longer than legs; tail long, reaches rear edge of tail membrane, which forms a point at tail tip.**
Species. There are four species: *L. aurita* is widespread and geographically variable; it is large, (FA = 47–54; WT = 12–22 g) and dark brown to rust. The other species have restricted distributions—*L. orinocensis* is small (FA = 41–44; WT = 7–10 g) and pale brown; *L. fernandezi* is small (FA = 42–43), with shorter ears (E = 19) and stiletto-shaped processes at base of noseleaf; and *L. marinkellei* is large (FA = 59), with large ears (E = 38).

Similar species. No other bats in the region have as large a noseleaf; hairy-nosed bats (*Mimon* spp.) have noseleaf about one-fourth to one-third shorter than ear and tail only half as long as the tail membrane.
Natural history. Feed on insects and some fruit. Common sword-nosed bats (*L. aurita*) roost in dense clusters of up to hundreds in caves and old tunnels, such as mine tunnels. Orinoco sword-nosed bats (*L. orinocensis*) have been found in large rock piles in hot grassland habitat. Three of the four species were only recently discovered and were first described between 1971 and 1982; these all have small known geographic ranges. Found in rainforest, deciduous forest, and plantations, pastures, and savanna.
Geographic range. Central and South America: *L. aurita* from S Mexico to N Bolivia and SE Brazil; *L. orinocensis* from the Colombian-Venezuelan Orinoco drainage; *L. fernandezi* and *L. marinkellei* from single localities in the Venezuelan and Colombian Amazon regions, respectively.
Status. Widespread to rare: *L. fernandezi* is currently known from a single cave.
References. Ochoa, J., and C. Ibañez. 1982. Nuevo murciélago del genero *Lonchorhina* (Chiroptera: Phyllostomidae). *Mem. Soc. Cien. Nat. La Salle* 42 (118): 145–59.

Long-legged Bat
Macrophyllum macrophyllum
Plate C, map 43
Identification. Measurements: HB = 40–53; T = 38–49; HF = 13–16; E = 17–20; FA = 34–40; WT = 7–11 g.
Upperparts plain medium brown; wings and tail membranes brown; fur medium length. **Ears quite large, pointed,** slightly hairy on upper rim, lower rim not prolonged downward and forward as a fold on head; **noseleaf a large, broad spear with prominent central rib,** horseshoe at base of noseleaf broad, with expanded lobe behind nostrils; eyes small. **Legs and tail membrane long, membrane with a distinctive row of about seven vertical lines of dots each leading to a bump on the straight, rear edge of the tail membrane; tail long, enclosed in membrane and reaching its edge, but not forming a longer point; calcar about two times as**

Map 43

Long-legged bat, *Macrophyllum macrophyllum*

long as foot. Foot with elongate toes and claws. Underparts about the same color as back.
Similar species. These are the only bats with parallel rows of dots on the edge of the tail membrane. Little big-eared bats (*Micronycteris* spp.), short-tailed and little fruit bats (*Carollia, Rhinophylla* spp.), and hairy-nosed bats (*Mimon* spp.) have tails much shorter than membrane, or no tail; sword-nosed bats (*Lonchorhina*) have greatly elongated noseleaf and tragus.
Natural history. Long-legged bats feed on insects, including water striders. They may use their long toes and tail membranes to skim insects from the water surface. These bats are usually found near water and roost singly or in small groups in wet tunnels, culverts, under bridges, and in caves and abandoned buildings. Chiefly found in rainforest and rainforest clearings, sometimes in deciduous forest.
Geographic range. Central and South America: extreme S Mexico south to NE Bolivia, Paraguay, and SE Brazil.
Status. Usually uncommon, but very widespread.
References. Harrison, D. L. 1975. *Macrophyllum macrophyllum*. Mammalian Species, no. 62.

Round-eared Bats
Tonatia spp.
Plate C, map 44
Identification. Measurements: HB = 54–95; T = 7–25; HF = 10–19; E = 19–38; FA = 33–59; WT = 8–40 g.
Upperparts usually gray-brown faintly

Map 44

▨　Round-eared bats, *Tonatia*

frosted with silver, or warm brown, rarely rusty or dark, ashy gray frosted white; fur on neck pale at base; fur long and soft, not very dense. **Ears very large, rounded,** usually hairy on base of back and upper edge; **noseleaf large, broad,** spear-shaped; **naked tip of chin with U of tiny round tubercules.** Thumb often hairy, **with second, claw-bearing phalange longer than first phalange. Tail membrane longer than legs; tail less than half the length of membrane,** within it except for extreme tip, which protrudes above. Feet and claws robust, not elongated. **Underparts often washed with silver, often with pale collar on neck.** Small to large size.

Species. There are 5–10 species; up to 5 can occur together in South America. The more distinctive species include: *T. bidens,* large (FA = 55–59), with hairy face, feet, thumb, forearm, and ears; *T. silvicola,* large (FA = 50–59), face naked, usually with a white throat, lays its ears entirely back against the crown when they are touched; *T. carrikeri,* large with pure white underparts; *T. schulzi,* medium, with distinctive warty skin on forearms; and *T. brasiliensis,* small (FA = 33–40), warm brown. The latter is divided by some into as many as 5 species.

Similar species. The small species are very similar to little big-eared bats (*Micronycteris* spp.), which have single chin pads in a V, not round tubercules, and first joint of the thumb equal to or longer than the second. Hairy-nosed bats (*Mimon* spp.) and spear-nosed bats (*Phyllostomus* spp.) have pointed ear tips.

Natural history. Round-eared bats feed on insects and perhaps occasional fruit. They seem to fly everywhere within the forest. Several species, including *T. silvicola* with its peculiar folding ears and the small *T. brasiliensis,* roost in small groups of up to 10 in holes in arboreal termite nests. *T. bidens* roosts in tree hollows. Surprisingly little is known of these common bats. Found in mature and secondary rainforest, or more rarely deciduous forest and plantations.

Geographic range. Central and South America: extreme S Mexico south to N Argentina and Paraguay.

Status. Common to extremely rare (*T. schulzi*).

References. Genoways, H. H., and S. L. Williams. 1980. Results of the Alcoa Foundation–Suriname expeditions. I. A new species of bat of the genus *Tonatia* (Mammalia: Phyllostomatidae). *Ann. Carnegie Mus.* 49:203–11.

Hairy-nosed Bats
Mimon spp.
Plates C, 5, map 45

Identification. Measurements: HB = 52–75; T = 14–29; HF = 9–18; E = 24–38; FA = 46–58; WT = 10–15 g.

Ears large and pointed, thickly haired with pale hair on back of base and with some long hairs on both sides of upper edge; tragus narrow and pointed; noseleaf very long, narrow, spear-shaped, with simple basal horseshoe; tail reaches about halfway down tail membrane, tip just perforates top side; tail membrane longer than legs, calcar longer than foot; foot with long toes and claws; medium size. There are two color patterns: (1) **upperparts dark, blackish brown with a pale stripe down midback from crown to rump, wings and tail membrane blackish; sides of noseleaf serrated, noseleaf thinly haired to tip; underparts frosted silvery buff,** washed orange around throat; and (2) **pale, warm brown, hairs pale at base, fur long and dense, wing and tail membrane brown; noseleaf not hairy, smooth on sides, underparts about the same as back,** slightly paler on neck (FA = 53–58; E = 35–38).

Species. There are three or four species: *M. crenulatum* has color pattern (1); it is

Map 45

Hairy-nosed bats, *Mimon*

Map 46

Spear-nosed bats, *Phyllostomus*

placed by some in a separate genus, *Anthorhina*. *M. bennettii* and a possibly conspecific form *cozumelae* have color pattern (2). Another species occurs at higher elevations in the Andes.
Similar species. There are no other bats with a long, hairy noseleaf and a striped back; only one species of little big-eared bat (*Micronycteris daviesi*) is as large as *M. bennettii*, and it is dark gray-brown; round-eared bats (*Tonatia* spp.) have round ears and quite short noseleaf.
Natural history. These bats feed on insects, spiders, and small lizards.
M. bennettii and *M. cozumelae* roost in small groups in wet caves. *M. crenulatum* roosts in tree hollows and crevices in rotten stumps and under bark. These roosts have a very strong batty odor. *M. crenulatum* is frequently mist-netted in male-female pairs, suggesting that pairs forage together. Found in rainforest, gallery forest, and elfin forest, rarely in deciduous forest, and in secondary forest and plantations.
Geographic range. Central and South America: S Mexico south to N Bolivia and SE Brazil. *M. bennettii* is only found along the coast of South America from Colombia to SE Brazil.
Status. *M. crenulatum* is uncommon and *M. bennettii* is rare; both are widespread.

Spear-nosed Bats
Phyllostomus spp.
Plate C, map 46
Identification. Measurements: HB = 76–150; T = 7–30; HF = 12–27; E = 21–36; FA = 55–94; WT = 33–140 g.

Upperparts dark blackish brown, rich chestnut red-brown, rusty, or gray brown; fur short, velvety, flat and smooth on back, usually with a silvery sheen; head, neck, and shoulders often paler than back, more grayish. **Ears broadly triangular, medium length; noseleaf a simple, broad, medium-length spear; horseshoe at base forms a complete semicircular fold free of lip** under nostrils; **head and muzzle broad and powerful;** naked tip of chin with V of small tubercules; eyes large; noseleaf and lips black or dark brown. **Tail short, about one-third the length of tail membrane;** calcar slightly shorter to longer than foot. Underparts slightly to distinctly frosted paler than back.
Species. There are four species: *P. hastatus* is very large (FA = 79–94; WT = 92–140 g); *P. elongatus* is large (FA = 61–68; WT = 38–57 g), usually has extreme wingtips pure white, a dark belly, and large ears (E = 28–31); *P. discolor* is slightly smaller (FA = 55–69; WT = 33–45 g) and has short ears (E = 21–26) and distinctive pale grayish underparts; *P. latifolius* is the smallest (FA = 56–60); has dark gray underparts, short ears (E = 27), and short legs.
Similar species. See pale-faced spear-nosed bat (*Phylloderma stenops*); only one other leaf-nosed bat is larger than *Phyllostomus hastatus*—false vampire bats (*Vampyrum spectrum*) are huge (FA > 100 mm), with large ears, cup-shaped horseshoe, and no tail; all other large- and medium-sized

spear-nosed bats (Phyllostominae) have long fur and long ears; large spear-nosed fruit-eating bats (*Artibeus*) have no tails; other very large bats have no noseleaf.

Sounds. These bats are social and very noisy; they squabble at the roost and often call with loud chattering squawks while they feed or fly in the forest at night.

Natural history. Feed on fruits, large insects, and pollen and nectar. *P. elongatus* and *P. latifolius* are little known. The former roosts in small groups in tree holes, while the other two species roost in large congregations of tight clusters in tree holes or caves. These roosting congregations are divided up into smaller permanent social units consisting of harems of a single male, and from a few (*P. discolor*) to several dozen females and their young (*P. hastatus*), and groups of nonbreeding bachelors. In *P. hastatus,* females of a harem have fixed feeding areas adjacent to each other but separate from those of other harems. The bats emerge from their roosts at complete darkness and forage for about two hours before returning to the roost. Several *P. discolor* fly single file to forage. Found in mature and secondary rainforest, gardens and plantations, gallery forest, dry forest, and cloud forest.

Geographic range. Central and South America: S Mexico to N Bolivia and SE Brazil. *P. elongatus* and *P. latifolius* in South America east of the Andes. To 1,300 m elevation.

Status. *P. hastatus* is locally common and widespread in the rainforest; the other species are less common or rare; *P. latifolius* is known from only a few individuals.

References. McCracken, G. F., and J. W. Bradbury. 1981. Social organization and kinship in the polygynous bat *Phyllostomus hastatus. Behav. Ecol. Sociobiol.* 8:11–34.

Bradbury, J. W. 1977. Social organization and communication. In W. A. Wimsatt, ed., *The biology of bats,* 3:1–72. New York: Academic Press.

Pale-faced Bat

Phylloderma stenops
Plate C, map 47

Identification. Measurements: HB = 82–115; T = 12–24; HF = 20–23; E = 25–32; FA = 66–73; WT = 41–65 g.
Upperparts faded brown to slightly red-

Map 47

Pale-faced bat, *Phylloderma stenops*

dish brown, often with faint frost or sheen; **sides of neck and front of shoulders gray-frosted, continuous with grayish underparts; fur very short, often sparse,** or with sparse patches. Wings and tail membrane blackish brown, **wingtips white.** Wings attach high on sides of body. Ears large; muzzle narrow; **noseleaf a large broad spear, horseshoe merging with upper lip at center, without any free fold at center below nostrils; lips, noseleaf, and skin in general pale pinkish mottled with tan.** Tail membrane long, calcar about the same length as foot. **Large, long-bodied bats; adults always seem to look scruffy with faded, sparse hair, mottled skin on face and wing membranes.**

Similar species. Very similar to the larger spear-nosed bats (*Phyllostomus* spp.), which all have horseshoe with continuous, free flange or fold around entire base over lip and blackish noseleaf, lips, and wings; round-eared (*Tonatia* spp.) and woolly false vampire bats (*Chrotopterus auritus*) have long, dense fur; false vampires (*Vampyrum spectrum*) are larger, with a cup-shaped noseleaf and no tail.

Natural history. Has been recorded feeding on the larvae and pupae from a wasp nest, and fruit. These bats are found around streams and swamps or marshes. The roost is undescribed. Rainforest, dry forest, and plantations.

Geographic range. Central and South America: S Mexico south to N Bolivia and SE Brazil. To 2,600 m elevation.

Status. Uncommon to rare, but widespread.

Map 48

Fringe-lipped bat, *Trachops cirrhosus*

References. Jeanne, R. L. 1970. Note on a bat (*Phylloderma stenops*) preying upon the brood of a social wasp. *J. Mammal.* 51:624–25.

Fringe-lipped Bat
Trachops cirrhosus
Plates C, 5, map 48
Identification. Measurements: HB = 71–92; T = 13–21; HF = 16–22; E = 30–40; FA = 57–64; WT = 28–45 g.
Upperparts smoky brown to tawny brown faintly frosted with paler brown; fur long, pale at base on neck; wings and tail membrane brown, **wingtips often white. Ears large, rounded, prominently hairy on both sides of upper edge; chin, lips, and muzzle around nose studded with long, projecting tubercules; noseleaf moderately large, spear-shaped, edges of both basal horseshoe and spear serrated with tubercules on edges;** muzzle narrow. Tail membrane slightly shorter than legs, tail short, about one-third as long as membrane. Toes long, claws robust and strongly hooked, calcar shorter than foot. Underparts frosted silvery.
Similar species. These are the only bats with large projecting tubercules on muzzle.
Natural history. Feeds on frogs and lizards, with some insects and occasional small mammals. Fringe-lipped bats are carnivores that appear to feed largely on frogs. Experiments have shown that they home in on the calls of frogs and learn to avoid poisonous species by recognizing their calls. The tubercules on the chin may be chemical sensors; the bat touches a frog's skin (where poisons are found) with

the tubercules as it is about to grasp it. Fringe-lipped bats fly low in the forest understory and are especially common around streams, swamps, and ponds. They roost in small groups, usually a dozen or fewer, in hollow trees, caves, culverts, under bridges, or in houses in the forest. Found in rainforest, gallery forest, deciduous forest, and plantations and secondary vegetation.
Geographic range. Central and South America: S Mexico south to Bolivia and SE Brazil.
Status. Widespread and common.
References. Tuttle, M. D., and M. J. Ryan. 1981. Bat predation and the evolution of frog vocalizations in the Neotropics. *Science* 214:677–78.

Woolly False Vampire Bat
Chrotopterus auritus
Plate C, map 49
Identification. Measurements: HB = 94–114; T = 0–15; HF = 23–28; E = 40–49; FA = 77–87; WT = 61–94 g.
Upperparts gray-brown, often faintly frosted with paler; fur very long, dense, and woolly; wings and tail membrane dark brown, **wingtips almost always pure white. Ears very large, rounded,** hairy on back of inner rim; eyes large; muzzle long and narrow; **noseleaf** moderately large, sword-shaped **with prominent central rib, horseshoe and spear with continuous edge forming a hollow cup around nostrils;** lips, ears, and noseleaf pale pink-brown. Tail membrane long, calcar slightly shorter than foot, **tail a tiny or short stub** within membrane, **sometimes absent.** Underparts pale gray, fur brown at base. **A very large woolly bat.**
Similar species. False vampires (*Vampyrum spectrum*) are similar, but larger, with shorter fur, often a stripe down back, and wingtips usually dark; no other bats have noseleaf and horseshoe that form a hollow cup around nostrils.
Natural history. Feeds on small vertebrates such as small mammals, birds, and lizards, and large insects. Woolly false vampires take prey of up to 70 g weight, but mostly feed on smaller vertebrates of 10–35 g, especially mice. Prey is sometimes brought back to the roosts, which are in caves and hollow logs. These bats roost in small groups of two or three that can

Map 49

Map 50

▨ Woolly false vampire bat, *Chrotopterus auritus*

▨ False vampire bat, *Vampyrum spectrum*

consist of a male-female adult pair and the young. Found in primary and secondary rainforest and plantations, and in dry forest and chaco.

Geographic range. Central and South America: S Mexico south to N Bolivia, Paraguay, and SE Brazil.

Status. Uncommon, but seems ubiquitous in rainforest; very widespread.

References. Medellín, R. 1988. Prey of *Chrotopterus auritus,* with notes on feeding behavior. *J. Mammal.* 69:841–44.

McCarthy, T. J. 1987. Additional mammalian prey of the carnivorous bats, *Chrotopterus auritus* and *Vampyrum spectrum. Bat Res. News* 28:1–3.

False Vampire Bat

Vampyrum spectrum
Plate C, map 50

Identification. Measurements: HB = 135–158; T = 0; HF = 29–38; E = 39–49; FA = 98–110; WT = 126–190 g.

Upperparts dark brown, chestnut-brown, or rust-orange; midback from crown to rump with a broad, pale stripe, often indistinct; fur medium length, woolly, often sparse; wings and tail membrane blackish brown; wingtips usually dark, sometimes partly bleached, but not pure white. **Ears very long, rounded; eyes large; muzzle long and narrow; canines large; noseleaf medium-sized, lance-shaped, horseshoe and spear with continuous rim raised to form hollow cup around nostrils.** Tail membrane long, calcar longer than foot; **tail absent.** Feet large, robust, with huge, curved claws. Underparts pale, dirty gray-brown to yellow-brown; fur much shorter than on back. **A huge bat with a wingspan of about 80–90 cm.**

Similar species. No other New World bats attain this size; see woolly false vampire (*Chrotopterus auritus*).

Natural history. Feeds on birds, bats, and rodents, possibly rarely takes insects or fruit. Birds brought by false vampires back to a roost were most often large species that weigh 20–150 g, such as cuckoos, anis, parakeets, trogons, and orioles. False vampires roost in small, tight groups in hollow trees. Groups seem to consist of a monogamous adult pair and one to three of their nonbreeding young. A member of the group that catches a large prey will sometimes bring it back to the roost and may share it. One member of the group seems to stay behind to guard the young while the other hunts. These bats often hunt in open areas such as over swamps, forest edges, and secondary vegetation, but they are also found deep in the forest.

Geographic range. Central and South America: S Mexico south to N Bolivia and the southern Amazon Basin. To 1,650 m elevation.

Status. Widespread. False vampires are the top predators of their order, and, like other top predators such as harpy eagles and jaguars, they are always relatively rare.

References. Vehrencamp, S. L., F. G. Stiles, and J. W. Bradbury. 1977. Observations on the foraging behavior and avian prey of the Neotropical carnivorous bat, *Vampyrum spectrum. J. Mammal.* 58:469–78.

Navarro, D., and D. E. Wilson. 1982. *Vampyrum spectrum.* Mammalian Species, no. 184.

Nectar-feeding or Long-tongued Bats (Subfamily Glossophaginae)

This subfamily includes bats that all feed on flower nectar. Most are small (7–20 g); they have elongated muzzles, often with the tip of the lower jaw overshot, projecting farther than the upper jaw; and a narrow tongue that can be greatly extended and usually has a brush of hairlike papillae at its tip. The lower incisor teeth are absent or reduced in size, and the lower lip has a concave pad or deep, V-shaped notch; these features provide a channel for the tongue. The noseleaf is a short triangle; there are fine whiskers, generally at least as long as the noseleaf, growing from around the nose and forward from the chin; the ears are relatively short and set low on the sides of the head. The tail is short, and the tail membrane is shorter than the legs. This is one of the most poorly known groups of bats: almost nothing has been recorded of the natural history of many species. These little bats are important or essential pollinators of many plants on whose nectar they feed. Such plants often have large, pale flowers that open at night, and musky odors. Bat-pollinated New World plants include balsa trees, ceiba (silk-cotton) trees, and jícaro (calabash), as well as desert plants such as saguaro cactus and some agaves. Most species of long-tongued bats also eat some insects and fruit. There are 13 genera in the subfamily, 8 in the rainforest region. Five genera of tiny species (5–10 g) are distinguished by differences in their skulls but are very similar externally and are difficult to identify with certainty in the hand. Four of the genera not described herein occupy arid New World habitats, and one is found on Caribbean islands.

Common Long-tongued Bats
Glossophaga spp.
Plates D, 6, map 51
Identification. Measurements: HB = 53–65; T = 3–10; HF = 9–13; E = 12–16; FA = 32–40; WT = 10–13 g.
Upperparts pale gray to pale brown, fur bicolored pale at base dark at tip, pale showing through on neck and shoulders; wings and tail membrane brown. **Muzzle slightly, not greatly, elongated, lower jaw about the same length as upper;** noseleaf a small spear, reaches close to corner of eye when laid back; **horseshoe around nostrils with no free flange on sides or bottom; chin tip a deep V with sides a ripple of tiny tubercules, no tubercules in center;** ears relatively short, slightly hairy on back of inner edge, tips rounded. **Tail very short; tail membrane shorter than legs;** calcar about half as long as foot. Young dark gray-brown.
Species. The five or six species are externally similar; two, *G. soricina* and *G. commissarisi,* occur mainly in rainforest and three occur more often in drier habitats, but two or three species may occur together in Central America and N South America. There is individual variation in color within a species: bats from dry habitats, and young bats, are gray.
Similar species. Chestnut long-tongued bats (*Lionycteris spurrelli*) have fur uni-

Map 51

Common long-tongued bats, *Glossophaga*

formly dark to base; all other long-tongued bats have lower jaw distinctly longer than upper; short-tailed fruit bats (*Carollia* spp.) have a free flange on the side of the horseshoe and a tubercule in the center of the chin notch; little fruit bats (*Rhinophylla*) have no tail and a flanged horseshoe.
Natural history. Feed on nectar, fruit, insects, and pollen. Common long-tongued bats are the least specialized of the long-tongued bats, and they have a broad diet. They roost in caves, buildings, tunnels, hollow trees, culverts, drains, or under bridges. They hang spaced apart in roosting colonies of from a half-dozen to many hundreds of individuals. Found throughout

the rainforest and in secondary forest, plantations, deciduous forest, and savanna. **Geographic range.** Central and South America: N Mexico south to SE Brazil, Paraguay, and N Argentina. *G. soricina,* from Mexico to Argentina, is the only species in the Amazon and Río Paraguay basins. *G. longirostris,* from Colombia, Venezuela, and Guyana; *G. alticola* on the Pacific coast of Central America from Mexico to El Salvador; *G. commissarisi* on the Pacific coast from Mexico to Peru. **Status.** Common and widespread. **References.** Webster, W. D., and J. K. Jones, Jr. 1980. Taxonomic and nomenclatorial notes on bats of the genus *Glossophaga* in North America, with description of a new species. *Occas. Pap. Mus. Texas Tech. Univ.* 7:1–12.

Chestnut Long-tongued Bat
Lionycteris spurrelli
Plate D, map 52
Identification. Measurements: HB = 50–60; T = 6–9; HF = 9–13; E = 12–15; FA = 32–37; WT = 7–11 g.
Upperparts rich, uniform chocolate to chestnut brown; fur unicolored to base, short to medium length, **velvety;** area between shoulder blades sometimes with some gray hairs; wings and tail membrane blackish brown. **Muzzle slightly, not greatly, elongated,** lower jaw about as long as upper; **chin with a single, large, concave V-shaped pad with smooth sides, not a deep groove;** noseleaf small, without horseshoe; ear short, tip rounded. Tail short, half the length of membrane; tail membrane short, about half the length of leg at center; calcar about half as long as foot. **Underparts about the same as back or only faintly frosted slightly paler.** Young like adults.
Variation. Color varies individually, within populations, from chestnut to brown.
Similar species. Common long-tongued bats (Glossophaga spp.) have fur distinctly pale at base and a deep groove on chin with tubercules on edges; see common long-tongued bats for other comparisons.
Natural history. Roosts in caves and crevices and is found in rainforest, gardens, and plantations.
Geographic range. Central and South America; barely enters Panama near the

Map 52

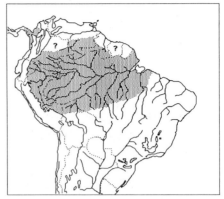

Chestnut long-tongued bat, *Lionycteris spurrelli*

Colombian border, south to N Bolivia and the southern Amazon Basin of Brazil, east to Belém. To 1,400 m elevation.
Status. Uncommon over large geographic range.
References. Williams, S. L., and H. H. Genoways. 1980. Results of the Alcoa Foundation–Suriname expeditions. II. Additional records of bats (Mammalia: Chiroptera) from Suriname. *Ann. Carnegie Mus.* 49:213–36.

Spear-nosed Long-tongued Bats
Lonchophylla spp.
Plate D, map 53
Identification. Measurements: HB = 46–78; T = 5–11; HF = 8–14; E = 12–18; FA = 30–45; WT = 5–21 g.
Upperparts pale orange, pale brown, smoky brown, or dark gray; fur bicolor with distinctly pale base, neck and shoulders often paler than back. Muzzle moderately, not extremely, long and narrow, lower jaw longer than upper; **noseleaf a well-developed spear,** no free horseshoe flange around base; chin tip a broad concave pad with a few tubercules on sides near lip, smooth sides farther down. Tail short; tail membrane short to quite long, but shorter than leg; calcar shorter than foot. Underparts the same as back or paler, grayish.
Species. Six of the about seven species occupy lowland rainforest: *L. thomasi* is small (5–6 g), dark smoky brown with fur distinctly pale at base; *L. mordax* is small

Map 53

Spear-nosed long-tongued bats, *Lonchophylla*

Map 54

Hairy-legged long-tongued bats, *Anoura*

(9 g), with fur barely paler at base than tip, usually with a dark belly washed with grayish (but one from Bahía has completely white underparts); *L. robusta* is large (14–17 g) and pale orange; *L. hesperia* is medium-sized and pale brown with pale, gray-brown underparts; *L. handleyi* is similar to *L. robusta*.

Similar species. The noseleaves on these bats are larger than on any similar species; common (*Glossophaga* spp.) and chestnut (*Lionycteris spurrelli*) long-tongued bats have lower jaws that do not protrude beyond upper; long-nosed (*Choeroniscus* spp.), Underwood's (*Hylonycteris underwoodi*), and Ega (*Scleronycteris ega*) long-tongued bats have extremely long, narrow, tubelike muzzles; dark long-tongued bats (*Lichonycteris* spp.) have tricolored fur; hairy-legged long-tongued bats (*Anoura* spp.) have no tail.

Natural history. Feed on nectar, insects, and probably fruit and pollen. *L. thomasi* roosts in small groups in hollow trees and small caves; *L. robusta* roosts in small groups in caves. These are the most common long-tongued bats in parts of the Amazon Basin. Found in rainforest, gardens, and plantations; some species occupy cloud forest and deciduous forest.

Geographic range. Central and South America: Nicaragua south to Bolivia and SE Brazil; *L. thomasi, L. mordax,* and *L. robusta* in Central and South America; *L. hesperia,* W coast of Peru and Ecuador; *L. handleyi* is from Ecuador and Peru; *L. bokermani* is from SE Brazil.

Status. Common to rare: *L. hesperia* is

known from only five individuals.

References. Hill, J. E. 1980. A note on *Lonchophylla* (Chiroptera: Phyllostomatidae) from Ecuador and Peru, with the description of a new species. *Bull. Brit. Mus. Nat. Hist. Zool.* 38:233–36.

Hairy-legged Long-tongued Bats
Anoura spp.
Plate D, map 54

Identification. Measurements: HB = 56–78; T = 0–7; HF = 10–14; E = 12–17; FA = 35–45; WT = 10–20 g.

Upperparts medium to dark brown, pale grayish brown, or blackish brown; fur quite short, bicolored, pale at base, often faintly frosted silver or pale; wings and tail membrane black or blackish brown. **Muzzle long and narrow, lower jaw longer than upper; noseleaf a small triangle,** whiskers around muzzle sometimes prominent; **ears short,** rounded, triangular. **Tail absent or a tiny, inconspicuous stub; tail membrane virtually absent, reduced to a narrow band down inside of hindleg to rump, this band hairy along edge;** calcar less than half length of foot. Underparts lightly frosted paler than back. Young like adults.

Species. The four species are all found in rainforest: *A. caudifera* is small (FA = 35–39; 10–13 g), brown, with a tiny tail, and a narrow, semicircular remnant tail membrane; *A. cultrata* is large (FA = 38–45; 14–23 g), blackish, often with a tiny tail; *A. geoffroyi* is large (FA = 41–45; 13–19 g), smoky brown, with no tail; *A. latidens* is large

(FA = 42–46), pale grayish brown, with pale throat and chest.
Similar species. Short-haired long-tongued bats (*Leptonycteris*) of arid habitats (not described in this book) have extremely short fur and are much larger; all other long-tongued bats have well-developed tail membranes; other bats with noseleaves and reduced or no tail membranes have short, broad muzzles.
Natural history. Feed on nectar, insects, fruit, and pollen. *A. cultrata* is restricted to montane areas and is known to roost only in caves and tunnels. The other species roost in small groups in caves or tree hollows and are found in lowland rainforest, deciduous forest, gardens, and plantations. Although they will use other roosts, these bats seem common only where there are caves, and they are rare or perhaps absent from lowland Amazonian forests lacking high ground, rocks, and caves.
Geographic range. Central and South America: Central Mexico (*A. geoffroyi*); or Costa Rica (*A. cultrata*) south to Peru, Bolivia, and SE Brazil. *A. caudifera* in South America east of the Andes only; *A. latidens* from Venezuela. To about 3,000 m elevation.
Status. Common to rare, restricted to widespread.
References. Tamsitt, J. R., and D. Nagorsen. 1982. *Anoura cultrata*. Mammalian Species, no. 179.
Handley, C. O., Jr. 1984. New species of mammals from northern South America: A long-tongued bat, genus *Anoura* Gray. *Proc. Biol. Soc. Washington* 97:513–21.

Dark Long-tongued Bats
Lichonycteris spp.
Plate D, map 55
Identification. Measurements: HB = 49–55; T = 6–10; HF = 8–11; E = 11–13; FA = 31–34; WT = 6–8 g.
Upperparts pale to dark brown; fur tricolor, dark at base and tip, pale brown in middle, dark base indistinct, looks like a shadow; elbows thinly furred above. Muzzle elongated, robust, not narrow and tubelike; lower jaw longer than upper; noseleaf small; whiskers prominent, chin whiskers as long as muzzle whiskers; ear very short, rounded. Tail short; tail membrane long; **calcar almost**

Map 55

Dark long-tongued bats, *Lichonycteris*

as long as foot. Underparts brown paler than back, about same color as central pale band of dorsal fur, bicolored faintly dark at base, paler at tip.
Species. Both species occupy rainforest: *L. obscura* is dark brown; *L. degener* is pale brown; some think the latter is a subspecies of the former.
Similar species. These are the only long-tongued bats with tricolored fur in South America; Underwood's long-tongued bats (*Hylonycteris underwoodi*, Central America) are similar but have a longer, narrower, tubelike muzzle and thickly furred elbows.
Natural history. Diet and roosts apparently unrecorded. Found in lowland and lower montane evergreen rainforest and plantations.
Geographic range. Central and South America: Guatemala south to Peru, NE Bolivia, and the Amazon Basin of Brazil. To about 1,000 m elevation.
Status. Rare, but widespread.

Ega Long-tongued Bat
Scleronycteris ega
Map 56
Identification. Measurements: HB ~ 57; T ~ 6; FA = 35.
Upperparts blackish brown; fur bicolor, paler brown at base, long; wings and tail membranes blackish brown. Muzzle greatly elongated, lower jaw projecting prominently forward and downward; a very long whisker grows from below front of ear. Distal thumb joint (with claw) distinctly long and thin, longer than first joint (in wing membrane). Tail short, tail

Map 56   Map 57

Underwood's long-tongued bat, *Hylonycteris underwoodi*

Ega long-tongued bat, *Scleronycteris ega*

Long-nosed long-tongued bats, *Choeroniscus*

membrane long, calcar about three-fourths length of foot; hindfoot tiny. Underparts brown, slightly paler than back, throat like back.

Similar species. Difficult to distinguish from long-nosed long-tongued bats (*Choeroniscus* spp.), which lack a long whisker in front of ear, have the distal thumb joint robust and about the same length as the first joint, and fur of medium length; dark long-tongued bats (*Lichonycteris* spp.) have tricolored fur; chestnut long-tongued bats (*Lionycteris spurrelli*) have unicolored fur.

Natural history. From lowland rainforest.

Geographic range. South America: central Amazon and upper Orinoco drainages of Brazil and Venezuela. Known from only four localities.

Status. Rare; known from only four individuals.

Long-nosed Long-tongued Bats
Choeroniscus spp.
Plate D, map 57

Identification. Measurements: HB = 51–66; T = 6–11; HF = 7–11; E = 11–13; FA = 31–41; WT = 5–8 g.

Upperparts dark brown, fur bicolor paler at base, sometimes indistinctly so; fur sometimes slightly frosted, medium length; elbows thinly furred; wings and tail membrane blackish brown. Muzzle greatly elongated into a long, narrow tube; lower jaw longer than upper; noseleaf small, triangular; line of nostrils close to horizontal; whiskers prominent, those from chin as long as those around nose; ears short, triangular, barely protruding above crown. Two thumb joints of about equal length, or second only slightly longer than first. Tail very short; tail membrane long, shorter than legs; calcar about three-fourths as long as foot. Underparts the same as back or slightly paler. Tiny, very long-nosed bats.

Species. The three to five species all occupy rainforest: *C. godmani* has fur pale at base; *C. minor* has fur almost unicolored; *C. periosus* is large (FA = 41); *C. intermedius* has medium brown fur pale at base and is small (FA = 33–35).

Similar species. Ega (*Scleronycteris ega*) and dark (*Lichonycteris* spp.) long-tongued bats are similar; all have the second thumb joint (with claw) distinctly longer than the first (in wing membrane). Dark long-tongued bats have tricolor fur and nostrils set at about 45 degrees. Ega long-tongued bats have a long whisker growing from near ear and long fur.

Natural history. The single described roost was a group of eight under a fallen log over a stream. Found in rainforest and openings in the forest such as marshes and plantations, and montane forests.

Geographic range. Central and South America: Mexico to N Bolivia and the Amazon Basin of Brazil.

Status. Uncommon to rare but widespread.

Underwood's Long-tongued Bat
Hylonycteris underwoodi
Map 56
Identification. Measurements: HB = 48–60; T = 4–10; HF = 7–11; E = 10–13; FA = 31–34; WT = 6–9 g.
Upperparts uniform dark brown; fur tricolor, dark at base and tip with paler brown center, dark band at base indistinct, looks like a dark shadow; wing and tail membrane blackish brown; elbows thickly furred above and below. Muzzle elongated, narrow, tubelike, lower jaw longer than upper; ear short, rounded, narrow across base; noseleaf short. Tail short; tail membrane long; calcar more than half as long as foot. Underparts brown, about the same as back or with more orange tints, fur bicolor dark at base. Tiny, dark, narrow-muzzled bats.
Variation. Some Underwood's bats from drier areas of Mexico have bicolored fur (some of these have tricolor fur on throat) and are paler; those from Costa Rica and Panama are very dark, almost blackish, with dark band at base of fur distinct.

Similar species. Dark long-tongued bats (*Lichonycteris* spp.) are the only other long-tongued bats with tricolored fur; they have robust muzzles and thinly haired elbows. Long-nosed long-tongued bats (*Choeroniscus* spp.) are difficult to distinguish from bicolor-furred individuals; Underwood's bats generally have hairier elbows and longer second thumb joint but probably cannot be distinguished externally with certainty.
Natural history. Feeds on pollen, fruit, insects, and probably nectar. Underwood's bats roost in small groups of two to eight in caves, hollow logs, tunnels, and under bridges. They may be chiefly montane. Found in rainforest, deciduous forest, and cloud forest.
Geographic range. Central America: SW Mexico to W Panama. To about 2,600 m elevation.
Status. Rare, geographic range patchy and much deforested.
References. Jones, J. K., Jr., and J. A. Homan. 1974. *Hylonycteris underwoodi.* Mammalian Species, no. 32.

Short-tailed and Little Fruit Bats (Subfamily Carolliinae)

This subfamily includes plain little bats without any striking distinguishing features. They have medium-sized muzzles, well-developed, medium-sized noseleaves and ears; short or absent tails; and moderate to greatly reduced tail membranes. They all seem to live in the rainforest understory, where they feed on small fruits, but there is no natural history information on several species. The two genera and seven species are all found in the Neotropical rainforest region.

Short-tailed Fruit Bats
Carollia spp.
Plates E, 6, map 58
Identification. Measurements: HB = 44–74; T = 6–12; HF = 9–18; E = 15–23; FA = 33–45; WT = 9–25 g.
Upperparts gray to warm brown, rarely rusty; fur moderately long and soft, parts easily to show strongly to indistinctly three- to four-banded hairs, dark at base, often looks variegated; wings and tail membranes gray-brown. Muzzle short and narrow, noseleaf quite large, reaches to eye when flattened, spear-shaped, horseshoe around nostrils with free flange on side; chin with a V-shaped line of small round warts surrounding a large central wart; ears medium, triangular, pointed,

Map 58

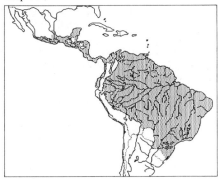

Short-tailed fruit bats, *Carollia*

reach about to nose when flattened forward. **Tail short, tip sticks up through tail membrane; tail membrane long,** but shorter than legs, calcar shorter than foot. Underparts uniform, about the same color as back, or frosted. Young gray.
Species. The four species all occupy lowland rainforest: *C. perspicillata,*
C. brevicauda, and *C. subrufa* are all usually gray-brown, with distinctly banded variegated-looking fur. *C. perspicillata* is large (FA = 40–45; 10–23 g) and in some populations there are rusty-tinted individuals (not seen in the other gray-brown species). *C. castanea* is small (FA = 33–38; 10–13 g) and uniform chestnut, with bands on fur indistinct. Two to three species occur together in most of the rainforest region.
Similar species. Common long-tongued bats (*Glossophaga* spp.) have fur pale at base and no wart in center of chin V; little fruit bats (*Rhinophylla* spp.) have no tail.
Natural history. Feed on fruit and insects, supplemented by nectar in the dry season. These are often the most numerous bats in lowland rainforests, and they seem most common in disturbed habitats. They use the understory vegetation levels, where they concentrate their feeding on the fruits of shrubs and treelets, especially the slender, green, candlelike fruits of plants of the genus *Piper.* Because of their high numbers, short-tailed fruit bats are probably one of the most important seed dispersers for *Piper* and many other plants with small fruits. They have been shown to be instrumental in carrying seeds into cleared areas and thus stimulating forest regeneration after a small clear-cut. They usually fly about a kilometer from their roost to feed. Short-tailed fruit bats roost in small groups, with individuals separated, in tree hollows, caves, overhanging banks, tunnels, culverts, or abandoned buildings. Roosting groups consist of small harems of females with one male, or bachelor males. Found in mature and disturbed rainforest, gardens and plantations, deciduous forest and gallery forest.
Geographic range. Central and South America: S Mexico to Bolivia, Paraguay, and SE Brazil. To at least 2,400 m elevation.
Status. Common and widespread.

Map 59

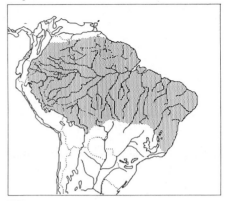

▨ Little fruit bats, *Rhinophylla*

References. Fleming, T. H. 1988. *The short-tailed fruit bat: A study in plant-animal interactions.* Chicago: University of Chicago Press.
De Foresta, H., P. Charles-Dominique, C. Erard, and M. F. Prévost. 1984. Zoochorie et premiers stades de la régénération naturelle après coupe en forêt guyanaise. *Rev. Ecol.* 39:369–400.

Little Fruit Bats
Rhinophylla spp.
Plate E, map 59
Identification. Measurements: HB = 43–68; T = 0; HF = 8–12; E = 13–17; FA = 31–38; WT = 7–13 g.
Upperparts pale gray, pale brown, or blackish brown; fur unicolored gray or brown to base except tips of hairs slightly darker, no distinct pale basal bands on hair; fur dense. **Muzzle short,** medium width; **noseleaf large, spear-shaped, when flattened reaches well behind eye to center of forehead, horseshoe well developed, with free flange on sides;** chin with a large, smooth, sculptured pad, no tiny warts; **ears triangular, medium,** reach to front of eye when flattened forward. **Tail absent; tail membrane rudimentary, either a deeply notched V or reduced to a narrow band on inside of leg** from calcar to rump; this band with or without hairy fringe; calcar half the length of foot, or less. **Underparts about the same as back** or frosted, fur unicolor gray or brown to base. Young like adult.

Species. The three species all occupy rainforest: *R. pumilio* is gray, with tail membrane V-shaped and hairless, with a band across rump; *R. fischerae* is beige, with tail membrane a hairy band down leg, not forming a band between legs; *R. alethina* is blackish, with tail membrane a band down leg only, legs and feet very hairy.

Similar species. Short-tailed fruit bats (*Carollia* spp.) have tails; hairy-legged long-tongued bats (*Anoura* spp.) have long muzzles, short noseleaf and ears, and pale hair base; yellow-shouldered fruit bats (*Sturnira* spp.) are larger ($>$ 15 g), have broad muzzles and short, broad noseleaf that does not reach behind eye; fruit-eating bats (*Artibeus* spp.) have broad muzzles and often stripes on face.

Natural history. Feeds on fruit of understory shrubs. Found only in lowland rainforest or gardens and plantations in forest; rare in deciduous forest.

Geographic range. South America: *R. pumilio* in the Amazon Basin and the Guianas; *R. fischerae* along the main Amazon and its upper western tributaries in Colombia, Ecuador, and Peru; *R. alethina* is known only from one locality on the west coast of Colombia. To 1,400 m elevation.

Status. *R. alethina* apparently rare; *R. fischerae* and *R. pumilio* locally common to uncommon.

References. Tuttle, M. 1970. Distribution and zoogeography of Peruvian bats, with comments on natural history. *Univ. Kansas Sci. Bull.* 49:45–86.

Neotropical Fruit Bats (Subfamily Stenodermatinae)

Dental formulas are highly variable. Stenodermatines are small to large; they have short, broad muzzles, large eyes, and no tails; tail membranes much shorter than the legs or sometimes absent; medium-length plain triangular ears; and all but two species have medium-sized, fleshy, spear-shaped noseleaves. Many species have striking pale stripes; the basic stripe pattern is a single stripe down the middle of the back from neck to rump and four stripes on the face, one above and one below each eye. The development of each of these stripes varies from species to species, from bright to faint to none. These bats all feed on fruit, supplemented by flower nectar in the dry season when fruit is scarce. To feed they pick a fruit from a tree and fly away, carrying it in the mouth to a feeding roost, which is usually a thin horizontal stem beneath the shelter of foliage. A favored spot is under a horizontal palm leaf. There the bat chews up the fruit and sucks the juices from indigestible fibers, which are spit out on the ground in small wads, along with any large seeds. A bat will use the same roost many times, and one can find these throughout the forest: a pile of seeds, feces, and wads of fibers beneath a large leaf (which may be several meters above the ground or above the roost). Many stenodermatines roost among foliage, where their white stripes probably act as camouflage. Bats of several genera in this subfamily make "tents" of large leaves, usually palms or palmettos, by biting the sides and ribs of a leaf in a characteristic pattern (forming a row of jagged holes, unlike the rows of round holes drilled by insect larvae through rolled leaves), so that the sides of the leaf drop down to form a shelter. Tent-making species can be found during the day by quietly approaching a strangely drooping leaf with bitten sides and looking under it without touching it (the bats are wary and will fly off if too much noise is made or the roost is touched; beware of wasps—they often use an old bat tent to shelter their own nest). Neotropical fruit bats are the main seed dispersers for many plants. They are particularly important in carrying seeds for early secondary or successional growth into gaps in the forest caused by natural disturbances or cutting by man, thus restoring new forest, and they are important for maintaining the plant species richness of the forest. This is the largest subfamily of leaf-nosed bats, with about 16 genera.

Map 60

Yellow-shouldered fruit bats, *Sturnira*

Yellow-shouldered or Hairy-legged Fruit Bats
Sturnira spp.
Plate E, map 60
Identification. Measurements: HB = 54–101; T = 0; HF = 7–21; E = 15–22; FA = 39–62; WT = 13–67 g.
Upperparts gray, pale brown, or yellow-brown; fur conspicuously tricolored, frosted with dark hair tips overlying silver or pale beige, hair base dark; shoulders and sometimes neck of adults stained yellow to orange with secretion from glands, hair sometimes stuck together with thick, dark secretion; yellow shoulders usually most prominent on large males. **Muzzle short, broad;** noseleaf a short, broad spear, not reaching beyond eye when flattened against muzzle; ear short, triangular; eye quite large. **Tail absent; tail membrane virtually absent, reduced to a narrow band; feet and inner side of legs very hairy;** thumb long and robust. Underparts gray, or gray-brown, like pale part of dorsal fur. **Medium to large robust, roly-poly woolly looking bats with large heads, thick necks, burly bodies, and short, strong, hairy legs.**
Species. There are about 12 species; 6 occupy lowland and montane rainforest; the others are mostly montane in the N Andes. Species are mainly distinguished by features of the teeth and are difficult to identify in the hand.
Similar species. Vampires (*Desmodus* spp., *Diphylla*) do not have spear-shaped noseleaves; other hairy-legged bats with no

tail membranes have facial stripes or long, narrow muzzles.
Natural history. Feed on fruit and nectar; especially fond of the nectar of banana flowers. These bats are often common, but little is known about them. They seem most numerous in open spaces in the forest such as gardens, clearings, streams, and edges. Their normal roosts are unknown and are probably somewhere high in the forest canopy (or they would have been discovered). Rarely they have been found in caves, hollow trees, tunnels, or culverts. The advantage of lack of a tail membrane is seen in a bat in the hand, which can twist and turn its legs and body like a contortionist. It seems likely that yellow-shouldered bats are adapted for clambering around quadrupedally among flowers and fruits. Found in lowland and montane rainforest, deciduous forest, cloud forest, gardens, and plantations.
Geographic range. Central and South America: central Mexico south to Uruguay. To at least 3,500 m elevation.
Status. Common to very rare (*S. nana*); widespread (*S. lilium*) to restricted (*S. aratathomasi, nana*).
References. Tamsitt, J. R., and C. Haüser. 1985. *Sturnira magna*. Mammalian Species, no. 240.
Soriano, P. J., and J. Molinari. 1987. *Sturnira aratathomasi*. Mammalian Species, no. 284.
Molinari, J., and P. J. Soriano. 1987. *Sturnira bidens*. Mammalian Species, no. 276.

Tent-making Bats
Uroderma spp.
Plates E, 6, map 61
Identification. Measurements: HB = 55–68; T = 0; HF = 10–15; E = 14–20; FA = 39–47; WT = 14–21 g.
Upperparts dark to pale gray or gray-brown or beige; fur bicolor, pale at base; midback with prominent white stripe. Face with faint to prominent white stripes above and below eye; muzzle short and broad; **noseleaf a large fleshy spear,** reaching to eye when pressed back, **horseshoe well developed, with a large free flange all the way around base,** pale colored on sides where facial stripe ends; ears triangular, cream-colored along side of rim

Map 61

Map 62

Tent-making bats, *Uroderma*

White-lined bats, *Vampyrops*

by facial stripes, tragus pale; **lips and tongue blackish. Tail membrane naked, a deeply notched V.** Underparts gray or gray-brown slightly paler or more gray than back.

Species. There are two species, both in rainforest: *U. bilobatum* is dark gray-brown, with strong facial stripes; *U. magnirostrum* is beige, with faint facial stripes, head, neck and shoulders paler than back, and edges of wings translucent. The black tongue fades in preserved specimens.

Similar species. White-lined (*Vampyrops* spp.) and great stripe-faced (*Vampyrodes* spp.) bats have pink tongues and usually a hairy fringe on the tail membrane; yellow-eared bats (*Vampyressa* spp.) are smaller (FA = 30–37); big-eyed bats (*Chiroderma* spp.) have fur with a dark band at base.

Natural history. Feed on fruit, nectar, and insects. These bats roost under "tents" which they make by cutting a line down each side of the midrib of a large leaf, such as banana, so that the sides droop downward. A small (2–52) group of bats hang in a tight cluster from the midrib of the leaf. Found in mature and disturbed rainforest, gardens and plantations, and deciduous forest.

Geographic range. Central and South America: S Mexico south to N Bolivia and SE Brazil. To about 1,000 m elevation.

Status. Common (*U. bilobatum*) or uncommon (*U. magnirostrum*) and widespread.

References. Baker, R. J., and C. L.

Clark. 1987. *Uroderma bilobatum.* Mammalian Species, no. 279.

White-lined Fruit Bats
Vampyrops spp.
Plate E, map 62

Identification. Measurements: HB = 53–103; T = 0; HF = 10–19; E = 15–25; FA = 35–63; WT = 13–63 g.

Upperparts blackish brown to warm brown; midback with bright white stripe; shoulders but not top of neck often paler than back; elbows often densely furred; fur without dark band at base. Muzzle short; **facial stripes above eyes bright to indistinct, often fading out forward of rear edge of ear; stripes below eye less distinct, often narrow or fainter;** noseleaf a large fleshy spear, horseshoe at base broad and with free flange on sides and usually front, sides of horseshoe usually dark, sometimes cream, not bright yellow; ear sometimes with pale rim, usually not; **tongue pink. Tail membrane a deeply notched V, very short at center, always fringed with hair;** calcar less than half the length of foot. Underparts the same as back, or slightly frosted paler gray or brown. Young like adults. **Small to large, robust bats.**

Species. There are about 10 species; most are found in lowland rainforest. One of the most common and widespread throughout the region is the smallest, *V. helleri* (FA = 35–40; WT = 13–16 g); it is pale brown. Up to six species can occur in the same area; some are mostly montane.

Similar species. Great stripe-faced bats

(*Vampyrodes* spp.; FA = 45–56) are warm brown, have yellowish sides of horseshoe and lower ears, all facial stripes broad and bright, upper stripes sharp to behind ear, back stripe sharp to base of tail membrane, and elbows thinly haired, but are difficult to distinguish in the field from some species. Most yellow-eared bats (*Vampyressa* spp.) are smaller, with no fringe of hairs on tail membrane. Big-eyed bats (*Chiroderma* spp.) have a dark band at base of fur and usually indistinct white stripes. Tent-making bats (*Uroderma* spp.) have dark tongues.

Natural history. Feed on fruit, nectar, and insects; diets of some species are unknown. These bats roost in small groups of 6–20, probably harems, in caves and in tents they make from palm leaves. Found in mature and disturbed lowland rainforest, gardens, clearings and plantations, cloud forest, deciduous forest, and caatinga.

Geographic range. Central and South America: S Mexico south to Bolivia, Paraguay, Uruguay, and SE Brazil.

Status. CITES Appendix III for *V. lineatus* (Ur). Common to uncommon, widespread. To about 2,500 m elevation.

References. Willig, M. R., and R. H. Hollander. 1987. *Vampyrops lineatus.* Mammalian Species, no. 275.

Gardner, A. L., and D. C. Carter. 1972. A review of the Peruvian species of *Vampyrops* (Chiroptera, Phyllostomatidae). *J. Mammal.* 53:72–84.

Great Stripe-faced Bats
Vampyrodes spp.
Plate E, map 63

Identification. Measurements: HB = 73–80; T = 0; HF = 12–17; E = 20–25; FA = 45–57; WT = 28–38 g.

Upperparts warm cinnamon-brown, neck and shoulders paler than back, midback with sharp, bright white line from between ears to end of rump; elbows thinly furred; fur pale at base. Muzzle very broad, U-shaped; noseleaf a large fleshy spear; **white stripes above and below eye broad and distinct,** expanding posteriorly to almost coalesce behind ear, stripes reach rear edge of ear; **ear rim, tragus, and side of noseleaf pale to bright yellow; lips and tongue pink. Tail membrane a deeply notched V, very narrow at center of**

Map 63

Great stripe-faced bats, *Vampyrodes*

body, fringed with hair; calcar less than half the length of foot. Underparts gray-brown. Young like adults. **Large, brightly striped, brown bats.**

Species. Two species are usually recognized: *V. caraccioli* from South America and *V. major* from Central America; they are very similar and perhaps conspecific.

Similar species. White-lined bats (*Vampyrops* spp.), if brown, are usually smaller, with stripes less distinct, rims of ear and horseshoe whitish or gray, elbows often thickly furred; yellow-eared bats (*Vampyressa* spp.) are all much smaller; big-eyed bats (*Chiroderma* spp.) have dark band at base of fur and usually indistinct stripes; tent-making bats (*Uroderma* spp.) have no hairy fringe on tail membrane and black tongues; other Neotropical fruit bats do not have back stripe.

Natural history. Feed on fruit, especially figs. Great stripe-faced bats roost in small groups of two to four in foliage and under palm leaves; the leaves are not modified to form a tent, but few roosts have been reported. Found in lowland rainforest and in gardens and plantations.

Geographic range. Central and South America: S Mexico south to N Bolivia and the S Amazon Basin of Brazil. To about 1,000 m elevation.

Status. Sometimes common, widespread.

Big-eyed Bats
Chiroderma spp.
Plate E, map 64

Identification. Measurements: HB = 53–82; T = 0; HF = 10–15; E = 16–21;

Map 64

Map 65

▨ Big-eyed bats, *Chiroderma*

▨ Yellow-eared bats, *Vampyressa*

FA = 38–52; WT = 13–26 g.
**Upperparts brown to gray brown; fur
with distinct dark band at base especially
on lower back; midback with sharp,
faint, or no white line.** Muzzle short and
very broad, noseleaf with wide, flared-
out sides in leaf shape, horseshoe with
free flange around all of base; **middle upper incisor teeth long, narrow, with simple points;** eyes large; ears with rims or
base and sides of noseleaf yellow or
brown; **facial stripes strong to indistinct,**
most ending forward of ear, rarely to middle of ear; tongue pink. **Tail membrane**
slightly hairy above, without hairy fringe,
quite long at center, calcar about three-
fourths of length of foot. Underparts gray-
brown. Young like adults.
Species. Three of the five species are
found in the rainforest region; two or three
can occur together: *C. trinitatum* is the
smallest (FA = 38–43; WT = 13–15 g)
and has a middorsal stripe; *C. salvini* is
large (FA = 46–52) and has large, bright
facial stripes; *C. villosum* is large
(FA = 44–50) and has indistinct facial
stripes; *C. doriae* occupies deciduous forest
in SE Brazil.
Similar species. Yellow-shouldered bats
have no tail membrane and no stripes; no
other striped bats of this size have a dark
band at base of hair (see yellow-eared
bats).
Natural history. Feed on fruit, especially
figs. *C. villosum* roosts in tree hollows;
roosts of the other species are apparently
unrecorded. Found in mature and disturbed
rainforest, gardens, and plantations.

Geographic range. Central and South
America: S Mexico south to Bolivia and
the S Amazon Basin of Brazil. *C. trinitatum* and *C. villosum* from Central America
to South America east of the Andes;
C. salvini to west of the Andes. To 2,200
m elevation.
Status. All species common to uncommon
and widespread except *C. doriae* and
C. improvisum (Guadeloupe), which are
geographically restricted and rare.

Yellow-eared Bats
Vampyressa spp.
Map 65
Identification. Measurements: HB = 45–
62; T = 0; HF = 7–12; E = 11–18;
FA = 29–39; WT = 7–16 g.
**Upperparts gray-brown to tawny or cin-
namon-brown, sometimes slightly frosted;**
neck and shoulders paler; **fur with faint
(like a shadow) to distinct dark band at
base; midback with or without white
stripe; wings and tail membrane brown,**
usually pale on forward edge. Muzzle short
and broad; noseleaf a large fleshy spear,
horseshoe with complete free flange around
base; **rims and base of ears, tragus, and
sides of noseleaf bright yellow; white facial stripes distinct, stripes above eyes
broad, sometimes coalescing above nose,**
often extending to rear of ear, stripes below eyes fainter; tongue pink; central upper
incisors bilobed, much longer than outer
incisors. **Tail membrane** deeply notched,
lightly haired above near body, **without
hairy fringe in bats with middorsal
stripe, with slight hairy fringe in center**

in bats without stripe; calcar less than half the length of foot. Underparts gray-brown little paler than back. Young gray or gray-brown. **Small, chunky bats.**
Species. Four of the five species occupy lowland rainforest; the other is montane. Up to three species might occur together. *V. pusilla* is the only lowland species with no back stripe; it is the smallest (FA = 29–33; WT = 7–8 g) and is tawny brown. The other lowland species are gray-brown, with a back stripe and no hairy fringe on tail membrane (*V. brocki, V. nymphaea, V. bidens*).
Similar species. White-lined bats (*Vampyrops* spp.) have a bright middorsal stripe, hairy fringe on tail membrane, and fur with no dark band at base; tent-making bats (*Uroderma* spp.) have blackish tongue and fur with no dark band at base; big-eyed bats (*Chiroderma* spp.) of this size have facial stripes ending at base of ear; Macconnell's bat (*Mesophylla macconnelli*) has no facial stripes; dwarf fruit-eating bats (*Artibeus* spp.) have no middorsal stripe and short facial stripes that end at base or middle of ear.
Natural history. Feed on fruit, especially figs; the diet of most species is unrecorded. *V. pusilla* has been found roosting singly in a tent made from a *Philodendron* leaf. This is apparently the only record of a roost site for the genus. Found in mature and disturbed rainforest, gardens, and plantations.
Geographic range. Central and South America: S Mexico south to S Peru, the S Amazon Basin of Brazil, and Paraguay and SE Brazil.
Status. Most species widespread, uncommon.
References. Peterson, R. L. 1968. A new bat of the genus *Vampyressa* from Guyana, South America, with a brief systematic review of the genus. Roy. Ont. Mus. Life Sci. Contrib., no. 73.
  Lewis, S. E., and D. E. Wilson. 1987. *Vampyressa pusilla.* Mammalian Species, no. 292.

Macconnell's Bat
Mesophylla macconnelli
Map 66
Identification. Measurements: HB = 44–50; T = 0; HF = 9–11; E = 12–15; FA = 29–33; WT = 7–9 g.

Map 66

Macconell's bat, *Mesophylla macconnelli*

Head and shoulders pale whitish beige, back pale beige; no white stripes on head or body. Wings gray-brown, **wrist, distal forearm, and thumb yellow; skin on forearm with tiny pale speckles. Noseleaf yellow,** a narrow spear, horseshoe with complete free flange at base; **ears all yellow** or pale beige toward tip; eyes large, sometimes with dark smudge in front of corner. **Tail membrane short, semitransparent, naked,** calcar less than half the length of foot. Underparts pale whitish beige, palest on throat. Tiny, delicate bats.
Variation. This species is close to and may belong in the genus *Vampyressa;* it is sometimes placed in the genus *Ectophylla,* to which it does not seem closely related.
Similar species. Little yellow-eared bats (*Vampyressa pusilla*) are similar but have facial stripes, gray wrists and thumbs, and a slight fringe of hair in center edge of tail membrane; dwarf fruit-eating bats (*Artibeus* [*Dermanura*]) usually have facial stripes and gray thumb and are usually uniform gray or brown.
Natural history. Macconnell's bats have been found roosting in a hollow tree, and a group of three was roosting under a tent made from a single, broad palm leaf about 150 cm high. The leaf had been cut in a broad V from its opposite edges to the midrib. Found in mature and disturbed lowland rainforest.
Geographic range. Central and South America: Costa Rica south to N Bolivia and the S Amazon Basin of Brazil. To about 1,000 m elevation.
Status. Uncommon and widespread.

Map 67

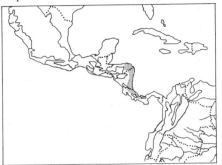

Honduran white bat, *Ectophylla alba*

Map 68

Fruit-eating bats, *Artibeus*

References. Gardner, A. L. 1977. Chromosomal variation in *Vampyressa* and a review of chromosomal evolution of the Phyllostomidae (Chiroptera). *Syst. Zool.* 26:300–318.

Honduran White Bat
Ectophylla alba
Plate 6, map 67
Identification. Measurements: HB = 37–47; T = 0; HF = 9–10; E = 14–15; FA = 26–30; WT = 4–6 g.
Head and shoulders white, grading to gray-white on midback and pale brownish white on lower back; wings blackish brown except forearms and thumbs yellow, propatagium translucent. Face with dark smudge from corner of eye to nose; **noseleaf bright yellow,** a narrow, large spear, horseshoe with free flange entirely around base; **ears bright yellow;** eyes large, blackish. Tail membrane naked, semitransparent, short; calcar less than half as long as foot. Throat white; belly pale gray or gray-brown. **Tiny, delicate bats.**
Similar species. Ghost bats (*Diclidurus* spp.) have white ears and no noseleaf and are larger; there are no other white bats in the region.
Natural history. Roosts in groups of one to six under tents made from large, horizontal, bananalike leaves of *platanillos* (*Heliconia* spp). The bats bite the leaf along each side of the midrib until the sides fold down to form a shelter. Found in mature, disturbed, and secondary rainforest and plantations of trees.
Geographic range. Central America: Honduras and Nicaragua south to Panama, mostly in the Caribbean lowlands only.
Status. Unknown; rare and potentially threatened by deforestation.
References. Timm, R. M. 1982. *Ectophylla alba*. Mammalian Species, no. 166.
Timm, R. M., and J. Mortimer. 1976. Selection of roost sites by Honduran white bats, *Ectophylla alba* (Chiroptera: Phyllostomatidae). *Ecology* 57:385–89.

Large Fruit-eating Bats
Artibeus spp.
Plates E, 6, map 68
Identification. Measurements: HB = 70–130; T = 0; HF = 12–21; E = 18–28; FA = 48–75; WT = 25–86 g.
Upperparts almost black to gray or brown; fur pale at base or unicolor; wings and tail membrane black in gray or black animals, brown in brown; **no white stripe on midback. Head large and powerful; four pale facial stripes usually present,** sometimes absent, stripes below eye often indistinct or absent, **upper stripes end at level of ear; muzzle short, broad; upper tooth rows horseshoe-shaped; tongue black; noseleaf a broad fleshy spear, reaches to rear corner of eye or beyond when pressed back;** horseshoe either with free flange all around base or only at sides, usually blackish, with pale edges in brown species; eyes large, dark brown; **ears medium-sized, triangular, tragus and lower rims pale or whole ear dark gray;** chin with V-shaped row of small tubercules with pad in center; central incisors short, bilobed. Tail membrane usu-

ally short to very short at center, U- or V-shaped. **Thickset, large, muscular bats** that bite fiercely when handled.
Species. There are about nine species, all but one in lowland rainforest (the dozen small species are described in the following account). Up to four species can occur together. The most common include: *A. jamaicensis,* large (50–60 g) dark brown to dark gray, fur slightly frosted, with distinct pale band at base, facial stripes distinct but faint; *A. lituratus,* very large (50–86 g) brown, wings brown, facial stripes broad and bright (these two species are among the most common bats throughout the Neotropical rainforest); *A. obscurus,* large (40 g), rich blackish to dark chocolate, fur long, only slightly paler at base, or unicolor, facial stripes faint; *A. hirsutus,* medium (24–30 g) pale gray with long fur, faint facial stripes, hairy tail membrane. There is disagreement about how many species are represented in this group and what to call them.
Similar species. Big-eyed bats (*Chiroderma* spp.) have a distinct dark band at base of fur, long, narrow, single-lobed middle upper incisors, a pink tongue, and usually a middorsal stripe; spear-nosed bats (*Phyllostomus* spp.) have tails.
Natural history. Feed mainly on fruit, especially figs, nectar, and insects. Large fruit-eating bats are some of the most important dispersers of seeds into early successional forest. Jamaican fruit bats (*A. jamaicensis*) are often the most common bats in the forest, along with common short-tailed fruit bats (*Carollia* spp.). They fly low in the understory and use trails as flyways. When encountered as they fly to a feeding roost carrying a fruit, they may drop the fruit in surprise, seeming to throw it toward the observer as they veer sharply to avoid collision. Fruit-eating bats roost in small groups in hollow trees and branches, among dense foliage, under palm leaves, in caves, or occasionally in buildings, and at least one species sometimes uses tents made from leaves. *A. jamaicensis* groups consist of harems of one male and several females that prefer to roost in tiny tree hollows, and groups of bachelor males that roost in foliage or tents. In caves, many harems may roost close together. Found in mature and secondary rainforest and deciduous forest, gardens, and plantations.
Geographic range. Central and South America: central Mexico south to Paraguay and N Argentina.
Status. Widespread and common.
References. Owen, R. D. (1987). Phylogenetic analyses of the bat subfamily Stenodermatinae (Mammalia: Chiroptera). Spec. Pub. Mus. Texas Tech., no. 26.
  Morrison, D. W. 1978. Foraging ecology and energetics of the frugivorous bat *Artibeus jamaicensis. Ecology* 59:716–23.

Dwarf Fruit-eating Bats
Artibeus (Dermanura) spp.
Map 68
Identification. Measurements: HB = 47–69; T = 0; HF = 9–14; E = 16–20; FA = 34–52; WT = 10–22 g.
Upperparts usually gray or gray-brown, sometimes pale brown or dark chocolate-brown, fur either pale or faintly dark at base; **wings and tail membrane blackish. Facial stripes present or absent, usually end at upper edge of ear, rarely at middle of ear; no midback stripe.** Noseleaf spear-shaped, often pale on sides of base; ears triangular, rims sometimes cream or yellow; four upper incisors of about equal size, center incisors bilobed; muzzle broad and powerful, upper tooth row horseshoe-shaped. Tail membrane with or without fringe of hairs on edge, V-shaped, often short; if very short, then edge with hairy fringe; if longer, without hairy fringe.
Species. There are about a dozen species; all probably occur in lowland rainforest. Most cannot reliably be distinguished in the hand. Species include: *A. hartii* (formerly *Enchisthenes*), medium (13–17 g), dark chocolate-brown, blackish on head, tail membrane short and hairy, facial stripes narrow pale brown; *A. glaucus,* small (12 g), uniform smoky brown or gray, fur uniformly dark to base, facial stripes bright, rims of ears, tragus, and sides of noseleaf pale yellow, wings black; *A. watsoni,* small (13 g), pale beige to gray, stripes prominent, tail membrane naked. Some place this group in a separate genus.
Similar species. Yellow-eared bats (*Vampyressa*) have brown wings, facial stripes to middle or rear of ear; Macconnell's bats (*Mesophylla macconnelli*) have no stripes,

yellow thumb and wrist, semitransparent tail membrane.

Natural history. Feed on small fruits and insects. At least two species roost in tents made from palm leaves, and another species has been found roosting under a banana leaf. Found in mature and disturbed lowland and montane rainforest, plantations and gardens, and cloud forest and deciduous forest.

Geographic range. Central and South America: central Mexico south to N Bolivia and the S Amazon Basin of Brazil. To at least 2,400 m elevation.

Status. Uncommon to rare, most species widespread.

References. Handley, C. O., Jr. 1987. New species from northern South America: Fruit-eating bats, genus *Artibeus* Leach. *Fieldiana Zool.* 39:163–72.

Map 69

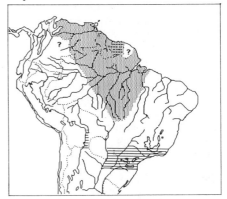

▨ Little white-shouldered bat, *Ametrida centurio*
≡ Ipanema or double-lipped bat, *Pygoderma bilabiatum*

Ipanema or Double-lipped Bat
Pygoderma bilabiatum
Plate E, map 69

Identification. Measurements: HB = 60–85; T = 0; HF = 12–15; E = 17–21; FA = 36–41; WT = 15–22 g; females strikingly larger than males.

Upperparts warm brown; forequarters paler than rump; fur long, tricolor with dark bands at base and tip; shoulder at junction of wing with pure white spot; wings and tail membrane brown. Head broad and short; **mouth wide, upper lip with a fold from base of noseleaf to corner of mouth forming a "double" lip; insides of upper and lower lips at corner of mouth with dense fringe of long papillae; noseleaf a large, narrow, fleshy spear; chin bald, without tubercules;** ears triangular, broad at base, with a partial membrane across head from back of upper side, tragus yellow; eyes large, chocolate-brown. Tail membrane short, V-shaped, well haired, with hairy fringe. **Chest thinly haired,** almost naked in males, hairer in females; belly medium brown, frosted.

Similar species. No similar bats have a double lip; little white-shouldered bats (*Ametrida*) are smaller and have a white spot below ear; the two other bats with shoulder spots do not have large spear-shaped noseleaves.

Natural history. Apparently feeds on soft, pulpy fruits. Found in mature and second-ary forests and plantations with fruit trees, in rainforest and deciduous forest.

Geographic range. South America: known from three disjunct areas—Surinam; south-central Bolivia and adjacent Argentina; and E Paraguay to SE Brazil.

Status. Locally common but patchy, rare in many areas; widespread.

References. Webster, W. D., and R. D. Owen. 1984. *Pygoderma bilabiatum.* Mammalian Species, no. 220.

Myers, P. 1981. Observations on *Pygoderma bilabiatum* (Wagner). *Z. Saugetierk.* 46:146–51.

Little White-shouldered Bat
Ametrida centurio
Map 69

Identification. Measurements: HB = 38–55; T = 0; HF = 10–11; E = 13–16; FA = 25–33; WT = 8–12 g; females larger than males.

Upperparts warm pale brown to cinnamon, males more dusky, with grayer tints than females; forequarters paler than hindquarters; fur tricolor, dark at base and tip; shoulder at angle of wing with pure white spot; neck below ear with smaller, fainter white spot; wings and tail membrane brown. **Face short and broad; mouth wide and monkeylike, muzzle naked,** lips with fringe of papillae inside; **noseleaf a short broad spear flattened back against base of eyes; eyes large and bulging, males with a swollen pad below.**

Ears triangular, broad at base, brown with yellow base and tragus. Tail membrane short, V-shaped, hairy, with hairy fringe. *Similar species.* See double-lipped bat. *Natural history.* Found in lowland rainforest and clearings and edges derived from it. *Geographic range.* Central and South America: one individual is known from the Canal Zone of Panama; all others are from Venezuela, the Guianas, and the eastern Amazon Basin of Brazil. *Status.* Rare in a large geographic range.

Visored Bat
Sphaeronycteris toxophyllum
Plate E, map 70

Identification. Measurements: HB = 52–63; T = 0; HF = 11–13; E = 13–17; FA = 37–42; WT = 18 g; females larger than males.

Upperparts cinnamon-brown, males sometimes tinted gray; fur tricolored, dark band at base and tip; **shoulder at angle of wing and neck below ear with pure white spots;** wings and tail membrane brown. Face short and broad, mouth wide, monkeylike, smiling; forehead with large, fleshy, horizontal fold (visor), much larger in males than females, males appear to be wearing a baseball cap; noseleaf ∩-shaped, pushed forward by bottom of visor; muzzle and chin naked; eyes large, bulging, metallic golden brown; ears triangular, base yellow, tip brown, tragus yellow; **chin with fold of skin that can be pulled up over face, fold has translucent spots over eyes when pulled up.** Tail membrane short, V-shaped, hairy. Underparts like back.
Similar species. No other bats have a face like this (see double-lipped bat [*Pygoderma bilabiatum*]); wrinkle-faced bats (*Centurio senex*) have no forward-projecting visor or noseleaf.
Natural history. Found in many habitats, including mature and secondary rainforest, cloud forest, and deciduous forest, and in gardens, plantations, and pastures. *Geographic range.* South America: Venezuela and the western Amazon Basin in Colombia, Peru, Brazil, and N Bolivia. To 2,200 m elevation.
Status. Rare to uncommon in a large geographic range.

Map 70

▨ Wrinkle-faced bat, *Centurio senex*
▤ Visored bat, *Sphaeronycteris toxophyllum*

Wrinkle-faced Bat
Centurio senex
Plate E, map 70

Identification. Measurements: HB = 53–67; T = 0; HF = 10–15; E = 13–19; FA = 42–47; WT = 16–25 g.
Upperparts warm pale brown to beige; fur tricolor, dark at base and tip; **shoulder at angle of wing with pure white spot;** wing brown, **membranes between third–fourth and fourth–fifth fingers (nearest to body) with striking ladderlike pattern** of alternating translucent and opaque bars, most pronounced in males; forearms very hairy. **Head and face a round ball, face naked, with deep, complex skin-folds and channels between eyes and around nostrils; mouth very wide;** eyes large and bulging; **no free noseleaf; chin with loose fold of skin that can be pulled up over eyes,** translucent spots where fold covers eyes, this mask well developed in males, less so in females; males only have additional large folds or lappets on the chin; **ears long, narrow, thin, naked, rounded,** brown distally, yellow basally, **upper edge of ears with large distinctive horizontal lobes covering part of brow.** Tail membrane narrow, V-shaped, hairy. Throat whitish, belly colored like back. Neck very short, head appears stuck onto large, square, squat body. Males have a musky odor.
Variation. Males have more extremely developed facial folds than females.
Similar species. No other bats have a face

like this or laddered wings; visored bats (*Sphaeronycteris toxophyllum*) have a nose-leaf and projecting visor on forehead. *Natural history.* Feeds on fruits, from which it extracts the juice. Wrinkle-faced bats roost among foliage. Males roost singly or in groups of two or three; a number may be dispersed around the same tree. Females roost in dense clusters. While roosting, the mask of skin is pulled up tightly completely over the face and forehead and covers the horizontal flaps of the ears; a ridge across the crown holds it in place. Wrinkle-faced bats fly rapidly, like big, heavy beetles, with a wobbly motion, sometimes with the body vertical to the ground. Found in mature and secondary rainforest, deciduous forest, gallery forest, plantations, and gardens. Apparently adapts well to extremely disturbed habitats and can live in city parks and scrubby forest near cane fields.
Geographic range. Central and South America: coastal Mexico south through Panama, coastal N Colombia, Venezuela, and Trinidad. To about 1,400 m elevation.
Status. Rare, but widespread and with wide habitat range.
References. Snow, J. L., J. K. Jones, Jr., and W. D. Webster. 1980. *Centurio senex.* Mammalian Species, no. 138.

Goodwin, G. G. and A. M. Greenhall. 1961. A review of the bats of Trinidad and Tobago. *Bull. Amer. Mus. Nat. Hist.* 122:187–302.

Vampire Bats (Subfamily Desmodontinae)

Dental formula: I1/2 or 2/2; C1/1; P1/2; M1/1; 2/1, or 2/2 = 20 or 26. These are the only true vampire bats in the world. They are highly specialized for feeding on the blood of birds and mammals. They have small folds above the nose and no spear-shaped structures; long, razor-sharp, caninelike, forward-pointing upper incisor teeth; and large, bladelike canine teeth. There is no tail, and the tail membrane is reduced. The legs and forearms are long, and the body is slender. The thumbs are long and thickened or have pads and are used as feet to walk, hop, or climb quadrupedally along the ground or on branches, to approach prey quietly. These bats never suck blood from their prey, but lick up the drops flowing from a small scoop they bite in the skin. The razor-sharp teeth make a quick and painless wound, without waking up sleeping prey. Vampires are usually feared by local people, who often have never seen one at close range and thus do not discriminate between vampires and other species of bats. There is a tendency to believe that larger bats are all vampires. People therefore kill bats at random (much as they behave toward snakes)—a practice to be discouraged because bats are beneficial. Vampires are an extraordinary example of complex evolutionary adaptation and should be objects of wonder rather than fear. They are related to the leaf-nosed bats, but they are sometimes placed in a separate family of their own. The two genera and three species are restricted to the New World tropics and subtropics.

Common Vampire Bats
Desmodus spp.
Plates F, 6, map 71
Identification. Measurements: HB = 72–93; T = O; HF = 14–22; E = 16–21; FA = 49–63; WT = 26–48 g.
Upperparts rich dark brown, rarely orange; fur short and straight, with a strong sheen, often silvery; hair on neck and shoulders sometimes whitish, with a prominent pale base; fur on back often sparse, especially on forequarters; wings blackish, wingtips black or white. Muzzle short, nostrils in small M-shaped fold, with another, ∩-shaped fold behind this, no free noseleaf, no horseshoe under nostrils; ears triangular with pointed tips, tragus hairy; chintip a deeply notched V, undersurface of chin naked, with dot of a gland at base of throat; middle upper incisors larger than canines, triangular and forward-pointing, canines triangular, very sharp. Thumbs long, either with distinct pads under joints or with entire length thickened; legs long, feet with long toes and claws. Tail membrane brown, a narrow U-shaped band, hairy. Underparts sometimes sharply demarcated from sides,

Map 71

 Common vampire bats, *Desmodus*

silvery gray or white with sheen, brown hair base showing through.
Species. There are two species, both in rainforest: the common vampire, *D. rotundus,* has stiff fur, with bright tips producing a coarse sheen, and black wingtips (except in Chile, where they are white); the white-winged vampire, *D. youngi,* sometimes placed in the genus *Diaemus,* has silky fur with a fine shine and sharply demarcated pure white wingtips.
Similar species. Hairy-legged vampires (*Diphylla ecaudata*) have long, dense fur and no tail membrane; wrinkle-faced (*Centurio senex*) and visored (*Sphaeronycteris toxophyllum*) bats have white shoulder-spots; all other tailless bats have noseleaves (except some bats restricted to Caribbean islands and not described in this book).
Natural history. Feed on blood. Common vampires feed on mammals, including humans, nowadays often on domestic ungulates, especially cattle and horses. They feed by silently approaching an animal (these bats fly extremely quietly), landing near it or on it, making a small scoop in the skin, and lapping up flowing blood with the tongue. White-winged vampires prefer to feed on the blood of birds (but if starving in captivity will attack a mammal); they are not known to attack cattle or humans. Vampires are agile quadrupedal runners and will readily run and hop on all fours, using the thickened thumb as a foot. Both species roost in caves and hollow trees. Common vampires have a complex social organization in which females form stable groups of 8–12 individuals that

roost together, feed in the same home range, and cooperate by sharing food with each other by regurgitation. Males form more temporary associations and compete for dominance of a group of females. These bats seem naturally rare in deep rainforest, but large concentrations of prey in the form of livestock introduced by man have caused them to become common in some areas, where they can transmit rabies, cause wounds susceptible to screwfly invasion, and bite humans. These species should be protected in forest, and only populations that feed on livestock should be controlled. Found in rainforest, deciduous forest, secondary vegetation, gallery forest, cerrado, chaco, pastures, and gardens.
Geographic range. Central and South America: N coastal Mexico, S Mexico south along both sides of the Andes to central Chile and S Argentina and Uruguay. To 1,500 m elevation.
Status. Common to rare (common vampire), or rare (white-winged vampire); both species widespread.
Local names. Vampiro (Span, Br); vampir (Su).
References. Greenhall, A. M., and U. Schmidt. 1988. *Natural history of vampire bats.* Boca Raton, Fla.: CRC Press.
Wilkinson, G. S. 1985. The social organization of the common vampire bat. *Behav. Ecol. Sociobiol.* 17:111–21.

Hairy-legged Vampire Bat
Diphylla ecaudata
Plate F, map 72
Identification. Measurements: HB = 72–86; T = O; HF = 13–19; E = 15–19; FA = 50–56; WT = 23–33 g.
Upperparts rich brown, pale to almost blackish, or near chestnut; fur soft and dense, with strong sheen, hair on shoulders white at base; forearms hairy. **Noseleaf very short, consisting of two inconspicuous, ∩-shaped folds, one behind the other; face attractive, mouth with a permanent pleasant smile; eyes large; ears short and rounded, lower edge prolonged forward in a small fold, forms a line under eye to nose.** Incisors as in common vampires. **Tail membrane almost absent, a narrow band down side of leg, hairy above and below; calcar forms a spur**

Map 72

▨ Hairy-legged vampire bat, *Diphylla ecaudata*

with free tip. Thumb long and thick. Chin naked; underparts gray-brown with sheen.
Variation. Paler animals seem to come from drier habitats.
Similar species. Other vampires (*Desmodus* spp.) have short fur and tail membrane that forms a band between legs; see common vampires for other comparisons.
Natural history. Feeds on blood, chiefly or exclusively of birds. Hairy-legged vampires

roost singly or in small colonies in caves, tunnels, and hollow trees. These vampires do not seem to be as agile at walking quadupedally on flat surfaces as are the common vampires; they approach their prey by crawling upside down along the bottom of a branch. They feed like the common vampires, by biting a small incision and licking up the drops of blood that flow from the wound. They sometimes attack chickens or other domestic poultry but do not usually seem to do them much harm. Found in rainforest, deciduous forest, dry forest, caatinga, and gardens and plantantions.
Geographic range. North, Central, and South America: Texas (one record) south along the Caribbean coast of Mexico, all countries of Central America, south to Bolivia and SE Brazil. To about 1,500 m elevation.
Status. Rare or uncommon, but extremely widespread.
References. Greenhall, A. M., U. Schmidt, and G. Joermann. 1984. *Diphylla ecaudata*. Mammalian Species, no. 227.

Funnel-eared, Thumbless, and Sucker-footed Bats (Natalidae, Furipteridae, Thyropteridae)

Dental formula: I2/3, C1/1, P3/3, M3/3 = 38. These are three small families of tiny insectivorous bats. The funnel-eared bats (Natalidae) includes one genus and four species. The thumbless bats (Furipteridae) includes two genera, each with one species. The sucker-footed bats (Thyropteridae) includes one genus with two species. The ecology of all species is poorly known; they all have a fluttering style of flight and may be specialized for hunting insects within dense vegetation. All are restricted to the New World tropics and subtropics.

Funnel-eared Bats
Natalus spp.
Plates F, 7, map 73
Identification. Measurements: HB = 45–64; T = 45–60; HF = 7–11; E = 12–16; FA = 36–42; WT = 4–7 g.
Upperparts pale beige, tawny yellow-brown, gold, or bright rust-orange; fur long, lax, slightly wavy; wings and tail membrane pale brown. **Muzzle slightly elongate** and narrow, **forehead rising abruptly above; nose and chin simple, with no leaf or folds, mouth somewhat funnel-shaped; ears triangular, broad at

base, funnel-shaped, pointing forward; eyes small,** inconspicuous. **Tail very long, about equal to head and body,** entirely enclosed within membrane; **tail membrane slightly longer than legs, comes to a point at tail tip;** calcar longer than foot; **hindlegs very long, leg to tip of claws about the same length as head and body;** thumbs tiny. Underparts the same color as back or slightly paler. **Tiny bats** with a large expanse of delicate, crinkled membrane.
Species. Two of the four species are found in lowland, continental rainforest; the other

Map 73

Funnel-eared bats, *Natalus*

Map 74

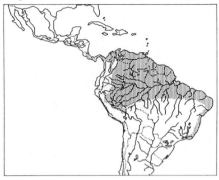

Thumbless bat, *Furipterus horrens*

two occupy Caribbean islands: *N. stramineus* is slightly smaller (HB = 45–48), with the tail usually slightly longer than the head and body; *N. tumidirostris* is larger (HB = 50–64), with the tail shorter than the head and body; color varies individually from beige to orange in both species.
Similar species. Thumbless bats (*Furipterus horrens*) have no thumb claw, and the tail is much shorter than tail membrane; in all vespertilionid bats the tail and tail membrane are much longer than the legs, and held curled forward in a coil at rest. No other bats with no noseleaf have a long tail entirely within the tail membrane, which forms a point at its edge.
Natural history. Feed on insects. Funnel-eared bats roost in the dark recesses of humid caves, where they hang singly or in groups of well-separated individuals, sometimes in colonies of thousands. They may be restricted to regions with caves (which are absent in much of the Amazon Basin). They fly with a fluttering flight. Found in rainforest, but more often in more arid habitats of deciduous or dry forest and gardens and plantations.
Geographic range. Central and South America: both coasts of Mexico almost to the U.S. border, south through Central America and along the north and east coast of South America to SE Brazil, and the lower Amazon Basin.
Status. Widespread but generally patchy in distribution, numerous in a few caves.

Thumbless Bat
Furipterus horrens
Plate F, map 74
Identification. Measurements: HB = 33–43; T = 20–27; HF = 6–9; E = 9–12; FA = 30–40; WT = 3 g.
Upperparts dark smoky gray, slate blue-gray, or dark gray-brown; **fur long, dense, and lax,** wings and tail membrane dark brown. Muzzle short, narrow, **nose and chin plain,** without folds or leaf, **forehead large, rising abruptly at right angle to muzzle;** ear short, broad at base; eye tiny, lost in fur. **Tail about half as long as tail membrane,** entirely enclosed within it; tail membrane longer than leg, entirely filled with pattern of close-set, fine, parallel lines at right angles to tail; **thumb a rudimentary stub, with no claw;** legs long, almost as long as head and body. Underparts slate gray. **Tiny bats like long-legged balls of gray fluff.**
Similar species. All other bats in rainforest have well-developed thumbs with claws.
Natural history. Feeds on insects. Thumbless bats roost in small clusters in colonies of up to at least 60 in caves, horizontal fallen logs, and deep cracks between rocks; one such roost was among large boulders in a riverbed, which were exposed during the dry season. They forage near the forest floor, with a slow, fluttering, mothlike flight. Found in lowland rainforest.
Geographic range. Central and South America: Costa Rica south to the southern Amazon basins of Peru and Brazil, and to SE coastal Brazil.

Map 75

Sucker-footed bats, *Thyroptera*

Status. Rare, but widespread.
References. LaVal, R. 1977. Notes on some Costa Rican bats. *Brenesia* 10–11:77–83.

Sucker-footed Bats
Thyroptera spp.
Plates F, 7, map 75
Identification. Measurements: HB = 37–48; T = 24–37; HF = 4–7; E = 11–14; FA = 31–38; WT = 3–5 g.
Upperparts rich chocolate-brown to chestnut; fur long and soft. Face short, muzzle narrow, without noseleaf; ears triangular, broad at base, sharply pointed; eyes tiny. **Ankle- and thumb-bases with moist, fleshy sucker disks. Tail membrane pointed, longer than legs; tail** longer than membrane, within it to posterior edge, **tip protrudes free of membrane for a short distance beyond it. Hindfoot tiny, with third and fourth digits fused into one thick digit.** Tiny, delicate bats.
Species. The two species are found in rainforest: *T. tricolor* has sharply contrasting white or cream underparts and two

small bumps on rear edge of calcar; *T. discifera* has underparts the same as back or only slightly paler, and one bump on rear of calcar.
Similar species. No other bats in region have suckers on feet.
Natural history. Feed on insects. *T. tricolor* roost in small groups of 1–9 inside the rolled new leaves of *platanillos* (*Heliconia* spp.) and bananas, at the stage when the leaves are just unrolling to open. Bats must therefore move from leaf to leaf every day or two. The bats in a group stay together as they move from roost to roost. *Platanillos* tend to grow on relatively good soil in forest openings, such as treefall gaps, and on riversides, and that is where these bats can be found, but they choose roosts in the shade. When the females have young, they can move to "maternity" roosts in hollow logs on the forest floor. *T. discifera* is poorly known, but it has been found roosting beneath open, dead banana leaves. Unlike most other bats, sucker-footed bats roost head upward, attached to the flat surface of a leaf by their suckers. Their local distribution seems to be patchy; they are apparently limited to areas where they find numerous appropriate roosts. Found only in lowland rainforest and gardens and plantations derived from it.
Geographic range. Central and South America: SE Mexico south through Central America to the southern Amazon Basin and the Atlantic forests of SE Brazil.
Status. *T. discifera* is rare; *T. tricolor* is uncommon and apparently patchy in a huge geographic range.
References. Wilson, D. E., and J. S. Findley. 1977. *Thyroptera tricolor*. Mammalian Species, no. 71.
    Wilson, D. E. 1978. *Thyroptera discifera*. Mammalian Species, no. 104.

Vespertilionid Bats (Vespertilionidae)

Dental formula: I1/3 or 2/3; C1/1; P1/2, 2/2, or 3/3; M3/3. This is a large family of small- to medium-sized bats that feed on insects. They have plain faces with no noseleaves, small eyes, and long, wide tail membranes which come to a point at the tip of the long tail. At rest the tail and membrane are curled forward under the body. During flight at the moment of prey capture, the large tail membrane is brought forward and used as a scoop to help trap flying insects. Most species are agile and rapid flyers. In northern regions, most species hibernate in caves during the winter, or both migrate and hibernate, or migrate far enough south to be able to remain active. Vespertilionids have one or, occasionally, two to four young, which are left behind in the roost while the mother forages. The natural history and social behavior of Neotropical species are poorly known. This is the most geographically widespread family of bats; members are found on all continents, with about 37 genera and over 300 species worldwide.

Little Brown Bats
Myotis spp.
Plate F, map 76

Identification. Measurements: about HB = 36–54; T = 30–42; HF = 7–10; E = 11–17; FA = 29–44; WT = 4–7 g.
Upperparts blackish brown, smoky brown, chestnut, reddish, or orangish; fur short or medium length, unicolor entirely dark to base or with tips frosted; wings and tail membrane usually blackish, sometimes brown. **Head triangular, broad between ears, coming to point at nose, without noseleaf** or other folds; **eyes tiny; ears triangular, pointed; first tooth behind canine tiny and peglike, creating a large gap between canine and first large tooth behind it. Tail membrane much longer than legs, coming to a point at tip of tail, tail entirely within membrane, usually curled forward under body when not in flight; calcar shorter than foot.** Thumb with long distal joint. Underparts paler than back with hairs frosted pale at tips, dark at base, rear of abdomen usually paler than chest.
Species. This is the largest and most widespread genus of bats in the world, with about 88 species. About 8 of the 35 New World species are found in rainforest; 3 or 4 can occur together at one locality. They are often quite similar. Some of the most common and widespread in rainforest include: *M. nigricans,* almost black (FA = 33–36); *M. riparius,* smoky gray or brown (FA = 32–35); *M. albescens,* fur sometimes frosted with pale tips (FA = 34–37).
Similar species. Most big brown bats (*Eptesicus* spp.) are larger (10 g); they have even fur that is very dark at the base, often blackish, evenly tipped over the entire upperparts with slightly to distinctly paler brown, which gives the appearance of a rich, smooth sheen; the first tooth behind the canine is large, leaving no gap between the canine and the first large tooth behind it. Little yellow bats (*Rhogeessa* spp.) have fur pale at base; sheath-tailed bats (Emballonuridae) have a short tail whose tip sticks

Map 76

▨ Little brown bats, *Myotis*

out above the tail membrane partway along its length; thumbless bats (*Furipterus* spp.) have a short tail and no thumb claw; funnel-eared (*Natalus* spp.) bats are pale, with funnel-shaped ears and mouth.
Natural history. Feed on tiny insects. Little brown bats seem to favor flying in open areas or gaps in the forest, such as those formed by treefalls, streams, trails, roads, swamps or pools, and gardens, or in clearings around houses. They fly swiftly, with rapid wing beats, making frequent dives and turns to catch their prey. They may be high or low in the forest. Forest species roost in tree hollows, rock crevices, or buildings. Most species of this genus roost in close contact in tight clusters. In all lowland habitats.
Geographic range. Worldwide; subarctic to equatorial; in the entire rainforest region. To at least 2,400 m elevation.
Status. Rainforest species are common to rare; most appear widespread.
References. LaVal, R. K. 1973. A revision of the Neotropical bats of the genus *Myotis. Los Angeles County Nat. Hist. Mus. Sci. Bull.* 15.
Wilson, D. E., and R. K. LaVal. 1974. *Myotis nigricans.* Mammalian Species, no. 39.

Map 77

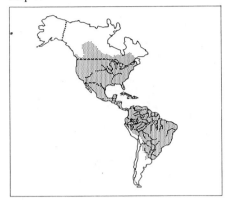

Big brown bats, *Eptesicus*

Big Brown Bats
Eptesicus spp.
Plates F, 7, map 77
Identification. Measurements: HB = 48–84; T = 33–49; HF = 8–13; E = 11–19; FA = 37–54; WT = 5–16 g.
Upperparts black-brown to tawny yellow-ocher; fur deep blackish at base with tip always paler, from slightly paler, rich brown, to much paler tawny, on dark animals paler tips impart a sheen; **fur medium-long, of even length, often silky; wings and tail membranes usually black,** sometimes dark brown, strong and leathery. **Face plain, pointed; without noseleaf; ears triangular, pointed; eyes small; first tooth behind upper canine large, leaving little gap between canine and first large tooth. Tail shorter than head and body length (40–90%), entirely within membrane, tail membrane comes to a point at tail tip, tail held curled forward under body at rest;** calcar longer than foot. Underparts yellowish, paler than back, hair blackish at base, usually tawny to gold at tips.
Species. There are about 30 species worldwide; 4 of the about 6 New World species can occupy lowland rainforest habitats: *E. fuscus* is large (FA = 48–54; WT ~ 15 g) and tawny yellow-ocher; *E. brasiliensis* is dark brown to reddish and smaller (FA = 40–46; WT = 9–12 g); *E. furinalis* is dark brown to orange-tinged brown, to dark tawny (FA = 37–43; WT = 7–10 g); and *E. andinus* is blackish (FA = 42–

48; WT = 11 g) and largely montane.
Similar species. Little brown bats (*Myotis* spp.) have a tiny first tooth behind the canine, leaving an apparent gap; usually have unicolor fur except that *M. albescens,* which are very similar to *E. furinalis* but smaller (FA = 34–37), have underparts frosted with more whitish or grayish, whereas in *E. furinalis* they are yellow or tawny.
Natural history. Feed on insects. Big brown bats roost in tree hollows, holes in snags in lagoons, buildings, and caves. *E. fuscus* flies in a steady, straight course, with occasional dives to catch prey, and it favors unencumbered areas. These bats are often in and around buildings and country towns. Found in rainforest, cloud forest, gallery forest, deserts, gardens, plantations, and fields.
Geographic range. Worldwide including the Americas from Canada to Argentina and Uruguay.
Status. Rare in rainforest; common in other habitats.
References. Hill, J. E., and D. L. Harrison. 1987. The baculum in the Vespertilioninae (Chiroptera: Vespertilionidae) with a systematic review, a synopsis of *Pipistrellus* and *Eptesicus* and description of a new genus and subgenus. *Bull. Brit. Mus. Nat. Hist.,* zool. ser. 52, no. 7.

Black-winged Little Yellow Bat
Rhogeessa tumida
Plate F, map 78
Identification. Measurements: HB = 36–49; T = 26–31; HF = 6–8; E = 12–14; FA = 25–33; WT = 3–5 g.
Upperparts pale yellow lightly to heavily frosted or overlaid with brown; fur bicolored pale yellow at base with brown tips; wings blackish or brown. Face triangular, nose pointed, plain, no noseleaf; eyes tiny; **ears triangular,** narrow at base, pale at base with dark tips; tragus a long, narrow blade. Tail membrane much longer than legs, coming to point at tip of tail, tail entirely within it; tail shorter than head and body. Underparts pale yellow or gold. Miniscule bats.
Variation. Varies geographically from pale in Central America to dark brown–frosted in South America. Five other species occupy dry habitats.

Map 78

Map 79

 Black-winged little yellow bat, *Rhogeessa tumida*

Big-eared brown bats, *Histiotus*

Similar species. Little brown bats (*Myotis* spp.) have fur dark at base; funnel-eared bats (*Natalus* spp.) have pale membranes, funnel-shaped ears, tail as long as head and body.
Natural history. Feeds on insects. Little yellow bats roost in colonies in hollow trees. They start to forage at dusk, sallying out from the shelter of trees with a more rapid flight than sheath-tailed bats. This genus includes the smallest New World bats. Found in a wide range of habitats from rainforest to dry forest.
Geographic range. Central and South America: E Mexico south through all of Central America to W and N Colombia and Ecuador, Venezuela, the Guianas, Trinidad, N Bolivia, and the Amazon Basin of Brazil to coastal Bahía.
Status. Apparently rare but very widespread.
References. LaVal, R. K. 1973. Systematics of the genus *Rhogeessa* (Chiroptera: Vespertilionidae). Occas. Pap. Mus. Nat. Hist. Univ. Kansas, 19:1–47.

Big-eared Brown Bats
Histiotus spp.
Plate F, map 79
Identification. Measurements: HB = 54–68; T = 47–58; HF = 8–10; E = 27–35; FA = 45–51; WT = 14–16 g.
Upperparts dark brown, tawny brown, or beige; fur long and lax, very dark brown at base with paler tips. Face triangular, nose plain and sharply pointed,

without noseleaf; ears enormous, broad and long, inner edges almost meet over brow; tragus long, lance-shaped. Tail membrane much longer than legs, tail within it to edge, then tip extends free beyond it for a short distance. Underparts frosted white or beige, hair dark brown at base.
Species. There are about four species; they are mostly montane or temperate, but *H. velatus* may be found in Atlantic coastal rainforests and the cloud forest species *H. montanus* may occasionally be found below 1,000 m.
Similar species. There are no other small bats with enormous ears and no noseleaf on the South American continent.
Natural history. Feed on insects. *H. velatus* has been found roosting in small clusters in the roofs and attics of buildings. Found in mature and disturbed premontane and montane humid forests and in temperate forests.
Geographic range. South America: the entire Andean chain and subtropical and temperate South America, mountains of Venezuela, greater Rio São Francisco drainage of SE Brazil, and Paraguay. To above 2,000 m elevation.
Status. All species apparently rare; widespread to restricted in range.
References. Mumford, R. E., and D. M. Knudson. 1978. Ecology of bats at Vicosa, Brazil. *Proc. 4th Inter. Bat Res. Conf.*, 287–95.

Map 80

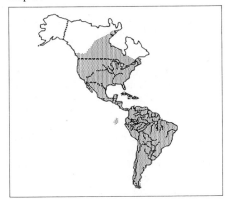

▨ Hoary or hairy-tailed bats, *Lasiurus*

Hoary or Hairy-tailed Bats
Lasiurus spp.
Plates F, 7, map 80
Identification. Measurements: HB = 52–90; T = 41–61; HF = 7–14; E = 10–20; FA = 36–57; WT = 6–21 g.
Upperparts rust-red, yellow, or brown frosted white, often frosted or variegated; fur dense, hairs with several color bands, dark band at base; forearm hairy below and with patches of hair near thumbs; **tail membrane long, wide, and thick, entirely filling space between legs, thickly furred above for half or more of its length;** wing membranes with pale patches surrounding bones. **Head short, muzzle blunt,** nostrils often widely separated, no noseleaf; **ears thick, short, broad, rounded, scarcely protruding above crown;** tragus a narrow panel displaced on a stalk in front of ear. Tail long, entirely within membrane, **at rest tail is bent forward and upward so that hairy tail membrane covers lower belly; posture at rest characteristic, curled forward in a tuck.** Underparts dirty yellow or red, hair dark at base. **Medium-sized, often beautifully patterned and brightly colored bats.**

Species. Four of the seven species can be found in lowland rainforest: *L. cinereus* is large (FA = 50–57), either brown (Central America) or pale yellow (Venezuela), strongly variegated or frosted with white; *L. borealis* is small (FA = 36–42), bright rust-orange on lower back, forequarters frosted with red over yellow, underparts yellowish; *L. ega* is medium-sized (FA = 47–51), entirely dirty yellow, without frosting; *L. egregius* is rust-red over yellow, with red underparts.
Similar species. No other bats in region have the combination of short blunt head and ears, no noseleaf, and long, thickly furred tail membrane.
Natural history. Feed on insects, especially moths and other large kinds, and have been reported preying on tiny species of bats. Hoary bats often fly high, with a strong, fast flight and slow wing beats. They sometimes forage around streetlights in towns. They roost singly among foliage, especially along forest edges, in places well sheltered from above, and their variegated colors probably serve as camouflage against the leaves. They can have litters of up to four young. Hoary bats include the most widespread species of New World bats. In North America they are seasonally migratory, but their behavior in the tropics is poorly known. Found in a wide variety of habitats from deep undisturbed Amazon rainforest to many kinds of disturbed forest, dry forest, and temperate and boreal forests.
Geographic range. North, Central, and South America: Canada to Chile. To at least 1,500 m elevation.
Status. Uncommon in rainforest; all rainforest species widespread except probably *L. egregius*.
References. Shump, K. A., Jr., and A. U. Shump. 1982. *Lasiurus borealis*. Mammalian Species, no. 183.
———. 1982. *Lasiurus cinereus*. Mammalian Species, no. 185.

Free-tailed or Mastiff Bats (Molossidae)

This is one of the most distinctive families of bats. They are small- to large-sized bats with long, thick, naked tails that extend free for about half their length beyond the edge of the tail membrane. They have short, velvety fur; leathery wings and tail membranes; short, broad hairy feet; rectangular, flattened bodies; very narrow wings; doglike faces with no noseleaf; wide mouths; and medium-sized black eyes. Their ears are complexly folded and distinctive: they are divided by a longitudinal fold into two open compartments, one facing outward and the other inward, over the eye; the forward or upper edge of the ear extends as a thick fold across the brow; the lower edge of the ear forms a fold that curves around forward under the ear, and this fold has a separate lobe called the antitragus that stands up in front of the ear opening and hides the true tragus, which lies behind it. The precise shape of both antitragus and upper edges of the ear is an important character for identification of genera and species. Bats of this family chiefly feed on large insects such as moths and beetles. They can be recognized by their strong, swift, erratic flight and narrow wings. Some species have echolocation calls in the frequency range audible to humans, and they can be clearly heard uttering musical single chirps as they cruise and bursts of sound as they dive to capture prey. The flattened bodies of these bats allow them to roost in narrow crevices in caves, between rocks, under bark, or in tree holes. Worldwide they have become commensal with man, roosting by the hundreds tightly packed in tiny spaces in the roofs of buildings. They are muscular and agile at crawling or scuttling quadrupedally. During the day in the roost they become torpid, with body temperature near ambient levels. They are able to open their mouths extremely wide, and when threatened they draw back their lips and open their mouths in a ferocious-looking threat. These are the common larger bats that fly swiftly over the rooftops of tropical towns at dusk. Large populations in caves feed on enormous numbers of insects and produce guano and are particularly beneficial to man. There are about 14 genera and 90 species worldwide, with 10 genera in the New World.

Dog-faced Bats
Molossops spp.
Plate G, map 81
Identification. Measurements: HB = 43–66; T = 21–29; HF = 5–11; E = 12–15; FA = 30–38; WT = 5–16 g.
Upperparts dark brown, fur pale at base. Fur on forehead diminishes gradually down face between eyes. Muzzle pointed; lips without folds, upper lip angles back steeply from nose to undershot mouth; nostrils set in slightly raised pads with hard upper rim; center of muzzle flat, without ridge. **Ears simple, triangular, pointed, upper edges widely separated on crown,** reach forward of eye when flattened, **division into two longitudinal compartments rudimentary, accordianlike slight flexible fold where upper edge of ear meets head, lower edge of ear not folded back. Tail about 40–50% of the length of head and body, free from membrane for less than half its length.** Round gland on throat present in males. Underparts like back or frosted pale, or gray-brown or cinnamon, throat sometimes white.
Species. The two species are found in lowland rainforest: *M. neglectus* is larger (FA = 36–38; WT = 15–16 g) and has dark underparts; *M. temminckii* is tiny (FA = 30–32; WT = 4–9 g) and has underparts frosted or paler than back.

Map 81

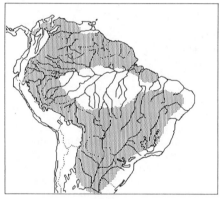

▨ Dog-faced bats, *Molossops*

Similar species. Flat-headed bats (*Neoplatymops mattogrossensis*) have bumps on the forearm and highly flattened body and head; doglike bats (*Cynomops* spp.) have no extra fold at upper edge of ear, that folds down strongly, forming separate compartment over eye, and hair of crown that ends abruptly on a line; mastiff bats (*Molossus* spp.) have the top of the muzzle raised in a ridge.
Natural history. *M. temminckii* roost in groups of two or three in hollow trees, rock crevices, buildings, and fence stakes.

They choose holes 1–3 m from the ground, with tiny entrances, often slits less than 2 cm wide or holes barely large enough for the bats to squeeze through. They are silent in the roost, so their presence is difficult to detect. They fly out at dusk for short foraging bouts, after which they return and rest in the roost. They frequent mature and highly disturbed rainforest, dry forest, areas around human habitations, and agricultural areas.

Geographic range. South America: Guyana, Venezuela, Colombia, Peru, Bolivia, Argentina, Paraguay, and Brazil south of the Amazon Basin (*M. temminckii*), or in the Amazon Basin (*M. neglectus*).

Status. Locally common to rare; *M. neglectus* is known from only 10 individuals, both species widespread.

References. Vizotto, L. D., and V. A. Taddei. 1976. Notas sobre *Molossops temminckii temminckii* e *Molossops planirostris* (Chiroptera-Molossidae). *Naturalia* 2:47–59.

Myers, P., and R. M. Wetzel. 1983. Systematics and zoogeography of the bats of the Chaco Boreal. Misc. Pub. Mus. Zool. Univ. Mich., no. 165.

Doglike Bats
Cynomops spp.
Plate G, map 82

Identification. Measurements: HB = 44–89; T = 22–40; HF = 8–13; E = 14–20; FA = 30–48; WT = 10–42 g. Upperparts blackish brown, chestnut, or orange-brown, fur short, dark or pale at base; **hair of crown ends abruptly, usually on line between ears. Muzzle broad, chin broad, rounded in profile; upper lip without vertical wrinkles, slightly puckered in to meet lower lip;** nostrils like holes punched in surface, not on raised pad, no sculptured surface around or between nostrils, top of muzzle between eyes and nostrils flat, without raised central ridge. **Ear rounded, folded longitudinally to form compartment over eye, when flattened forward reaches midway from eye to nose; upper edges of ears do not meet in center of forehead; rear edge of ear slightly folded backward,** antitragus almost rectangular, tragus a narrow spike. Throat gland present. Tail usually 40–50% of head and body length. Underparts dark,

Map 82

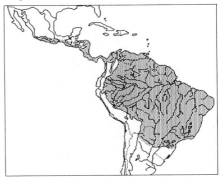

Doglike bats, *Cynomops*

slightly grayer and paler than back or throat and belly whitish.

Species. There are about five species; three occur in lowland rainforest: *C. abrasus,* medium-sized, often chestnut; *C. greenhalli,* small and blackish; *C. planirostris,* medium-sized, with pale fur base and whitish underparts. These species are sometimes placed in the genus *Molossops*.

Similar species. Dog-faced bats (*Molossops* spp.) have pointed ears and fur decreasing gradually down face between eyes; mastiff bats (*Molossus* spp.) have a raised ridge down center of face.

Natural history. Feed on insects. Doglike bats roost in small groups of up to eight in small holes in rotting trees in swamps, hollow branches, fence posts, and buildings. *M. planirostris* roosts are 1.5–5 m above the ground; the bats are noisy in the roost, twittering and scrambling around, especially just before they emerge at night. They often seem to forage around or over ponds or swamps, or in clearings. Rainforest, gallery forest, llanos, and savannas.

Geographic range. Central and South America: W coast of Mexico south through all Central and South American countries to Paraguay.

Status. Common to rare, widespread.

References. Vizotto, L. D., and V. A. Taddei. 1976. Notas sobre *Molossops temminckii temminckii* e *Molossops planirostris* (Chiroptera-Molossidae). *Naturalia* 2:47–59.

Map 83

Flat-headed bat, *Neoplatymops mattogrossensis*

Map 84

Brazilian free-tailed bat, *Tadarida brasiliensis*

Flat-headed Bat
Neoplatymops mattogrossensis
Plate G, map 83
Identification. Measurements: HB = 46–
54; T = 22–29; HF = 7–8; E = 13–15,
27–31; WT = 7–9 g.
**Upperparts dark brown variegated over
pale yellow,** hairs yellow with brown tips;
fur short, sparse; **forearm skin above
sprinkled with small bumps. Head flat,
muzzle long, pointed, upper lip under
nose angled steeply back to undershot
mouth; ears sharply directed forward,
upper edges do not meet in center of
crown. Body flattened. Throat and chest
pale orange to dirty yellow, forming a V
on belly sharply demarcated from brown
sides under wings.** Base of throat with
round gland in both sexes. Tiny, flat bats
with forward-pointing faces.
Variation. This species is sometimes
placed in the genus *Molossops*.
Similar species. These are the only tiny
free-tailed bats with bumps on forearm.
Natural history. Roosts under slabs of
rock and in rock crevices. The geographic
distribution of flat-headed bats may be in
part determined by the presence of appropriate roost sites. Found in arid and caatinga habitats, but also on some rare rock
mountains in rainforest.
Geographic range. South America: Venezuela, Guyana, and Brazil in Rio Branco
and from Matto Grosso and Bahía; distribution probably poorly known.
Status. Common to rare.

References. Willig, M. R., and J. K.
Jones, Jr. 1985. *Neoplatymops mattogrossensis.* Mammalian Species, no. 244.

Brazilian Free-tailed Bat
Tadarida brasiliensis
Plates G, 7, map 84
Identification. Measurements: HB = 49–
66; T = 29–43; HF = 8–11; E = 16–
20; FA = 36–46; WT = 11–13 g.
**Upperparts dark brown to gray-brown,
fur usually uniform to base,** sometimes
base paler. Muzzle upturned, pointed; **upper lip with deep vertical folds, hangs
over lower lip; chin and jaw flat,** face and
chin with short stiff, blunt bristles; nose
backed by hard flange. **Ears meet, or almost meet, on center of crown but are
not joined;** rear edge of ear not folded
back, **ear when flattened reaches tip of
nose.** Tail 50–70% as long as head and
body (usually about 60%). Underparts
lightly frosted slightly paler than back,
throat darker than abdomen.
Variation. There is some geographic variation in size.
Similar species. See broad-eared free-
tailed bats (*Nyctinomops* spp.).
Natural history. Feeds mainly on moths,
beetles, flies, and hymenopterans. These
bats form renowned colonies of millions in
caves in the SW United States, but in the
tropics they roost in more modest numbers
in tree holes, caves, rock crevices, and the
roofs of houses. When foraging, they fly
high and swiftly in open spaces. This is

not a rainforest species, but it is found in dry or montane habitats on the fringes of the entire rainforest region, and some records appear to be from within it, perhaps from cleared areas or savannas.

Geographic range. North, Central, and South America: the southern half of the United States to N Argentina.

Status. Locally common and widespread. Some populations are threatened and declining from overuse of toxic pesticides and disturbance of roosting caves by guano miners.

References. Constantine, D. G. 1967. Activity patterns of the Mexican free-tailed bat. Univ. N. Mex. Pub. Biol., no. 7.

Broad-eared Free-tailed Bats
Nyctinomops spp.
Plate G, map 85

Identification. Measurements: HB = 52–87; T = 34–56; HF = 8–13; E = 13–30; FA = 40–64; WT = 10–30 g.

Upperparts dark brown to reddish, fur either pale at base or uniform; **wings and tail membrane semitranslucent.** Muzzle pointed, upturned, **upper lip deeply wrinkled with vertical folds, overhangs lower lip; chin and jaw flat,** face and chin with a few long, slender bristles; nose upturned, nostrils raised on little tubes, backed by hard ridge. **Ears large, upper edges join on forehead on fleshy stalk,** rear edge of ear not folded back, ears **when flattened forward reach tip of nose.** Tail about 50–80% of head and body length. **Throat usually brown, darker than slightly frosted abdomen.**

Species. Three of the four species occur in lowland rainforest: *N. aurispinosus* has FA > 47; *N. laticaudatus* has FA = 40–46; *N. macrotis* has long, lax fur and huge ears and is the largest (FA = 58–64). These species are placed by some in the genus *Tadarida*.

Similar species. Brazilian free-tailed bats (*Tadarida brasiliensis*) have ears that meet but do not join on forehead; other free-tailed bats do not have wrinkled lips.

Sounds. Foraging *T. macrotis* often emit a loud piercing chatter. It is unclear whether these are audible echolocation pulses or communication calls.

Natural history. Feed on insects, mainly moths. Broad-eared free-tailed bats roost in

Map 85

▨ Broad-eared free-tailed bats, *Nyctinomops*

colonies of up to scores in caves, rock crevices, hollow trees, crevices between palm leaves, and buildings. They occupy many habitats from rainforest to arid scrub.

Geographic range. North, Central, and South America: SW United States to N Argentina and Uruguay.

Status. Apparently uncommon, but widespread.

References. Barbour, R. W., and W. H. Davis. 1969. *Bats of America.* Lexington: University Press of Kentucky.

Bonneted Bats
Eumops spp.
Plate G, map 86

Identification. Measurements: HB = 60–125; T = 27–68; HF = 9–20; E = 18–44; FA = 37–83; WT = 10–60 g.

Upperparts blackish, pale brown, cinnamon, gray-brown, or chestnut; fur usually pale at base. **Muzzle pointed, upper lip without vertical folds, slanting sharply back from nose; nostrils bounded behind by horny, raised, semicircular ridges that join to form a central keel between nostrils; no raised ridge down center of face; ears large, flattened forward like a hat pulled down over eyes, upper edges joining in midbrow;** highly developed deep longitudinal fold or keel separating the two compartments of ear; ear flattened forward reaches more than midway from eye to nose, sometimes to tip of nose; antitragus large, semiovate, broadest at base. Underparts like back or frosted beige; throat with

Map 86

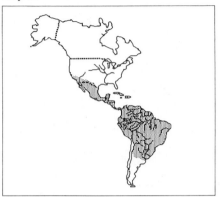

Bonneted bats, *Eumops*

Map 87

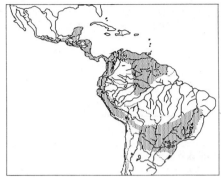

Crested mastiff bats, *Promops*

large round gland in males. Small to very large bats.

Species. Seven of the nine species have been found in lowland rainforest. Two species are noteworthy for their very large size—*E. perotis* (FA = 75–83) and *E. underwoodi* (FA ~ 71).

Similar species. Broad-eared free-tailed bats (*Nyctinomops* spp.) have wrinkled lips; no other free-tailed bats have ears broadly connected on midbrow and reaching forward of eye when flattened forward.

Sounds. Some, if not all, species emit audible echolocation chirps.

Natural history. Feed on large insects, for which the audible echolocation is an adaptation; *E. perotis* specializes on butterflies and moths. Bonneted bats roost in small groups in tree holes, cliffs, and roofs of buildings. Because of their large size and narrow wings, the largest species must fly swiftly to remain airborne, and they apparently are unable to take flight from a level surface. They roost high enough up to allow them to gain speed as they drop from the roost. These bats forage at great heights. Found in many habitats from rainforest to arid scrub and in towns.

Geographic range. North, Central, and South America: SW United States to central Argentina.

Status. Most species are rarely encountered; they may be rare or simply difficult to capture because of their high flying and roosting behaviors.

References. Eger, J. L. 1977. Systematics

of the genus *Eumops* (Chiroptera: Molossidae). Life Sci. Contrib. Roy. Ontario Mus., no. 110.

Vaughn, T. A. 1959. Functional morphology of three bats; *Eumops, Myotis, Macrotis. Univ. Kansas Pub. Mus. Nat. Hist.* 12:1–153.

Crested Mastiff Bats
Promops spp.
Plate G, map 87

Identification. Measurements: HB = 68–89; T = 50–55; HF = 12–13; E = 13–18; FA = 46–57; WT = 14–25 g.
Upperparts dark brown to chestnut, head and/or neck usually darker than back, blackish; fur velvety. **Head short, broad, crown with slightly longer tuft of hair between ears; ears reach midway from eye to ear when laid forward,** rounded, inner edges meet with fold near horizontal; **lower edge of ear slightly folded back; lips with no vertical wrinkles; nostrils backed by slight horny ridge** that does not descend between nostrils; **top of muzzle from between ears to nose with raised central ridge;** chin rounded; **palate between the upper teeth deeply concave.** Throat with round gland in both sexes, more highly developed in males. **Four lower incisors. Tail 60–75% of the length of head and body; tail membrane covers more than half of tail. Underparts paler than back, uniform brown, not frosted, throat and chest darker than abdomen.**
Species. The two species are both found in

rainforest: *P. centralis* is larger (FA = 51–57), and *P. nasutus* is smaller (FA = 46–50).

Similar species. Difficult to distinguish from mastiff bats (*Molossus* spp.), which have throat paler than abdomen, two lower incisors, and do not have deeply concave palate or a crest of slightly longer hairs on crown.

Natural history. Crested mastiff bats roost in small groups of about six in holes in rotting trees, houses (*P. nasutus*), or under palm leaves (*P. centralis*). Found in openings in rainforest, or in dry forest.

Geographic range. Central and South America: S Mexico to N Argentina. To at least 1,000 m elevation.

Status. Both species apparently rare but widespread.

References. Sazima, I., and W. Uieda. 1977. O morcego *Promops nasutus* no sudeste Brasileiro (Chiroptera, Molossidae). *Ciência e Cultura* 29:312–14.

Mastiff Bats

Molossus spp.

Plate G, map 88

Identification. Measurements: HB = 59–89; T = 29–51; HF = 8–15; E = 12–19; FA = 36–53; WT = 9–38 g.

Upperparts black, dark brown, or chestnut; fur often paler at base than tip; fur short and velvety; wings and tail membrane black. **Muzzle broad, with large rounded chin; lips without deep vertical wrinkles, slightly puckered so that upper does not overhang lower, upper lip wide below nose, not steeply angled backward;** nostrils simple, without upstanding hard ridge behind; **center of muzzle between ears and nose raised in a ridge. Ears short, rounded, upper edges almost but do not quite meet in center of crown at horizontal angle;** rear edge of ear folded back; **ears when flattened reach midway from eye to nose.** Two lower incisors. Tail usually about 50–60% of head and body length. **Throat with round gland at base in both sexes; underparts like back, not frosted, throat often slightly paler than belly,** with hairs paler at base.

Species. All of the about seven species occur in lowland rainforest. The most widespread and common are: *M. rufus* (formerly called *M. ater*), the largest (FA = 47–

Map 88

Mastiff bats, *Molossus*

53; WT = 25–38 g), black or red, with a naked tail membrane; *M. molossus* (FA = 36–41; WT = 12–15 g), pale base of hairs visible through fur. There has been some confusion about the names and status of several species.

Similar species. Difficult to distinguish in the hand from crested mastiff bats (*Promops* spp.), which have deeply concave palate, tail usually > 60% of head and body length, a crest of longer hairs on crown, and four lower incisors; no other free-tailed bats have the midline of the muzzle raised in a ridge.

Sounds. Colonies in roofs become active with much twittering, bumping, and scrambling about at nightfall (but most bats twitter and squawk in the roost).

Natural history. Feed on insects, chiefly moths, beetles, and flying ants. Mastiff bats roost in tree holes, rotting trees, rock piles, and buildings. They are often found roosting in large colonies of hundreds in narrow, closed spaces under the roofs of houses in towns; often both *M. rufus* and *M. molossus* in the same roost. These are the common, narrow-winged bats that fly rapidly like swifts over the rooftops of tropical cities and towns at dusk. *M. rufus* leave the roost at nightfall and forage for about an hour before returning; only a few leave the roost again to forage for a second time before dawn. Found in rainforest and many types of drier habitats, and in towns and cities. These bats rarely carry rabies, and colonies in roofs are more of a nuisance than a threat to health. Their consumption of large quantities of insects is beneficial.

Geographic range. Central to South America: N Mexico to N Argentina and Uruguay.

Status. Common to uncommon, widespread.

References. Marques, S. A. 1986. Activity cycle, feeding and reproduction of *Molossus ater* (Chiroptera: Molossidae) in Brazil. *Bol. Mus. Paraense Emilio Goeldi, Zool.* 2:159–79.

Freeman, P. W. 1981. A multivariate study of the family Molossidae (Mammalia, Chiroptera): Morphpology, ecology, evolution. *Fieldiana Zool.*, n.s., no. 7.

Monkeys (Primates)

The three families of New World monkeys form a distinct group (Platyrrhini) that differs in many features from the Old World monkeys and apes (Catarrhini). The most obvious external difference is in the structure of the nose, which has close-set, downward-pointing nostrils in Old World primates (including ourselves), and wide-apart, sideward-pointing nostrils in New World monkeys. New World monkeys have short muzzles and flat, naked faces; large, forward-facing eyes; short ear pinnae; short necks and torsos; long hindlimbs; long, prehensile digits; plantigrade stance; and long tails in all but two species. All are primarily arboreal, and they descend to the ground only to cross an open space. A few species will forage on the ground when food is scarce in the trees. Neotropical monkeys show great geographical variation in color patterns. This poses problems when we try to sort them into species (see Appendix C, section 2), and there is still much controversy about the classification of species in certain genera. This book presents the conservative view for a number of genera, where, despite recent revisions, the taxonomy still appears unclear. We expect that many species-level changes will be made in future classifications of the monkeys. All New World monkeys are listed under CITES Appendix II.

Marmosets and Tamarins (Callitrichidae)
Goeldi's Monkey (Callimiconidae)

Dental formula: I2/2, C1/1, P3/3, M2/2 = 32. The marmosets (*Callithrix, Cebuella*) and tamarins (*Saguinus, Leontopithecus*) are distinguished from other New World monkeys by tiny size (about 100–600 g; the largest weighs as much as a large squirrel); long, nonprehensile tails that can be tightly coiled; heads decorated with a variety of ear tufts, tassels, ruffs, manes, mustaches, or mantles of long hair; hands and feet with claws instead of nails; frequent twins instead of a single young; and a variety of internal features. There are two major groups of callitrichids—the marmosets (*Callithrix* spp., *Cebuella*) and the tamarins (*Saguinus* spp., *Leontopithecus* spp.). The main difference between them is that the former have specialized lower jaws and teeth (small procumbent canines) that they use to dig holes in the bark of trees and vines. Marmosets feed extensively on the plant exudates that collect in these holes (sap, gum, resin). Tamarins have large, straight lower canines and do not gouge holes in bark, although they will feed on exudates they find. Members of both groups also feed on fruits and extensively on insects. All callitrichids have quite similar behavior when seen in the field: they run lightly along branches, rapidly springing between trunks or stems with "vertical cling and leap" jumps—taking off and landing with the head and body held vertical. While active, they constantly bob and turn their heads, peering in every direction. They call with similar birdlike whistles and soft chirping sounds. These monkeys are usually wary and stealthy, hiding quickly behind trunks, but they are often heard and can easily be found if their calls are recognized. Those species that have been studied live in small families consisting of a single breeding female, one or more males (they can be polyandrous or monogamous), two to four subadults, and carried juveniles. Young are usually carried on the back of a male. Females may have young twice a year. Most species favor dense vegetation with festoons of vines, where their insect prey is numerous; they thus adapt well to secondary and disturbed forest. Because they are too small to be prey for human hunters and thrive in secondary vegetation, callitrichids are usually the most common monkeys near villages or towns; some are even found in city parks. There are 4 genera and about 17 species of Callitrichidae, all in the Neotropics.

Goeldi's monkeys (Callimiconidae) are much like tamarins in size and external appearance but differ from them in many internal anatomical features. They have only one young at a time. The family contains only one species. The major reference for all members of these two families is P. Hershkovitz, *Living New World monkeys (Platyrrhini)*, vol. 1 (Chicago: University of Chicago Press, 1977).

Map 89

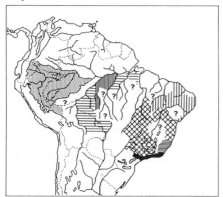

▦ Pygmy marmoset, *Cebuella pygmaea*
▤ Silvery marmoset, *Callithrix argentata*
▥ Tufted-ear marmoset, *C. jacchus jacchus*
▨ Tassel-ear marmoset, *C. humeralifer*
▦ Tufted-ear marmoset, *C. j. geoffroyi*
▩ Tufted-ear marmoset, *C. j. penicillata*
■ Tufted-ear marmoset, *C. j. aurita*

Pygmy Marmoset
Cebuella pygmaea
Plate 9, map 89

Identification. Measurements: HB = 117–152; T = 172–229; HF = 36–46; E = 15–22; WT = 107–141 g.

Head and forequarters tawny gold-gray, with fine striations visible at close range; back and **hindquarters coarsely variegated with tawny buff and gray.** Back of head, neck, and shoulders with a mane of longer hair. Face thinly haired, with a pale stripe from between eyes to nose, muzzle sometimes grizzled whitish; eyes large. Hands and feet tawny. **Tail slender, tapered at the tip, tawny, faintly banded with black above.** Throat and chest tawny buff, belly yellowish white or buff, genital area black. Young like adults but without pale face stripe or black genital tuft. **A tiny monkey with a large head, from a distance looks pale buffy gray.** The world's smallest true monkey.

Similar species. Tamarins (*Saguinus* spp.) and Goeldi's monkeys (*Callimico goeldii*) in the same region are either mostly red and/or black or have striking white mustaches or muzzles; small squirrels are brown-olivaceous, with short legs and bushy tails.

Sounds. Alarm a soft raspy chirp like the sound of a katydid. Several monkeys usually call together.

Natural history. Diurnal; arboreal; small groups of two to six. Feed on tree sap and gum, insects, and occasionally fruits. Pygmy marmosets bite many small, round-to-oblong, 8–15 cm holes in the bark of a tree or vine and return day after day to feed on the sap or gum that seeps into the wounds. Each group has a few such trees, some with hundreds of old and currently used feeding holes. Only certain species of trees are used; these differ in different areas. Pygmy marmosets stay low in the forest, usually below 20 m. They move quietly and are shy and difficult to see. The surest way to find them is to wait discreetly in early morning or late afternoon near a feeding tree with freshly made holes. Tamarins may also come to "steal" sap, especially in the dry season. Throughout most of their geographic range these marmosets seem restricted to seasonally flooded forest, river margins, floodplains, or streamsides; rarely they are seen in terra firme forest far from water. Found with a patchy distribution in mature and secondary lowland rainforest.

Geographic range. South America: east of the Andes in Colombia, Ecuador, Peru, and Brazil (Acre), from the base of the Andes east to the Rio Purus. Limits of range not precisely known.

Status. CITES Appendix II. Locally common to rare; patchy. Not hunted, but potentially at local risk from selective deforestation of floodplain forests.

Local names. Saguí, leaõzinho (Br); mono de bolsillo, leoncito (Pe, Ec, Co); chichíco (Co).

References. Ramirez, M. F., C. R. Freese, and J. Revilla C. 1977. Feeding ecology of the pygmy marmoset, *Cebuella pygmaea,* in northeastern Peru. In D. Kleiman, ed., *The biology and conservation of the Callitrichidae,* 91–104. Washington, D.C.: Smithsonian Institution Press.

Soini, P. 1982. Ecology and population dynamics of the pygmy marmoset, *Cebuella pygmaea. Folia Primatol.* 39:1–21.

Silvery Marmoset
Callithrix argentata
Plate 8, map 89
Identification. Measurements: HB = 180–280; T = 265–380; HF = 52–72; E = 25–31; WT = 420 g.

Upperparts silvery white, often with brownish or sooty tinge to lower back, face reddish or white, eye black; or pale brown with whitish or tan neck and shoulders, face brown, hindfeet brown, forefeet tan, thigh with large pale yellowish or whitish patch from hip to knee. Ears naked, without tufts, exposed. Tail black or brown. Underparts white to yellowish, tan in brown races. External genitalia naked, white.

Variation. The pale, silvery race is found in the northern part of the species range in Pará, Brazil. Animals from between the Rios Tapajós and Cupari may be completely white, with white, yellow, or brownish tails. Animals from Rondônia, Matto Grosso, and Bolivia are browner. A race from Rondônia has forequarters pale gray-yellow, crown blackish with white patch between and above eyes, hands blackish, feet blackish or rusty, rump darker brownish gray, thighs rusty, and tail pitch black. There may be another silvery marmoset species, *C. emiliae,* in the same geographic region, but its status is unclear given the large color variation within the species.

Similar species. Tassel-ear marmosets (*C. humeralifer*) have pale ear tufts, a banded tail, and mottled or grizzled hindquarters; golden-handed tamarins (*Saguinus midas*) are completely blackish where they occur with silvery marmosets.

Sounds. Soft trills, chirps, and whistles.
Natural history. Diurnal; arboreal; groups of three to eight. Probably feeds on fruit, insects, and plant exudates. The pale color of silvery marmosets makes them hard to see against the pale bark of deciduous trees, which are common in parts of their range. To move from one clump of trees to the next in savanna, these marmosets descend to the ground and scamper across open grassland. They occupy primary and secondary forest from evergreen rainforest to deciduous gallery forest and clumps of trees in savanna.

Geographic range. South America: Brazil, Bolivia, and Paraguay, south of the Amazon from the Rio Madeira east to the Toncantins and south to E Bolivia (Santa Cruz). SE limits poorly known.

Status. CITES Appendix II. Locally common and geographic range large.

Local names. Sauim (Br).

References. De Vivo, M. 1985. On some monkeys from Rondônia (Primates: Callithrichidae, Cebidae). *Pap. Avulsos Zool. Univ. São Paulo* 36:103–10.

Tassel-ear Marmoset
Callithrix humeralifer
Plate 8, map 89
Identification. Measurements: HB = 198–300; T = 298–398; HF = 53–75; E = 28–30; WT = 300–400 g.

The three geographic races are described individually below. General features of the species are: ear hidden by long, yellow-white tufts of hair growing from both in front of and behind pinna; tail banded with faint or distinct bands.

Callithrix h. humeralifer. Head blackish, face gray; forequarters silvery gray frosted with black; hindquarters variegated black and yellow-white; hands and feet blackish; tail gray with more or less distinct black bands; throat yellowish; underparts from chest to thigh orange; hip and upper thigh often with a pale spot of silvery buff.

C. h. chrysoleuca. Head and body pure yellow-white; limbs, feet, and tail golden above and below, tail faintly banded with darker gold. Fur silken and shiny.

C. h. intermedius. Forequarters silvery gray; hindquarters variegated with blackish; hindlimbs and hip patches orange; forelimbs paler orange; throat and chest whitish or pale buff; belly and inguinal region orange; tail reddish brown at base becoming gray distally, with dark bands or spots.

Similar species. Silvery marmosets (*C. argentata*) have no ear tassels and no banding on tail; tassel-ear and silvery marmosets may have no geographic range overlap, but this is still uncertain.

Sounds. Birdlike whistles and twitters; alarm call a repeated "tsik."

Natural history. Diurnal; arboreal; groups of 2–13. Feeds on fruits, insects, and tree exudates. Tassel-ear marmosets prefer or require areas of dense, viny vegetation such as those caused by second growth into natural or human-caused disturbance, and river edges. They make holes in trees to produce exudate flow, but seem to use exudates to a far lesser extent than tufted-ear or pygmy marmosets. Sometimes they follow army-ant swarms and forage for flushed insects. They have small home ranges of about 13 ha. At night they sleep in vine-covered trees or, rarely, in tree holes. Found in mature and secondary rainforest.

Geographic range. Brazil: south of the Amazon in a narrow strip between the Rio Tapajós and lower Rio Madeira and Rio Roosevelt. *C. h. humeralifer,* between the Rios Tapajós and Canumá, southern limit unknown; *C. h. chrysoleuca,* between the Rios Madeira, Aripuaná, and Canumá; *C. h. intermedius* between Rios Roosevelt and Aripuaná.

Status. CITES Appendix II.

Local names. Sauim (Br).

References. Rylands, A. B. 1981. Preliminary field observations on the marmoset *Callithrix humeralifer intermedius* (Hershkovitz, 1977) at Dardanelos, Rio Aripuaná, Matto Grosso. *Primates* 22:46–59.

———. 1986. Ranging behaviour and habitat preference of a wild marmoset group, *Callithrix humeralifer* (Callitrichidae, Primates). *J. Zool., Lond.* 210:489–514.

Tufted-ear Marmoset
Callithrix jacchus
Plate 8, figure 4, map 89
Identification. Measurements: HB = 190–248; T = 270–350; HF = 52–68; E = 21–31; WT = 261–323 g; females usually below 300 g, males heavier.
The distinct geographic races of this marmoset are individually described below. General features are: **ears with tufts or tassels; posterior half of back finely banded with gray or orange and black; tail banded with black, often sharply.** There is variation within, and apparently some intergradation among, the forms described below; these forms are sometimes considered distinct species.

C. j. jacchus. **Head dark brown dominated by enormous whitish tufts** surrounding ear; forehead with white spot; shoulders grizzled pale gray or gray and buff; hindquarters and tail banded gray and white; underparts and feet gray.

C. j. penicillata. As above, but **ear tufts black;** face, muzzle and forehead, or only forehead, white; hindlegs buff-gray.

C. j. geoffroyi. As above, but **face and crown pure white** sharply demarcated from pitch black hood; ear tufts black, growing from in front of ear; back banded gray and black variegated bright rust on lower back; underparts dark brown or black except throat white.

C. j. flaviceps. **Face with dark central mask surrounding nose and eyes; rest of head ochraceous,** including long tuft growing from inside of ear pinna; back banded orange and black; hands and feet yellowish orange; underparts buffy with black stripe down chest, belly, and inguinal region.

C. j. aurita. **Face, forehead, and long tuft growing from inside of ear pinna, whitish** to pale buff; **sides of head black;** crown buff; back blackish brown; tail grizzled yellowish gray-buff with sharp black bands; underparts blackish buff with black midline on chest and belly.

Similar species. Lion tamarins (*Leontopithecus rosalia*) are larger and red-gold and/ or black, with no grizzling or bands on body.

Sounds. High-pitched whistles, chirps, and twitters; mobbing alarm a repeated "tsik."

Natural history. Diurnal; arboreal; small groups of 2–13. Feeds on fruits, insects, and the exudates of trees. Tufted-ear marmosets bite holes in the bark of trees, then lick the exudate. Frequently used trees are riddled with hundreds of horizontal slitlike to round holes, 1–1.5 cm wide by 2–20 cm long. About a third of the total day's activity is spent feeding on the sap that seeps into these holes. Groups occupy small territories of 0.5–5 ha. At night members of a group sleep together in a dense tangle of vegetation or on a branch. Found in mature and secondary forests and plantations, even city parks. In evergreen and deciduous forests, gallery forests, and upland scrub.

Geographic range. South America: Brazil; E forests from Ceará to São Paulo. *C. j.* jacchus, in Ceará, Piauí, Pernambuco; *C. j. penicillata,* in dry and upland forest and scrub from Bahía and Goiás to São Paulo; *C. j. geoffroyi,* in coastal and

Figure 4. Head patterns of subspecies of **tufted-ear marmosets,** *Callithrix jacchus:* (*a*) *C. j. jacchus;* (*b*) *C. j. penicillata;* (*c*) *C. j. geoffroyi;* (*d*) *C. j. flaviceps.*

lowland Minas Gerais and Espírito Santo; *C. j. flaviceps,* at higher elevations in SE Minas Gerais; *C. j. aurita,* in coastal and lowland Minas Gerais and São Paulo. **Status.** CITES Appendix I, US-ESA endangered, *C. j. aurita* and *C. j. flaviceps* only; others CITES Appendix II. **Local names.** Sauim. **References.** Coimbra-Filho, A. F., and R. A. Mittermeier. 1977. Tree-gouging, exudate-eating and the "short-tusked" condition in *Callithrix and Cebuella.* In D. Kleiman, ed., *The biology and conservation of the Callitrichidae,* 105–15. Washington, D.C.: Smithsonian Institution Press.

Maier, W., C. Alonso, and A. Langguth. 1982. Field observations on *Callithrix jacchus jacchus* L. *Z. Säugetierk.* 47:334–36.

Saddleback Tamarin
Saguinus fuscicollis
Plate 9, map 90
Identification. Measurements: HB = 175–270; T = 250–383; HF = 55–77; E = 21–35; WT = 338–436 g.
Upperparts divided into three color zones: shoulders and forelegs pure black, dark brown, or mahogany red-brown; back from behind shoulders to hip **a "saddle" of variegated mixture of blackish and yellowish or orange striations; hindlegs dark brown to red-brown or deep red,** sometimes the same color as forelegs, sometimes much redder. **Head black, muzzle white, with short hairs,** some races with a white eyebrow band across forehead; eyes brown. Tail dark brown or blackish. Underparts dark reddish.
Variation. There are 13 subspecies; the 2 most distinctive are the following:
S. f. acrensis. Head and forequarters white; face with black skin; ears black; lower back variegated brown and orange; rump and hindlegs orange; underparts and tail yellow. There is evidence that this may be a distinct species.
S. f. melanoleucus. Entirely creamy white, sometimes washed with yellowish or buff, except ears black and face with dark skin. This is the only white tamarin.
Similar species. Black-mantle tamarins (*S. nigricollis*) have no distinct saddle and black forequarters (saddlebacks have red forequarters where the species overlap);

mustached tamarins (*S. mystax*) are black with long white mustaches; emperor tamarins (S. imperator) have long, drooping white mustaches; red-chested tamarins (*S. labiatus*) have sharply contrasting orange or red underparts; golden-mantle tamarins (*S. tripartitus*) have a golden ruff on neck; Goeldi's monkeys (*Callimico goeldii*) are entirely black, with a crown of long hairs.
Sounds. Birdlike twitters, trills, and whistles; loud whistles repeated 7–10 times.
Natural history. Diurnal; arboreal; groups of 2–12. Feeds on small fruits, nectar, and insects. Saddlebacks most often use the middle and lower levels of the forest, below the canopy, and are usually found in dense, viny habitats. They are usually wary and difficult to see, but often a soft trill betrays their presence nearby, or at a distance, loud whistles. They run actively through the vines and lower branches of larger trees, making many quick jumps across gaps, or they sit in a group, with their tail coiled forward or hanging straight down below. Saddlebacks spend most of their insect-foraging time investigating knotholes and crevices on the trunks of large trees, often near the ground. Groups consist of one breeding female, one or more adult males, and their young. They have large home ranges of 30–100 ha. At night they sleep in a dense festoon of vines surrounding a large tree, or in a tree hole. A group of this species sometimes travels and feeds with a group of another tamarin species that occupies the same territory (emperor, mustached, or red-chested, depending on the region). This is the most commonly seen tamarin in Peru, and it is the most widespread species; it overlaps the geographic ranges of five others. Mature, disturbed, and secondary lowland rainforests.
Geographic range. South America: east of the Andes in Colombia, Ecuador, Peru, Bolivia, and the upper Amazon Basin of Brazil, west of the Rio Madeira and south of the Caquetá/Japurá. *S. f. acrensis* and *S. f. melanoleucus* are from Acre, Brazil, at the headwaters of the Rio Juruá.
Status. CITES Appendix II. Widespread and common.
Local names. Sauim (Br); bebeleche

(Co); chichíco, leoncito (Ec); pichico (Pe).
References. Terborgh, J. 1983. *Five New World primates*. Princeton: Princeton University Press.

Terborgh, J., and A. W. Goldizen. 1985. On the mating system of the cooperatively breeding saddle-backed tamarin (*Saguinus fuscicollis*). *Behav. Ecol. Sociobiol.* 16:293–99.

Golden-mantle Tamarin
Saguinus tripartitus
Plate 9, map 90
Identification. Measurements: HB = 218–240; T = 316–341; E = 31–32. **Back variegated grayish and white or orange;** limbs orange. **Head black, with black collar of hair continuous under throat;** muzzle and sometimes face pure white, **neck with a ruff of bright golden to creamy fur dorsally, sharply contrasting with black crown.** Tail black above, orange on proximal part of undersurface. Feet and hands blackish orange. **Underparts orange.**
Variation. This tamarin is considered by some to be a subspecies of saddleback tamarin. However, it overlaps saddleback tamarins in geographic range and is highly distinct from all neighboring races of saddlebacks.
Similar species. See saddleback tamarin (*S. fuscicollis*).
Natural history. Found in lowland evergreen rainforest only. May occur in different habitat types than saddleback tamarin.
Geographic range. South America: east of the Andes in Ecuador and NE Peru; from Napo/Pastaza, Ecuador, and Loreto, Peru, apparently in the wedge between the Ríos Napo, Curaray, and perhaps Nanay. Range poorly known.
Status. CITES Appendix II. Unknown.
Local names. Leoncito, chichico (Ec).
References. Thorington, R. W., Jr. 1988. Taxonomic status of *Saguinus tripartitus* (Milne-Edwards, 1878). *Amer. J. Primatol.* 15:367–71.

Golden-handed or Midas Tamarin
Saguinus midas
Plate 9, map 90
Identification. Measurements: HB = 206–282; T = 316–440; HF = 57–80; E = 28–40; WT = 415–665 g.

Map 90

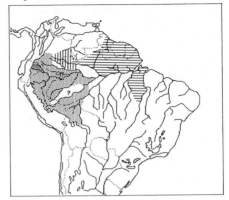

▨ Saddleback tamarin, *Saguinus fuscicollis*
▥ Mottled-face tamarin, *S. inustus*
▤ Midas tamarin, *S. midas*
▬ Golden-mantle tamarin, *S. tripartitus*

Face, head, and forequarters entirely black; back black variegated with yellowish or buff. Hands and feet bright golden yellow or orange, or black. Tail and underparts black. Young may have pale areas around eyes and mouth.
Variation. Animals from north of the Rio Amazonas have golden paws and yellowish variegation on back, and usually weigh 500–600 g; those from south of the Amazonas have black paws and backs frosted with buff (*S. m. niger*), larger ears, and may be smaller (usually < 500 g).
Similar species. Yellow-handed titis (*Callicebus torquatus*) are larger, with a pale collar on throat; bare-face tamarins (*Saguinus bicolor*) are whitish or yellowish with orange on the tail; silvery marmosets (*C. argentata*) have whitish forequarters. This is the only callitrichid in most of its geographic range. It is not clear whether silvery marmosets occupy the same habitats; there are a few collections from the same localities.
Sounds. Birdlike whistles; mild alarm a chorus of soft, cricketlike, froglike, or birdlike chirps.
Natural history. Diurnal; arboreal; groups of two to six. Feed on fruits and insects. These tamarins are generally in smaller groups than other callitrichids. Unlike other species they tend to frequent open, high forest formations, where they forage at heights of 5–25 m, and they often travel in

the canopy. They also favor dense, viny habitats such as swamps and streamsides. Found in mature and secondary terra firme rainforest.
Geographic range. South America: Brazil and the Guianas, the eastern Amazon Basin east of the Rio Negro and the lower Rio Xingu.
Status. CITES Appendix II. Widespread and common.
Local names. Sauim (Br); sagoewijntje, sagoewenki (Su).
References. Thorington, R. W. 1968. Observations of the tamarin *Saguinus midas.* *Folia Primatol.* 9:95–98.

Mittermeier, R. A., R. C. Bailey, and A. F. Coimbra-Filho. 1977. Conservation status of the Callitrichidae in Brazilian Amazonia, Surinam, and French Guiana. In D. Kleiman, ed., *The biology and conservation of the Callitrichidae,* 137–46. Washington, D.C.: Smithsonian Institution Press.

Black-mantle Tamarin
Saguinus nigricollis
Plate 9, map 91
Identification. Measurements: HB = 210–251; T = 308–361; HF = 63–72; E = 26–29; WT = 419–505 g.
Forequarters black, lower back and hindlimbs dark red; or forequarters blackish grizzled olivaceous and hindquarters blackish yellow-olivaceous (agouti).
Crown black, sides of head dark to pale brown; face thinly haired, black; **muzzle grizzled white,** with short hairs; ears black, inconspicuous. Hands, feet, and tail black. Underparts dark brown or black, including inguinal region. From a distance looks entirely grayish or black.
Variation. Animals from west of Iquitos are grizzled olivaceous; those from east of Iquitos (north of the Amazon/Solimões) are black with dark red hindquarters.
Similar species. Where they overlap, saddleback tamarins (*S. fuscicollis*) have dark red or orange forequarters and hindquarters and a grizzled gray saddle on the middle of the back; they are difficult to distinguish in poor light in the field; pygmy marmosets (*Cebuella pygmaea*) are much smaller and pale colored; Goeldi's monkeys (*Callimico goeldii*) are entirely black, with black muzzle and a ruff of long hair.

Map 91

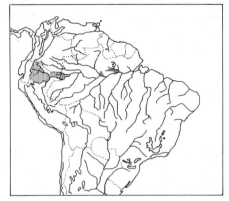

▦ Black-mantle tamarin, *Saguinus nigricollis*
▤ Brazilian bare-face tamarin, *S. bicolor*

Sounds. Birdlike chirps, whistles, and twitters.
Natural history. Diurnal; arboreal; groups of 4–8 which may join together to form large, temporary troops of up to 40. Feeds largely on insects, especially big orthopterans, and on fruit, nectar, and plant gum. Black-mantle tamarins favor dense, viny vegetation. They are the only callitrichids known to sometimes travel in large groups. The larger groups are noisy and resemble squirrel monkey troops as they jump and rummage through the forest searching for insects in a broad band from near the ground to the canopy. When in smaller groups, they usually forage in the lower levels of the forest. Found in mature, disturbed, and dense secondary rainforest.
Geographic range. South America: east of the Andes in Ecuador, Colombia, and N Peru; between the Ríos Putumayo, Napo, and the headwaters of the Putumayo. Southeastern limits of range unknown. Apparently absent from the wedge between the Ríos Napo and Curaray. To 1,000 m elevation.
Status. CITES Appendix II. Little hunted and locally common.
Local names: Leoncito, chichico (Ec); bebeleche (Co); pichico (Pe).
References. Izawa, K. 1978. A field study of the ecology and behavior of the black-mantle tamarin (*Saguinus nigricollis*). *Primates* 19:241–74.

Map 92

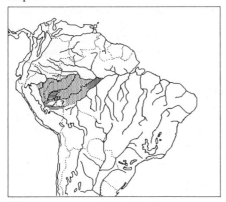

▓ Black-chested mustached tamarin, *Saguinus mystax*
▥ Red-chested mustached tamarin, *S. labiatus*
▤ Emperor tamarin, *S. imperator*

Red-chested Mustached Tamarin
Saguinus labiatus
Plate 9, map 92
Identification. Measurements: HB = 234–300; T = 345–410; HF = 69–78; WT = 500–650 g.
Back and hindlegs black heavily frosted or variegated with silvery gray. Head and face black with white nose and lips in a prominent ∧; ear with pale stripe of white, bare skin in front; crown sometimes with a narrow, white or copper stripe down center; **nape with white triangular patch.** Tail, hands, and feet black. Chin and sometimes throat black, rest of **underparts a sharply contrasting, bright, pure rust-red.**
Variation. Populations just south of the Rio Solimões have a rusty red or orange patch on the crown of head; animals from north of the river have the upper part of the chest black.
Similar species. See black-chested mustached tamarin.
Sounds. Birdlike whistles and chirps.
Natural history. Diurnal; arboreal; groups of 2–13. Feeds on fruits and insects. Red-chested tamarins chiefly use the middle layers of the forest, where they forage for small insects on twigs and leaves. They often travel in mixed groups with saddleback tamarins. Found in mature and secondary evergreen and deciduous forest.
Geographic range. South America: east of

the Andes in W Brazil, NE Bolivia, lowlands between Rios Purus and Madeira, and possibly north of the Rio Solimões. Range limits poorly known.
Status. CITES Appendix II. Locally common.
Local names. Sauim (Br).
References. Yoneda, M. 1984. Comparative studies on vertical separation, foraging behavior and traveling mode of saddlebacked tamarins (*Saguinus fuscicollis*) and red-chested moustached tamarins (*Saguinus labiatus*) in northern Bolivia. *Primates* 25:414–22.
    Pook, A. G., and G. Pook. 1982. Polyspecific association between *Saguinus fuscicollis, Saguinus labiatus, Callimico goeldii* and other primates in northwestern Bolivia. *Folia Primatol.* 38:196–216.

Black-chested Mustached Tamarin
Saguinus mystax
Plate 9, map 92
Identification. Measurements: HB = 235–280; T = 365–435; HF = 66–79; E = 26–31; WT = 536–700 g.
Forequarters black or dark, blackish brown; back and hindquarters blackish brown frosted or grizzled with gray or dark buff. **Head black with white nose and region around mouth, short** tufts of longer **white** hair forming **mustaches at sides of mouth** that part away from chin; spot of white, bare skin in front of ear; **crown black or bright, dark rust-red.** Hands, feet, and tail black. **Underparts dark brown,** in some populations with white genital area.
Variation. S. m. pileatus, with red crown and gray-frosted back, is from between the lower Rios Juruá and Purus.
Similar species. Red-chested mustached tamarins (*S. labiatus*) have red underparts; emperor tamarins (*S. imperator*) have long, downcurling white mustaches and orange tail; saddleback tamarins (*S. fuscicollis*) are smaller and have reddish hindquarters, a saddle of grizzled hair, less prominent whitish muzzle; Goeldi's monkeys (*Callimico goeldii*) are entirely black.
Sounds. Birdlike whistles and chirps.
Natural history. Diurnal; arboreal; groups of two to six. Feeds on fruits and insects. These tamarins sometimes travel in mixed groups with saddleback tamarins; they use a higher level of the forest than the saddle-

backs. They tend to run along branches, rarely using the "vertical cling and leap" mode of locomotion that is typical of saddlebacks. The mother is reported to carry the young, unlike other tamarins and marmosets, where males transport young. Found chiefly in mature rainforest, occasionally in old secondary forest.

Geographic range. South America: east of the Andes in Peru, Brazil, and perhaps Colombia, west of the Rios Purus and upper Madeira, and south of the Japurá. Range limits poorly known, especially north of the Solimões.

Status. CITES Appendix II. Locally common; seems less adaptable to disturbed habitat than other tamarins and, because of its larger size, is sometimes hunted for meat.

Local names. Sauim (Br); pichico (Pe).

References. Castro, R., and P. Soini. 1977. Field studies on *Saguinus mystax* and other callitrichids in Amazonian Peru. In D. Kleiman, ed., *The biology and conservation of the Callitrichidae,* 73–78. Washington, D.C.: Smithsonian Institution Press.

Ramirez, M. 1984. Ecology and conservation of the moustached tamarin in Peru. *Primate Conservation* 4:22–23.

Emperor Tamarin
Saguinus imperator
Plate 10, map 92

Identification. Measurements: HB = 230–258; T = 350–415; HF = 68–72; E = 28; WT ~ 400 g.

Forequarters, back, and hindquarters entirely grizzled pale gray washed with buff on sides. Head and ears blackish, **face dominated by long, downcurling white mustaches extending far below chin;** brow with a whitish band above eyes, eyes brown. **Tail bright rusty orange,** rust extending onto rump around tail base, tip of tail blackish; **or tail black above, orange below, with tip entirely black.** Hands and feet blackish. Throat and chest grizzled gray or brown, belly and insides of limbs bright orange or brown. Young have short mustaches.

Variation. Animals from the upper Río Juruá have gray underparts; those from Peru and the upper Río Juruá have a thin beard of long white hairs as well as mustaches.

Similar species. No other Neotropical mammals have long white mustaches. See saddleback tamarin (*S. fuscicollis*).

Sounds. Birdlike chirps, quavering whistles, and whistles in a long, descending series.

Natural history. Diurnal; arboreal, groups of 2–10. Feeds on fruit, insects, and flower nectar in the dry season. Emperor tamarins mainly forage for insects on leafy foliage and twigs, where they catch active prey with catlike stealth. They often form a mixed troop with a group of saddleback tamarins, with which they may defend a common territory. Emperor tamarins are shy and secretive and difficult to observe; they travel quietly. They prefer dense, viny vegetation, where they spend several hours a day resting inconspicuously. A quavering whistle or chirp sometimes reveals their location in a thicket. At night they sleep in large, vine-draped trees, especially ones isolated from the surrounding forest. They are rarer in their habitat than are saddleback tamarins. Found in mature and disturbed lowland rainforest, especially riversides and viny habitats.

Geographic range. South America: east of the Andes in Brazil, Peru, and Bolivia, SW Acre, Brazil, Pando, Bolivia, and north of the Río Madre de Dios, Peru.

Status. CITES Appendix II. Generally uncommon and apparently patchy.

Local names. Pichico, chico chico (Pe); sauim (Br).

References. Terborgh, J. 1983. *Five New World primates.* Princeton: Princeton University Press.

Brazilian Bare-face Tamarin
Saguinus bicolor
Plate 10, map 91

Identification. Measurements: HB = 208–283; T = 335–420; HF = 63–83; E = 25–31.

There are three distinctive subspecies. General features are: **face and head to level of ears naked, tail bicolored blackish or brown above, orange below; large, naked ears.**

Pied bare-face tamarin, *Saguinus bicolor bicolor.* **Forequarters white; back and hindquarters pale brown;** naked skin of face and head entirely black, or mottled with white patches; tail black above, rust-gold below; throat and chest white; belly brown mixed with buff; inner thighs and lower abdomen bright rust-red.

Martin's bare-face tamarin, *S. b. martinsi.* **Sides and hindlegs grizzled cinnamon; forearms orange;** naked skin of face black; crown to base of tail brown, in a broad band; tail black above, orange below; underparts orange.

Ochraceous bare-face tamarin, *S. b. ochraceous.* **Upperparts entirely pale brown** with buff tinge, paler on forequarters; legs like back or more orange; naked skin of face and head blackish; tail dark brown above, gold below; underparts entirely orange to golden brown.

Similar species. Midas tamarins (*S. midas*) are blackish with gold paws.

Sounds. Birdlike chirps and whistles.

Natural history. Diurnal; arboreal; in small family groups. Feeds on fruits and insects. Pied bare-face tamarins favor dense viny vegetation of secondary forests and forest edges. Small populations live in isolated secondary forest plots in the city of Manaus, including the grounds of the Instituto Nacional de Pesquisas da Amazonia (INPA) and the Hotel Tropica. In these places it is the only tamarin present, but it is said to occur together with the midas tamarin in a forest outside the city. Found in disturbed and secondary lowland rainforest.

Geographic range. Brazil: north of the Rio Amazonas in a narrow, discontinuous strip near the river from the Rio Negro east to the lower Cuminá. Pied bare-faced tamarins are restricted to a 30 km radius of Manaus; the other two subspecies probably range between the lower Rios Uatumá and Cuminá (Erepecuru). Range limits not precisely known.

Status. CITES Appendix I, US-ESA endangered. Poorly known, perhaps at risk. This species has the smallest geographic range of any Amazon Basin primate and lives in a densely inhabited region. Although it thrives in secondary habitats, populations may already be highly fragmented.

Local names. Sauim.

Map 93

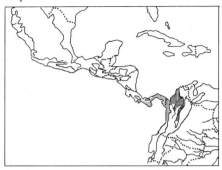

▦ Geoffroy's tamarin, *Saguinus geoffroyi*
▥ Cotton-top tamarin, *S. oedipus*
▤ Silvery-brown bare-face tamarin, *S. leucopus*

Silvery-brown Bare-face Tamarin
Saguinus leucopus
Plate 10, map 93

Identification. Measurements: HB = 223–263; T = 347–417; HF = 68–80; E = 24–30.

Upperparts pale silvery brown, or yellowish white variegated with brown. Face almost naked, thinly haired with white; **brow and crown to ears whitish; neck and top of head from behind ears with ruff of sharply contrasting darker brown;** ears thinly haired, dark. **Tail brown,** distinctly darker than body, frosted silver toward tip, **extreme tip usually white.** Legs and feet whitish or silvery; ankle and sometimes foot with red-brown spot(s). **Underparts dark to bright rusty orange.**

Similar species. Cotton-top tamarins (*S. oedipus*) have long plumes of white hair on head, sharply bicolored body with blackish back and white limbs, and pale underparts.

Sounds. Birdlike warbles, chirps, and trill; a shrill "tee-tee."

Natural history. Diurnal; arboreal; groups of up to 15. Probably feeds on fruits, particularly undergrowth berries, and insects. This species uses all heights of the forest and favors edge habitats such as streamsides. It thrives in second-growth vegetation, where it is the only remaining primate in much of its geographic range. Habits reported to be similar to those of cotton-top tamarin. Found in disturbed lowland and lower montane rainforest.

Geographic range. South America: Colombia east of the Andes, between the eastern bank of the lower Río Cauca and the western bank of the middle Río Magdalena, and foothills of the central Andes. One of the smallest geographic ranges among tamarins.

Status. CITES Appendix I, US-ESA endangered. Most of the geographic range has been deforested.

Local names. Titi, titi gris.

References. Hernández-Camacho, J., and R. W. Cooper. 1976. The non-human primates of Colombia. In R. W. Thorington and P. G. Heltne, eds., *Neotropical primates,* 35–69. Washington, D.C.: National Academy of Science.

Green, K. M. 1978. Neotropical primate censusing in northern Colombia. *Primates* 19:537–50.

Mottled-face Tamarin
Saguinus inustus
Plate 10, map 90

Identification. Measurements: HB = 208– 259; T = 330–410; HF = 70–75.

Entirely blackish, except hindquarters dark red. Face almost naked; with pure white patch of skin below nose; cheeks and area around eyes naked, white, mottled with variable amounts of black; ears naked, black, sometimes with white patches. External genitalia naked, white.

Similar species. Midas tamarins (*S. midas*) have contrasting yellow hands and feet; saddleback tamarins (*S. fuscicollis*) have a saddle of tawny variegated fur in midback; Goeldi's monkeys (*Callimico goeldii*) have a completely black face.

Natural history. Undescribed.

Geographic range. South America: SE Colombia and NW Brazil, from a small, remote region between the Ríos Guaviare, Japurá, and upper Río Negro, range probably incompletely known.

Status. CITES Appendix II. Unknown.

Local names. Sauim (Br); mico diablo, diablito, tití diablo (Co).

Cotton-top Tamarin
Saguinus oedipus
Plate 10, map 93

Identification. Measurements: HB = 205– 255; T = 307–411; HF = 66–72; E = 20–24; WT = 350–510 g; males slightly larger than females.

Back brown, sometimes frosted silvery or yellowish; hair of back overhangs sides in a long fringe. Head dominated by crest of long white hairs growing in a narrow peak between areas of naked skin on crown, widening backward to cover entire top and sides of neck; face, chin, and sides of crown to behind ear almost naked, black; with thin white hairs especially in a fringe over eyebrows and from corner of eye to jowl. Tail dark chestnut-red for first third, chestnut extending onto area around tail base and hind thighs, distal two-thirds black. Limbs and feet cream to yellowish white. Underparts cream to yellowish white. Young have short white hairs on "crest."

Similar species. Silvery brown tamarins (*S. leucopus*) are uniform-colored, with a short brown crest; red-naped tamarins (*S. geoffroyi*) have a dark chestnut neck and white triangle of short hairs on crown (no long white plumes).

Sounds. Birdlike chirps, long, ascending whistles and modulated, descending, then ascending, whistles.

Natural history. Diurnal; arboreal; groups of 2–13. Feeds on small fruits, insects, and occasional buds and leaves. Cotton-top tamarins defend small home ranges or territories of as little as 8 ha. They arise an hour after dawn and, like other tamarins, retire well before dark. They travel 1–2 km a day, foraging for insects, feeding on fruit, and seeking favorite sleeping spots, which are high in the forks of large trees, dense vine tangles, or tufts of leafy vegetation. Found in evergreen and deciduous mature and secondary forests.

Geographic range. South America: N Colombia east of the Río Atrato and west of the Ríos Cauca and lower Magdalena. To 400 m elevation.

Status. CITES Appendix I, US-ESA endangered. More than three-fourths of the original habitat of the cotton-top has been deforested, much of it for cattle pasture. The species is now restricted to a few isolated forest fragments.

Local names. Titi, titi blanco, titi leoncito, titi pielroja.

References. Neyman, P. F. 1977. Aspects of the ecology and social organization of free-ranging cotton-top tamarins (*Saguinus oedipus*) and the conservation status of the species. In D. Kleiman, ed., *Biology of the Callitrichidae*, 39–71. Washington, D.C.: Smithsonian Institution Press.

Red-naped or Geoffroy's Tamarin
Saguinus geoffroyi
Plate 10, map 93
Identification. Measurements: HB = 200–287; T = 315–423; HF = 64–82; E = 20–31; WT = 453–520 g.
Back and upper thighs black variegated (almost spotted) with short stripes of pale dirty yellow; forelegs, chest, and feet yellow-white; lower hindlegs and feet yellowish gray; hair of back overhangs belly in long fringe. **Face and sides of head to ear, naked, black, thinly haired with white,** especially prominent on eyebrow; stripe from corner of eye to below ear; **crown topped with narrow triangle of short, white, flat-lying hair; neck to shoulder dark, chestnut-red. Tail dark chestnut-red at base, distal half black.** Underparts pale yellow with buff tinge.
Variation. Considered by some to be a subspecies of cotton-top tamarin.
Similar species. Cotton-top tamarins (*S. oedipus*) have long white plumes on head; Central American squirrel monkeys (*Saimiri oerstedii*) are yellowish with black crown and muzzle, pale mask around eyes.
Sounds. Birdlike chirps, twitters, and long, descending whistles repeated two to four times.
Natural history. Diurnal; arboreal; groups of up to 14. Feeds on fruits, insects (chiefly orthopterans), plant exudates, and small amounts of leaves and shoots. Red-naped tamarins are most common in areas of dense vegetation, such as river edges and disturbed or secondary forest, where they forage for insects on thin vines and branches in the low shrub layer of the understory. They often hang by their hindfeet to reach a prey item. At night they sleep in large emergent trees. Found in a wide range of habitats from pluvial rainforest to seasonal dry forest.
Geographic range. Central and South America: S Costa Rica, Panama, and Colombia west of the Andes to the Río San Juan.
Status. US-ESA endangered. Threatened by deforestation.
Local names. Titi (Co, Pn); bichichi (Co);

Map 94

Goeldi's monkey, *Callimico goeldii*

Golden-rump lion tamarin, *Leontopithecus rosalia chrysopygus*, former

Golden lion tamarin, *L. r. rosalia*, former

Gold-and-black lion tamarin, *L. r. chrysomelas*, former

Lion tamarins, present ranges

marmoseta, tamarín (CR).
References. Garber, P. A. 1984. Proposed nutritional importance of plant exudates in the diet of the Panamanian tamarin, *Saguinus oedipus geoffroyi*. *Inter. J. Primatol.* 5:1–15.

———. 1984. Use of habitat and positional behavior in a Neotropical primate, *Saguinus oedipus*. In J. G. H. Cant, and P. S. Rodman, eds., *Adaptations for foraging in non-human primates*. New York: Columbia University Press.

Lion Tamarin
Leontopithecus rosalia
Plate 8, map 94
Identification. Measurements: HB = 200–336; T = 315–400; HF = 71–86; E = 24–26; WT = 361–710 g.
There are three geographically disjunct subspecies that may be full species: Generally, **gold, or black and gold; head with a mane of long hair parted in the middle on the top and extending completely around the head under the chin;** hands and fingers exceptionally long and narrow. These are the largest callitrichids.
   Golden lion tamarin, *L. r. rosalia*. **Entirely golden** with silky, shiny fur; face naked, pinkish gray; tail sometimes faintly barred with darker gold, or white-tipped.

Feet and tail with occasional black patches.

Golden-rump lion tamarin, *L. r. chrysopygus.* **Entirely glossy black except rump, upper thighs and tail base golden or reddish,** the size of the golden patch variable; outer surfaces of legs reddish or yellowish brown.

Gold-and-black lion tamarin, *L. r. chrysomelas.* **Crown and fringe around face, forelimbs, and often hindfeet golden; rest of body black,** except tail may be gold at base, above or below, or all black; hindquarters may be reddish. This form has distinctive behavioral and morphological characteristics and should probably be considered a separate species.

Similar species. Most tufted-ear marmosets (*Callithrix jacchus*) are gray, with banded tails, striations on back, and usually white markings on head.

Sounds. A large variety of birdlike calls: descending frequency trills, whines, whistles, and clucks.

Natural history. Diurnal; arboreal; groups of 2–11. Feeds on fruit and insects. The long, narrow hands of lion tamarins are apparently used for reaching into tree hollows and crevices to probe for hidden insects. Lion tamarins are usually active at heights of 5–10 m; they favor areas of dense primary vegetation, especially those with many bromeliads. At night the whole group sleeps together in a tree hollow, generally one with a small entrance. Found in humid to seasonal Atlantic forests, where it seems best adapted to mature or primary vegetation. Lion tamarins do not show the adaptability of tufted-ear marmosets to disturbed or man-made habitats and may always have been uncommon or rare.

Geographic range. South America: SE Brazil, remnant forest fragments in the state of Rio de Janeiro (golden lion tamarin); two forest fragments in the state of São Paulo (golden-rump tamarin); forest fragments in the state of Bahía (gold-and-black tamarin).

Status. CITES Appendix I, US-ESA endangered. Close to extinction. Over 90–95% of the original habitat of this species has been deforested. Although remaining populations are mostly in secondary or degraded habitats, the species adapts poorly to disturbance. Each subspecies is now reduced to a few (two for the golden-rump)

isolated fragments of forest and has wild populations of fewer than three hundred. Efforts are being made to preserve remaining populations and reintroduce captive-bred individuals to the wild.

Local names. Sauim preto (golden-rump tamarin); sauim vermelho, mico-leaõ (golden lion tamarin).

References. Coimbra-Filho, A. 1977. Natural shelters of *Leontopithecus rosalia* and some ecological implications (Callitrichidae: Primates). In D. Kleiman, ed., *Biology and conservation of the Callitrichidae,* 79–89. Washington, D.C.: Smithsonian Institution Press.

Kleiman, D. G. 1981. *Leontopithecus rosalia.* Mammalian Species, no. 148.

Goeldi's Monkey
Callimico goeldii
Plate 8, map 94

Identification. Measurements: HB = 210–234; T = 255–324; HF = 68–78; E = 20–26; WT ⌣ 480 g.

Upperparts, head, legs, and tail entirely glossy black or with slight grizzling of reddish or yellowish on rump, thighs, and tail base; **fur long and silky, feathery on tail, looks disheveled.** Face thinly haired, almost naked, black; **head with long hair in two tiers: a velvety crown on top and sides, with long side-whiskers framing face, and a longer ruff behind crown, from top of head to below ear; ear naked, hidden in ruff.** Tail not prehensile but can be coiled ventrally. **Belly, chest, and insides of limbs thinly haired, almost naked, with pure white skin** visible in extended postures. Young may have grizzled rump and thighs.

Similar species. All tamarins that occur in the same geographic region have white or gray mustaches or muzzles and ears that protrude through the head hair.

Sounds. Alarm call a loud chuck; also gives two- to three-second trills.

Natural history. Diurnal; arboreal and terrestrial; groups of up to eight. Feeds on invertebrates and fruit. Goeldi's monkeys usually travel close to the ground, below 5 m, with frequent "vertical clinging and leaping" between upright undergrowth trunks. When alarmed, they keep close to the ground while they flee. Goeldi's monkeys sometimes travel in association with tamarins, which usually stay above 5 m. They seem to be habitat specialists, inhabiting scrubby upland (terra firme) forest away from river floodplains. These forests have a broken canopy, dense undergrowth, and much bamboo.

Geographic range. South America: east of the Andes in Colombia, Ecuador, Peru, Bolivia, and Brazil (Acre); in a narrow strip within 4–500 km of the base of the Andes. To 500 m elevation. Range poorly known from fewer than 25 confirmed localities.

Status. CITES Appendix I, US-ESA endangered. These monkeys seem everywhere rare, with thinly scattered, patchy populations, and they are often unknown to local people. Apparent rarity puts this species at potential risk, but its true status and even its geographic range are unknown.

References. Heltne, P. G., J. F. Wojcik, and A. G. Pook. 1981. Goeldi's monkey, genus *Callimico*. In A. F. Coimbra-Filho and R. A. Mittermeier, eds., *Ecology and behavior of Neotropical primates*, 169–209. Rio de Janeiro: Academia Brasiliera de Ciências.

Pook, A. G., and G. Pook. 1981. A field study of the socio-ecology of the Goeldi's monkey (*Callimico goeldii*) in northern Bolivia. *Folia Primatol.* 35:288–312.

Monkeys (Cebidae)

Dental formula: I2/2, C1/1, P3/3, M3/3 = 36. The cebid monkeys range from 0.8 to 15 kg. They have five toes on all feet, thumb vestigial in two genera; nails on all digits. The heaviest species all have prehensile tails; the lightest species do not. Monkeys with prehensile tails wrap the tip tightly around branches, hang supported by it, and often carry the tail with the tip in a tight downward coil when they walk along the top of a branch; only prehensile-tailed cebids carry their tail tips coiled (but see marmosets and tamarins). Cebids give birth to one young, which is carried by the mother in large species and small nonmonagamous species, and often by the father in small monogamous species. All species are arboreal (unlike Old World monkeys, which include many terrestrial species). All species eat fruit; some also eat many insects, and others eat more leaves than fruit. There is much geographic variation in color between populations of monkeys in different areas separated by large rivers or mountains, and it is still unclear how many species are represented in several groups. There are 11 genera and a minimum of 31 species. This book generally follows the taxonomy presented in A. F. Coimbra-Filho and R. A. Mittermeier, *Ecology and Behavior of Neotropical Primates*, vol. 1 (Rio de Janeiro: Academia Brasileira de Ciências, 1981). Members of a genus, such as howlers or sakis, are similar in diet, behavior, and body shape.

Small monkeys. There are three genera of small monkeys that all weigh about 1 kg: night monkeys, titis, and squirrel monkeys. They are not closely related and have completely different ecologies. They are about twice as heavy as the marmosets and tamarins and do not travel by vertical clinging and leaping. They do not have prehensile tails.

Medium monkeys. There are two groups of monkeys that weigh 2–4 kg. The sakis, bearded sakis, and uakaris (Pithecinae) are a closely related group of specialized monkeys. They have bushy, nonprehensile tails that they sometimes wag below a branch, long hair (shaggy in the sakis and uakaris), and large, divergent upper and lower canine teeth that bulge under the lips and pull the mouth into a permanent frown. They appear to use these teeth to split open the husks of unripe fruits, on whose seeds they feed. This group is restricted to the Amazon Basin and the Guianas.

The capuchin monkeys are brown or yellowish with dark caps (one has a black back), and they have prehensile tails that are more weakly developed than those of the larger monkeys: the tip of the tail is hairy below, and the monkeys rarely swing supported only by the tail. They feed on fruits, invertebrates, and small vertebrates.

Large monkeys. The spider, woolly, woolly spider, and howler monkeys (Atelinae) all weigh 4 kg or more. They have strongly prehensile tails with a naked gripping surface on the underside of the tip. They feed on fruits and leaves. Spider and woolly spider monkeys have long, thin arms and legs and often swing by their arms or tail below branches; woolly monkeys sometimes swing by their arms, but howler monkeys rarely do so. Because of their large size (which gives the highest return per round of ammunition), these are the monkeys most intensively hunted for meat, and they are rare in many inhabited areas.

Monkeys are the most important seed dispersers for hundreds of plant species, especially canopy trees and lianas, and they are therefore vital elements of rainforest ecosystems. If they are eliminated from the forests by hunting, the plant species composition of the forests will eventually change.

Night or Owl Monkeys
Aotus spp.
Plate 11, map 95

Map 95

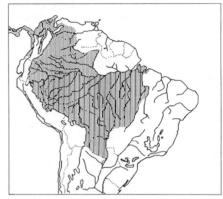

| | Night monkey, *Aotus* spp., gray-necked |
| | Night monkey, *Aotus* spp., red-necked |

Identification. Measurements: HB = 240–475; T = 220–418; HF = 89–96; E = 27–35; WT = 780–1,249 g.

Upperparts grizzled gray to brown; fur soft, dense, and woolly. Head round, with short fur, **face** slate brown **rimmed with white, brow patterned with three black stripes: a large central black spot between eyes, white patches above eyes, and two narrow black stripes from corner of eyes** converging backward and sometimes meeting on the rear of the crown; **eyes large,** round, brown, **set close together; eyeshine bright, reddish orange;** mouth with a slight, enigmatic smile. **Tail nonprehensile, never coiled, brown or dark rusty orange frosted with black for first half,** becoming **black at** slightly tufted **tip.** Fingers long and narrow, with expanded pads at tips. **Underparts pale yellow to bright orange;** these colors either extending onto insides of limbs and prominently onto sides of neck and throat, or not.

Species. Some authorities recognize as many as nine geographically separated species. These are outlined as follows: dull-colored animals with gray sides of neck, inner limbs and throat, and yellowish underparts, from Central America and the Amazon Basin north of the river, include *A. lemurinus, A. brumbacki, A. trivirgatus,* and *A. vociferans;* more brightly colored animals with orange neck and throat, and often orange inner limbs and tail base, are all from south of the Amazon River and include *A. miconax, A. nigriceps, A. nancymai, A. azarae,* and *A. infulatus.* The forms in each group are often difficult to

distinguish; in our opinion some of them are subspecies of a variable species, and the situation remains unclear. For reliable identification, a museum specimen is needed.

Similar species. These are the only monkeys with huge eyes and nocturnal activity. Kinkajous (*Potos flavus*) and olingos (*Bassaricyon* spp.) have no facial markings and have wide-set eyes with yellow-orange reflection.

Sounds. Alarm call soft metallic clicks often accompanied by low-pitched, resonating, tonal grunts; on moonlit nights (most often) a two- to four-syllable, low-pitched, owllike hoot.

Natural history. Nocturnal; arboreal; groups of two to five. Feeds on fruit, insects, and flower nectar. Usually night monkeys are seen in the upper half of the forest. They can be found throughout the forest, but they are most common in denser

areas with many vines, such as the edges
of rivers and open areas. These little monkeys travel quietly and are relatively sedentary. They may spend several hours in a
single fruit tree, moving around from time
to time or sitting quietly. The light of
flashlights disturbs them, and their eyeshine is often visible only for a few seconds before they turn their heads away or
flee. Their most evident feature at night is
the long, black-tipped tail hanging straight
down from a branch. They are most active
and easy to find on moonlit nights, when
their hooting betrays their location. By day
they sleep in a vine-draped tree, tree hollow, or dense festoon of vegetation. They
are monogamous, and the father often carries the young. In the temperate zone of
Paraguay where nights are cold, owl monkeys are sometimes active in the daytime.
These are one of the commonest monkeys
near human settlements. They occupy a
wide range of mature, disturbed, and secondary forested habitats from the wettest
rainforests to dry scrub in the Chaco.
Geographic range. Central and South
America: Panama to the Chaco of Argentina; except absent from the Guianas and
Brazil north of the Rio Amazonas and east
of the Rio Negro. To elevations of 3,200 m.
Status. CITES Appendix II. Widespread
and common. Little hunted except for the
laboratory animal trade; adapts well to disturbed forests.
Local names. Mono de noche (Span); ma-
caco-da-noite (Br); marta, mico dormilón,
mico de noche (Co); tutamono (Ec); mus-
muqui (Pe); jujaná, gná, marteja (Pn); mir-
ikiná (Gua). Note that some of the same
names are used for kinkajous, olingos, or
woolly opossums.
References. Wright, P. C. 1978. Home
range, activity pattern, and agonistic encounters of a group of night monkeys
(*Aotus trivirgatus*) in Peru. *Folia Primatol.*
29:43–55.

Hershkovitz, P. 1983. Two new species
of night monkeys, genus *Aotus* (Cebidae,
Platyrrhini): A preliminary report on *Aotus*
taxonomy. *Amer. J. Primatol.* 4:209–43.

Map 96

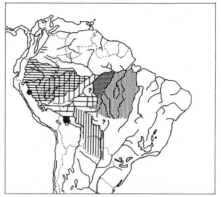

▦	Dusky titi monkey, *Callicebus moloch moloch*
▤	Dusky titi monkey, *C. m. cupreus*
▥	Dusky titi monkey, *C. m. donacophilus*
▥	Dusky titi monkey, *C. m. caligatus*
▤	Dusky titi monkey, *C. m. brunneus*
▦	Dusky titi monkey, *C. m. hoffmanni*
◉	Dusky titi monkey, *C. m. modestus* and *olallae*
◼	Dusky titi monkey, *C. m. oenanthe*

Dusky Titi Monkey
Callicebus moloch
Plate 11, map 96
Identification. Measurements: HB = 287–
360; T = 331–480; HF = 84–102; E =
26–36; WT = 928–1,400 g.
Upperparts dull brown, red-brown, or
finely grizzled **gray;** fur long and dense.
Head round, face small, flat, blackish or
gray; brow black or gray often contrasting
with crown; sides of head either contrasting
bright yellow or orange, or white around
ear, or same color as crown; ears inconspicuous. **Tail nonprehensile, thickly furred,
gray, brown, or blackish, tail tip often
pale,** dirty white or brown for up to half
its length, **usually hangs straight downward** or droops over a branch. Underparts
pale yellowish to bright or dark red-orange.
**Head held low, below level of shoulders;
usually sits with feet tucked forward under body close to hands**—a furry ball with
a dark face and pendant tail. Thick fur
hides the neck and makes the body, limbs,
and tail appear chunky, and the almost-na-
ked face look small.
Variation. Dusky titis are highly variable
both within and between populations.
Those from the dryer southern parts of the

range in Bolivia, Paraguay, and Matto Grosso are pale gray; those from south of the Amazon in Pará and Amazonas have bright orange or yellow cheeks. This taxon has recently been divided into as many as 10 species; their distributions are shown on the map.

Similar species. Tamarins (*Saguinus* spp.) are small and slender and jump with vertical-cling-and-leap locomotion; squirrel monkeys (*Saimiri* spp.) are slender, with short fur and slender tails; sakis (*Pithecia* spp.) are much larger and shaggier and have bushy tails.

Sounds. At dawn and occasionally at other times, pairs give extremely loud duets of prolonged, continuous, rapidly modulated whoops. Alarm call a soft grunt or, rarely, tamarinlike chirps and twitters.

Natural history. Diurnal; arboreal; groups of two to five. Feeds on leaves and fruits: the only largely folivorous small monkey. Dusky titis are most often found in areas of denser vegetation, especially on water edges and in swamp forest, where they use the middle and lower levels. They like to feed on the leaves of bamboo and are often found in bamboo thickets. Like most arboreal folivores, they are quite inactive and spend many hours each day just sitting, like blobs of fur on a branch. When running along a branch, they have a characteristic bouncing gait, bounding with both forefeet, then both hindfeet, moving together. They are monogamous, and the young is usually carried on the father's back. When disturbed, titis will often vanish into a dense tangle of vegetation, where they may remain for hours without betraying their presence by any sound or movement. Where hunted they are wary and stealthy and seldom seen, although they are usually common. Their presence in an area is always shown by their morning duets, often given from the middle of dense thickets. At night they sleep at medium heights (8–18 m) on branches sheltered by vegetation. Mature and disturbed lowland evergreen rainforest and gallery forest.

Geographic range. South America: east of the Andes in a strip along their base from Colombia to Bolivia and Paraguay, and Brazil south of the Amazon only. To 850 m elevation.

Status. CITES Appendix II. Widespread and common, not usually hunted for meat.

Local names. Zogue-zogue, uapusá (Br); zocay (Co); zocayo, saui (Ec); tocón (Pe); sauhí (TGua).

References. Hershkovitz, P. 1988. Origin, speciation, and distribution of South American titi monkeys, genus *Callicebus* (family Cebidae, Platyrrhini). *Proc. Acad. Nat. Sci. Philadelphia* 140:240–272.

Wright, P. 1986. Ecological correlates of monogamy in *Aotus* and *Callicebus*. In J. G. Else and P. C. Lee, eds., *Primate ecology and conservation*, 159–67. Proceedings of the 10th Congress of the International Primate Society, vol. 2.

Yellow-handed Titi Monkey
Callicebus torquatus
Plate 11, map 97

Identification. Measurements: HB = 310–375; T = 420–93; HF = 91–105; E = 23–33; WT = 0.8–1.5 kg.

Upperparts red-brown, dark brown, or almost black, thighs blackish to rust-red; fur long and dense. Face thinly haired with whitish; **neck with broad white to buff collar from ear to ear under throat; chin white. Hands bright yellow-white,** sharply demarcated from forearms, **or black. Tail dark brown or black.** Underparts red-brown to sooty brown. Locomotion and postures like dusky titi.

Variation. Animals from near the Río Orinoco are blackish. The degree of reddish color seems to vary individually within populations near the Río Amazonas. The black-handed race is from Colombia, between the upper Ríos Caquetá and Putumayo.

Similar species. Midas tamarins (*Saguinus midas*) have similar coloration, but no white collar, and are much smaller; no other monkeys are black/brown with yellow hands.

Sounds. Loud whooping duets at dawn; chucks and birdlike chirrups.

Natural history. Diurnal; arboreal; groups of two to five. Feeds mainly on fruit, with some insects. Yellow-handed titis are generally found in tall forest, where they use the middle and upper levels. They are quite active and spend most of the day moving. At night they sleep together, tails intertwined, on a large horizontal branch in the canopy. This species may be a habitat spe-

Map 97

▩ Yellow-handed titi monkey, *Callicebus torquatus*
▤ Masked titi monkey, *C. personatus*

cialist that is largely restricted to areas where the soil has been leached to almost pure white sand and stream waters are clear "black." In most regions within its range, it occurs patchily on pockets of such soil, but there may be some populations on other soils in Venezuela. Found only in mature rainforest.

Geographic range. South America: east of the Andes in the western Amazon Basin lowlands of Colombia, Venezuela, Peru, and Brazil.

Status. CITES Appendix II. Locally common but distribution patchy; too small to be hunted intensively except where larger game is scarce.

Local names. Japuça-de-coleira (Br); viduita, macaco caresebo (Co).

References. Kinzey, W. G. 1981. The titi monkeys, genus *Callicebus*. In R. Mittermeier and A. F. Coimbra-Filho, eds., *Ecology and behavior of Neotropical primates*, 1:241–76. Rio de Janeiro: Academia Brasileira de Ciências.

Kinzey, W. G., and A. H. Gentry. 1978. Habitat utilization in two species of *Callicebus*. In R. W. Sussman, ed., *Primate ecology*, 89–100. New York: John Wiley.

Masked Titi Monkey
Callicebus personatus
Plate 11, map 97
Identification: Measurements: HB = 325–550; T = 430–580; HF = 90–123; E = 30–35; WT = 1,050–1,650 g.

Upperparts **pale brown,** or frosted yellowish beige, **or grizzled gray** with brown lower back; **fur long, thick, dull colored.** Head small; **face round and flat, black; wide fringe of black hair encircling face** completely, **or not extending to chin, or black only around ear;** pale yellow crown of hair on top and side of head, or head grizzled gray; ears inconspicuous. **Tail nonprehensile, thickly haired,** dark orange, pale orange, yellowish brown, or dirty gray, sometimes with a pale tip. **Hands and feet black.** Throat and chest yellowish orange or brown, belly like back, but paler. Posture like dusky titi.

Variation. Titis from Minas Gerais and Espírito Santo are beige with sharply contrasting yellow and black crown; those from Rio de Janeiro and São Paulo are yellow-gray with black face and crown; and those from Bahía are grizzled gray with gray head and brown lower back.

Similar species. Marmosets (*Callithrix* spp.) and tamarins (*Leontopithecus rosalia*) are smaller, with relatively large heads and slender tails; all other monkeys in range are much larger.

Sounds. Duets with loud whooping calls at dawn.

Natural history. Diurnal; arboreal; groups of two to six. Masked titis feed on fruits supplemented by some leaves. They are quite inactive and spend about half the day resting. At night they sleep in a huddle, with tails intertwined, on a large, high branch in the canopy (25–40 m up). Mature, lowland Atlantic coastal rainforest and inland gallery forests.

Geographic range. South America: E Brazil, Bahía to São Paulo.

Status. Cites Appendix II. Threatened by deforestation of small geographic range.

Local names. Sahuí-guaçu.

References. Kinzey, W. G., and M. Becker. 1983. Activity pattern of the masked titi monkey, *Callicebus personatus*. *Primates* 24:337–43.

Common Squirrel Monkey
Saimiri sciureus
Plate 11, figure 5, map 98
Identification. Measurements: HB = 250–320; T = 370–435; HF = 78–90; E = 21–30; WT = 600–1,400 g; males heavier than females.

Upperparts gray-olivaceous; back faint to bright gold-olivaceous frosted with black; fur short. Head round, distinctly patterned, muzzle blackish, mask around eyes white; ears hairy, white with slight, pointed tufts; sides of neck behind ear white, crown a cap of black or gray. Females tend to have darker heads and cheeks than males. Tail nonprehensile, slightly tufted, olivaceous with black tip above, paler below. Forelimbs, hands, and feet bright gold to pale yellow- orange. Chin and throat whitish; chest and belly pale yellow to orange. Body slender and gracile.

Variation. Animals from Bolivia and SE Peru have black caps and bright gold limbs; those from S Pará and north of the Amazon have olivaceous caps; a dark, blackish form is found in a small region of islands at the mouth of the Rio Japurá. Authorities disagree on the number of squirrel monkey species; some recognize as many as six.

Similar species. Bare-eared squirrel monkeys (*S. ustus*) have naked ears, blue-gray caps, and gray forearms; these are the only small monkeys with black muzzle and white eye mask; marmosets (*Callithrix* spp.) and tamarins (*Saguinus* spp.) are smaller, with disheveled manes of long hair on the head, and they travel in small groups; capuchins (*Cebus* spp.) are larger and often walk with tail coiled.

Sounds. Very noisy, travel with constant

Map 98

 Common squirrel monkey, *Saimiri sciureus*
Bare-eared squirrel monkey, *S. ustus*
Central American squirrel monkey, *S. oerstedii*

squeals and high-pitched birdlike whistles and chirps, crashing and disturbance of vegetation.

Natural history. Diurnal; arboreal; large groups of 20 to more than 100. Feeds on insects, small ripe fruits such as figs, and nectar. Squirrel monkeys spend much of the day ranging through the forest searching for insects. Small, slender, and agile, they favor vegetation with many thin vines and branches, like that at river and lake edges. In constant motion, they run and jump through the vegetation, usually keeping to the middle and lower levels of the forest, sometimes even using the ground. They also climb to reach fruits in the high-

Figure 5. Head patterns of squirrel monkeys: (*a*) **common squirrel monkey,** *Saimiri sciureus;* (*b*) **bare-eared squirrel monkey,** *Saimiri ustus* (Drawings after P. Hershkovitz, 1984, *Amer. J. Primatol.*)

est treetops. They often associate with brown capuchin monkeys and travel with them for many hours. This species is most common in floodplain forests, especially on rich alluvial soils, and it adapts well to the dense vegetation of old secondary and disturbed forests. It is rare or absent in some areas far from rivers in poor-soil terra firme forests. Found in mature and disturbed lowland evergreen rainforest.

Geographic range. South America: the Amazon Basin and the Guianas. To 1,500 m elevation.

Status. CITES Appendix II. Widespread and often common. Too small to be hunted for food except where game is scarce, but formerly intensively exploited for the laboratory animal and pet trades.

Local names. Macaco-de-cheiro, boca preta (Br); titi (Co); fraile, frailecito, barizo (Co, Ec, Pe); eekhoornaap, monkimonki (Su).

References. Terborgh, J. 1983. *Five New World primates.* Princeton: Princeton University Press.

Hershkovitz, P. 1984. Taxonomy of squirrel monkeys genus *Saimiri* (Cebidae, Platyrrhini): A preliminary report with description of a hitherto unnamed form. *Amer. J. Primatol.* 7:155–210.

Bare-eared Squirrel Monkey
Saimiri ustus
Figure 5, map 98

Identification. Measurements: HB = 300–380; T = 410–470; HF = 65–80; E = 30; WT ∼ 1 kg.

Similar to the common squirrel monkey, except: ears naked; crown blue-gray; forearms usually gray, yellowish color of hands ending at wrists.

Variation. Animals from the lower Rio Tapajós have yellowish crown and forearms.

Similar species. See common squirrel monkey (*S. sciureus*): the two species apparently occur together in several localities.

Natural history. Probably similar to common squirrel monkey. Lowland evergreen rainforest only.

Geographic range. South America: Brazil, south of the Amazon between the Rios Purus and Tapajós.

Status. CITES Appendix II.

Local names. Macaco-de-cheiro, boca preta.

References. Thorington, R. W. 1985. The taxonomy and distribution of squirrel monkeys (*Saimiri*). In L. A. Rosenblum and C. L. Coe, eds., *Handbook of squirrel monkey research, 1–33.* New York: Plenum.

Central American Squirrel Monkey
Saimiri oerstedii
Plate 11, map 98

Identification. Measurements: HB = 266–291; T = 362–389; HF = 76–86; E = 27.

Back, lower legs, hands and feet orange-gold; shoulders and hips olivaceous; fur on back quite long at higher elevations. Head with black cap, black muzzle, white mask; cap and cheeks of males often paler than those of females; ear and sides of neck white. Tail olivaceous, distal half black, tip tufted. Chin and throat white; rest of underparts pale orange.

Variation. Males from the northern part of the range have olivaceous caps. Sometimes considered a subspecies of the common squirrel monkey.

Similar species. These are the only small monkeys in Central America with black muzzle and white mask.

Sounds and natural history. Similar to common squirrel monkey; group size is often small in remnant forest fragments. Because of its tiny, disjunct geographic range, unlike that of any other monkey, this species is thought to have been introduced by man to Central America. Found in lowland rainforest.

Geographic range. Central America: Costa Rica and Panama in a small area on the Pacific coast.

Status. CITES Appendix I, US-ESA endangered. Most of the former range has been deforested and only a few fragmented populations remain.

Local names. Tití, mono ardilla (Pn, CR).

References. Baldwin, J. D., and I. J. Baldwin. 1972. The ecology and behavior of squirrel monkeys (*Saimiri oerstedii*) in a natural forest in western Panama. *Folia Primatol.* 18:161–84.

Boinski, S. 1985. Status of the squirrel monkey *Saimiri oerstedi* in Costa Rica. *Primate Conservation* 6:15–16.

Brown Capuchin Monkey
Cebus apella
Plate 11, map 99

Map 99

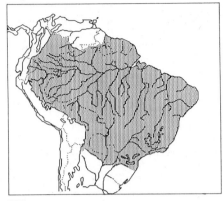

Brown capuchin monkey, *Cebus apella*

Identification. Measurements: HB = 350–488; T = 375–488; HF = 107–132; E = 28–43; WT = 1.7–4.5 kg; males larger than females.

Upperparts dark yellowish to slightly reddish **brown,** darkest middorsally; **shoulders paler than back,** mustard yellow or brown; sides and thighs frosted yellowish. **Head broad, crown completely covered with black or dark brown cap that extends down cheek as a distinct dark bar in front of ear;** hair of cap forming short tufts above ears, giving **top of head** a characteristic **flat, squared-off,** or **"eared"** frontal outline; **face flat, dark brown, pink, or brown mottled pink, fringed** completely **with yellowish** to white. **Tail prehensile, black or brown, darkest at tip, usually carried with tip in tight, downward coil. Hands, hindlimbs, and feet black or brown, darker than body.** Underparts brown or chest yellow. Body stocky and robust, especially adult males.

Variation. Animals from drier regions of Paraguay and Pernambuco, Brazil, are pale yellow-brown with yellow shoulders, yellow or reddish chests, and brown extremities; those from E Brazil are dark brown or blackish overall; those from the Amazon Basin have black hands, feet, caps, and tails. There is much individual variation between members of the same troop, especially in face color. Adult males are often darker than females.

Similar species. These are the only capuchin monkeys that occur with other capuchin species over large regions. They are distinguished from all others by dark bars in front of ears, brown face, flat or tufted top of head, and more robust build. No other yellowish brown, black-capped monkeys carry the tail coiled; woolly monkeys (*Lagothrix lagothricha*) have round heads, and black or dark faces with no markings or cap.

Sounds. Noisy. Emit frequent short, yipping whines (like hungry newborn puppies); alarm call a distinctive two-toned tonal clunking; plaintive birdlike whistles. Forage destructively by ripping apart and dropping vegetation, hammering nuts against branches, jumping noisily from tree to tree. Sounds of foraging can be heard for long distances.

Natural history. Diurnal; arboreal; groups of 5–20, usually about 10. Feeds on ripe fruits, palm nuts, arthropods and small vertebrates, and some nectar. Brown capuchins usually forage in the middle to lower levels of the forest but will feed on high-canopy fruit. These robust capuchins find animal foods by tearing apart dead vegetation; they will eat almost any small animal they find, including mouse opossums, bird's eggs or nestlings, and lizards. They also tear open well-protected buds or stems of palms and eat the young flowers or pith, or pry ripe palm fruits from a tight cluster. The stronger individuals break the hard nuts of *Astrocaryum* palms by banging them against branches (other capuchins also do this, but more rarely). They spend much foraging time in palm trees. Brown capuchins are one of the most commonly seen monkeys in Amazonia. Their noisy behavior makes them easy to detect, but where hunted they are difficult to approach and quickly give the alarm and flee. Squirrel monkeys will often forage for many hours with a group of brown capuchins. Found in many types of forest including mature rainforest, dry forest, gallery forest, and disturbed and secondary forest.

Geographic range. South America: east of the Andes from Colombia and Venezuela south to Paraguay and N Argentina. To 1,500 m elevation.

Status. CITES Appendix II. Widespread

and often common. Intensively hunted for meat in most parts of its range, but has greater reproductive potential, habitat flexibílity, and recuperative capacity than larger monkeys.

Local names. Macaco-prego (Br); mico maicero (Co); machin negro (Ec, Pe); mutsaap, bruine rolstaartaap, keskesi (Su); cai comun (Ar); caí (Gua).

References. Terborgh, J. 1983. *Five New World primates.* Princeton: Princeton University Press.

Janson, C. 1985. Aggressive competition and individual food consumption in wild brown capuchin monkeys. (*Cebus apella*). *Behav. Ecol. Sociobiol.* 18:125–38.

Wedge-capped or Weeping Capuchin Monkey

Cebus olivaceus

Plate 11, map 100

Identification. Measurements: HB = 374–460; T = 400–554; HF = 120–143; E = 35–51; WT = 2.3–4.2 kg.

Upperparts uniform, tawny brown sometimes frosted with dirty yellow; shoulders and upper arms sometimes silvery yellow; back of head and neck reddish; fur quite long, stands upright, giving the animals a bristly, robust look. Head brownish yellow with sharp, dark brown V-shaped cap, point of V connected to **a thin brown stripe down center of face to nose. Face pink, fringed to behind ears by yellowish white to silvery white.** Tail prehensile, dark brown, often carried with tip coiled. **Hands, wrists, and feet dark brown,** often contrasting with paler forearms. Chest pale brown, belly brown.

Variation. Animals from coastal Venezuela (Falcón) are entirely dark brown with red-tinged limbs and light brown shoulders and face. The name *C. nigrivitattus* is sometimes used for this species.

Similar species. Brown capuchins (*C. apella*) have dark bars in front of ears and no facial stripe; woolly monkeys (*Lagothrix lagothricha*) have completely dark head and face with no markings.

Sounds. Frequent short yips and tonal caws, barks like small dog, squeals and chatters. Sometimes hammers or taps nuts, rips vegetation apart noisily.

Natural history. Diurnal; arboreal; groups of 10–33. Feeds chiefly on ripe fruits, seeds and arthropods, with some vegetative plant parts. Wedge-capped capuchins mainly use the middle and lower levels of the forest, on occasion even descending to the ground. They spend several hours a day searching for invertebrates, often in palms, rummaging among leaves, breaking apart dead branches, and searching under foliage. Found in mature evergreen forest and gallery forest or, rarely, deciduous forest.

Geographic range. South America: Venezuela, the Guianas, and Brazil, east of the Rios Orinoco and Negro and north of the Rio Amazonas.

Status. CITES Appendix II. Common in some parts of Venezuela, where brown capuchins are absent, but over most of its geographic range is much less common than the brown capuchin.

Local names. Grijze capucijneraap, bergi-keskesi (Su); mono chuco (Ve).

References. Robinson, J. G. 1986. Seasonal variation in use of time and space by the wedge-capped capuchin monkey, *Cebus olivaceus:* Implications for foraging theory. Smithsonian Contrib. Zool., no. 431.

Mittermeier, R. A., and M. G. M. van Roosmalen. 1981. Preliminary observations on habitat utilization and diet in eight Surinam monkeys. *Folia Primatol.* 36:1–39.

Map 100

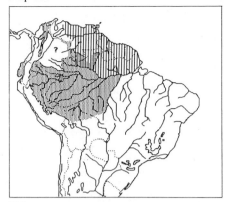

White-fronted capuchin monkey, *Cebus albifrons*

Wedge-capped or weeping capuchin monkey, *C. olivaceus*

White-fronted Capuchin Monkey
Cebus albifrons
Plate 11, map 100

Identification. Measurements: HB = 358–460; T = 401–475; HF = 112–136; E = 32–45; WT = 1.2–3.6 kg.

Upperparts pale smoky gray-brown, yellowish brown, or reddish brown; forearms and hindlegs yellow or rusty reddish. **Face pink fringed with silvery white; crown with sharp, wedge-shaped cap of dark smoky brown, generally extending forward as a thin stripe down center of brow. Tail prehensile, smoky silvery yellow, usually paler at tip than base, often carried with tip coiled.** Underparts yellowish. A gracile, slender, medium-sized monkey.

Variation. There is much individual color variation even within the same troop.

Similar species. Brown capuchins (*C. apella*) have dark bars in front of ears; both they and wedge-capped capuchins (*C. olivaceus*) have dark hands, feet, and tail tip and robust build; squirrel monkeys (*Saimiri* spp.) have black muzzles, white masks, and do not coil their tails; woolly monkeys (*Lagothrix lagothricha*) have dark faces with no markings.

Sounds. Calls often, with low whines and whistles. Travels and forages noisily.

Natural history. Diurnal; arboreal; groups of 7–30. Feeds on fruits, seeds, and arthropods, especially wasps. White-fronted capuchins use all levels of the forest, including occasionally the ground. They forage for hidden arthropods by manipulating the leaves and branches, but they do less destructive ripping and tearing than brown capuchins. Sometimes they hammer palm nuts to crack them. White-fronted capuchins are gracile and agile, moving more quickly and lightly than brown capuchins, and they are often more wary and difficult to approach. They form loose foraging associations with squirrel monkeys. Troops have large home ranges that overlap each other. Found in mature and disturbed rainforest.

Geographic range. South America: isolated areas in N Colombia and Venezuela and coastal Ecuador; and the middle and upper Amazon Basin of Colombia, Venezuela, Ecuador, Peru, Bolivia, and Brazil west of the Rios Negro and Tapajós. To 2,000 m elevation.

Map 101

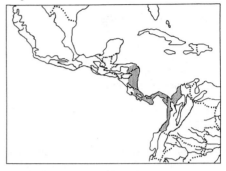

White-throated capuchin monkey,
Cebus capuchinus

Status. CITES Appendix II. Hunted for meat, but widespread in Amazon Basin; coastal populations may be threatened by deforestation and hunting.

Local names. Cairara (Co, Br); machin blanco (Ec, Pe); maicero cariblanco (Co).

References. Terborgh, J. 1983. *Five New World primates.* Princeton: Princeton University Press.

Hershkovitz, P. 1949. Mammals of northern Colombia. Preliminary report no. 4: Monkeys (Primates), with taxonomic revisions of some forms. *Proc. U.S. Nat. Mus.* 98:323–427.

White-throated Capuchin Monkey
Cebus capuchinus
Plate 11, map 101

Identification. Measurements: HB = 335–453; T = 350–551; HF = 110–150; E = 21–42; WT = 1.4–3.9 kg; males larger than females.

Upperparts black except sides of neck, shoulders and upper arms sharply contrasting pale yellow to whitish. Head yellow with black, V-shaped cap on crown; face pink with whitish hairs. Tail prehensile, black, sometimes brownish below, often carried with tip coiled. Hands and feet black. Throat and chest pale yellow; belly thinly haired, blackish.

Variation. Capuchins from Honduras have inner thighs frosted yellow.

Similar species. Squirrel monkeys (*Saimiri* spp.) have dark muzzles and do not hold their tails in a coil.

Sounds. Calls with yips, barks, whistles, and screams; forages noisily.

Natural history. Diurnal; arboreal; groups

of 2–24. Feeds on ripe fruits and arthropods. White-throated capuchins use all levels of the forest, sometimes including the ground. They forage for prey by examining leaves, picking through dead litter, splitting twigs, and pulling off bark. These are active monkeys that spend much of the day traveling and foraging. They may defend territories, unlike other capuchins. Found in a wide range of habitats including mature, disturbed, and secondary evergreen and deciduous forests.

Geographic range. Central and South America: Honduras south to N Colombia, west of the Andes to N Ecuador.

Status. CITES Appendix II. Many populations are probably threatened by deforestation.

Local names. Mico negro, carita blanca (Co); carablanca (CR); cariblanco (Pn).

References. Freese, C. H., and J. R. Oppenheimer. 1981. The capuchin monkeys, genus *Cebus*. In A. F. Coimbra-Filho and R. A. Mittermeier, eds., *Ecology and behavior of Neotropical primates*, 1:331–90. Rio de Janeiro: Academia Brasileira de Ciências.

Guianan Saki Monkey
Pithecia pithecia
Plate 12, map 102

Identification. Measurements: HB = 285–480; T = 328–455; HF = 107–131; E = 28–40; WT = 1.65–2.35 kg; males larger than females.

Male entirely glossy pitch black except brow and sides of face pure white or rusty orange, white cheeks sometimes grading to orange or pale buff beard; **face around mouth, nose, and eyes dark, framed by white. Female entirely grizzled gray or gray-brown;** face whitish, with a diagonal white stripe from below eye, down cheek beside muzzle. **Fur long and shaggy** on upperparts, hanging in a fringe on sides over thinly haired belly, makes the body appear much larger than it is; long hair grows forward in a mop over crown from rear of head. Head held low, below level of shoulders. **Tail nonprehensile, exceptionally bushy,** usually hangs straight downward when saki is sitting and is carried behind, not above, the body when traveling. Hands and feet black or gray; belly sometimes bright rusty orange.

Variation. On the N bank of the Rio Amazonas east of Manaus, males have orange faces and females have whitish crowns and orange bellies.

Similar species. Brown bearded sakis (*Chiropotes satanus*) have no white on face, are not shaggy over the shoulders, and often carry the tail raised in an arc over the back.

Sounds. Generally quiet; low grunts in alarm; clear whistles and trills.

Natural history. Diurnal; arboreal; groups of one to four, usually two or three. Feeds on fruit, seeds, and leaves. Guianan sakis seem to prefer tall, primary forest habitat, where they use the middle and lower levels most intensively. They are quiet and wary. Sakis characteristically travel by springing leaps, sometimes hopping bipedally without using their hands, and are locally known as ''flying monkeys.'' When a saki is sitting and apparently slightly alarmed, the plumelike tail is sometimes waggled back and forth. They are occasionally seen in apparent association with other species of monkeys. Found in a variety of lowland evergreen rainforest types, including terra firme, riverside, dense, and open, and although they can be found in disturbed or secondary vegetation, they are chiefly mature-forest monkeys.

Geographic range. South America: the Guianas, Venezuela, and Brazil north of the Rio Amazonas and east of the Rios Negro and Orinoco.

Map 102

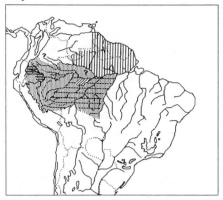

	Monk saki monkey, *Pithecia monachus*
	Guianan saki monkey, *P. pithecia*
	Equatorial saki monkey, *P. aequatorialis*
	Buffy saki monkey, *P. albicans*
	P. monachus irrorata

Status. CITES Appendix II. Usually uncommon or rare and occurs patchily, but widespread, and much of its geographic range is undisturbed by man.

Local names. Parauacu (Br); flying jack, breezy monkey (Gy); yarké (FG); witkopap, pluimstaartap, wanakoe. (Su).

References. Oliveira, J. M. S., M. G. Lima, C. Bonvíncíno, J. M. Ayres, and J. G. Fleagle. 1985. Preliminary notes on the ecology and behavior of the Guianan saki (*Pithecia pithecia*, Linnaeus 1776; Cebidae, Primate). *Acta Amazonica* 15:249–63.

Mittermeier, R. A., and M. G. M. van Roosmalen. 1983. A synecological study of Surinam monkeys. In A. G. Rhodin and K. Miyata, eds., *Advances in herpetology and evolutionary biology*, 521–34. Cambridge: Museum of Comparative Zoology, Harvard University.

Monk Saki Monkey
Pithecia monachus
Plate 12, map 102

Identification. Measurements: HB = 370–480; T = 404–500; HF = 116–147; E = 25–37; WT = 2.2–2.5 kg.

Upperparts black, or lightly to heavily frosted with dirty white, appears blackish or gray with black hair base showing below; **fur long and shaggy. Face** almost naked, **blackish** thinly haired with white, **white stripe diagonally down side of face from below eye; crown** almost bald, **lacking a mat of short hair; mop of long hair growing forward over brow; brow and cheeks the same color as back.** Hands and feet dirty white; legs sometimes whitish. Chin and throat naked; chest thinly haired whitish or pale buffy; belly thinly haired black with pale skin showing through. Tail, posture, and locomotion like Guianan saki. **Male and female alike.** Young almost black.

Variation. Blackish animals are from Rondônia (Brazil), Peru, Ecuador, and Colombia; those from east of the Rio Juruá are generally frosted with white and may have gray-rimmed faces: these have recently been recognized as a separate species, *P. irrorata.* Formerly sometimes called *P. hirsuta.*

Similar species. Equatorial sakis (*P. aequatorialis*) have contrasting white brow and cheeks, and sometimes red chest or beard; buffy sakis (*P. albicans*) have ruff, limbs, and underparts buff or orange.

Sounds. Generally quiet; calls with low-pitched whines, grunts, chirps.

Natural history. Diurnal; arboreal; groups of two to eight, usually three or four. Feeds on fruits, ripe and green seeds, leaves, and ants. Monk sakis use the middle and upper levels of the forest. They are shy and quiet, with behavior like that of equatorial sakis. At night they sleep on large branches in the upper canopy. There is only one breeding female in a group, and she gives birth every two to three years. Home ranges of groups overlap. Monk sakis are usually found in terra firme forest, where they frequent dense, leafy vegetation, but they can also be found in seasonally flooded forests. Found in lowland evergreen rainforest only.

Geographic range. South America: upper Amazon Basin of Colombia, Ecuador, Peru, Bolivia, and Brazil. To 600 m elevation.

Status. CITES Appendix II. Distribution patchy, locally common, uncommon, or rare, but widespread.

Local names. Mico volador (Co); parauacu (Br, Ec); saki (Ec); huapo negro, oso mono (Pe).

References. Happel, R. E. 1982. Ecology of *Pithecia hirsuta* in Peru. *J. Hum. Evol.* 11:581–90.

Soini, P. 1986. A synecological study of a primate community in the Pacaya-Samiria National Reserve, Peru. *Primate Conservation* 7:63–71.

Equatorial Saki Monkey
Pithecia aequatorialis
Plate 12, map 102

Identification. Measurements: HB = 392–440; T = 448–474; HF = 111–131; E = 28–40; WT = 2–2.5 kg.

Upperparts black sparsely frosted with dirty white; fur long, shaggy, and wavy, hangs in a long fringe on sides over nearly naked belly. **Face nearly naked, whitish, male with face surrounded by a broad, sharply contrasting border of short white hair,** or cheeks and beard red; **crown covered by a distinctive mat of short white or buffy hair. Female with cheeks and crown white like male, or distinctly**

darker, similar to back. **Feet dirty white. Throat, chest, and belly orange or dark brown.** Tail and body hair pattern, locomotion, and posture as in Guianan saki (*P. pithecia*).

Variation. This species was until recently confused with the monk saki; older accounts of saki natural history often do not adequately distinguish which species was studied.

Similar species. Monk sakis (*P. monachus*) have cheeks and crown the same color as upperparts or black, lack a mat of short dense hair on crown, and are generally grayish; titis (*Callicebus* spp.) do not have bushy tails.

Sounds. Usually silent; gives low grunts in alarm, also chirps and moaning calls.

Natural history. Diurnal; arboreal; groups of one to four. Diet probably fruit, seeds and leaves, similar to other sakis. In some remote areas, equatorial sakis can often be seen from a boat, from which they look like big black balls of fur with bushy tails, sitting in riverside trees. They seem to be most common in riverside, flooded, and swamp forests in poor-soil (blackwater) regions, but they are also found in terra firme forests away from water. Equatorial sakis are stealthy and quiet and will sit absolutely motionless in dense, leafy vegetation, their presence betrayed by long bushy tails hanging below a branch. If put to flight, they bound away acrobatically and suddenly disappear into hiding. Appear to be found mostly in lowland primary forests.

Geographic range. South America: in the upper Amazon Basin of Ecuador, Peru, and possibly Brazil.

Status. CITES Appendix II. Distribution appears patchy, rare or absent in many areas, locally common in a few.

Local names. Parauacu (Br, Ec); saki (Ec); huapo negro, oso mono (Pe).

References. Hershkovitz, P. 1987. The taxonomy of South American sakis, genus *Pithecia* (Cebidae, Platyrrhini): A preliminary report and critical review with the description of a new species and a new subspecies. *Amer. J. Primatol.* 12:387–468.

Buffy Saki Monkey
Pithecia albicans
Plate 12, map 102

Identification. Measurements: HB = 365–560; T = 405–570; HF = 120–132.

Back and tail blackish; ruff, arms, legs, fringe on sides, and underparts buff, orange, or reddish. Sexes similar. Long, shaggy hair of body and tail as in other sakis. Infants under 3 weeks old are chocolate brown.

Similar species. Monk sakis (*P. monachus*) are gray or black, with a dark belly.

Natural history. Diurnal; arboreal; groups of one to seven. Feeds on fruits, young leaves, seeds of unripe fruits, and insects. Buffy sakis use the middle and upper levels of tall forest. They move with frequent strong leaps, but they spend much of the day sitting motionless and hide from observers in the same way as equatorial sakis (above). Found in undisturbed and logged primary forests.

Geographic range. South America: Brazil, south of the Solimões between the lower Rios Purus and Juruá.

Status. CITES Appendix II. Locally common and not intensively hunted, but geographic range small.

Local names. Parauacu.

References. Johns, A. 1986. Notes on the ecology and current status of the buffy saki, *Pithecia albicans. Primate Conservation* 7:26–29.

Brown Bearded Saki Monkey
Chiropotes satanus
Plate 12, map 103

Identification. Measurements: HB = 327–440; T = 300–465; HF = 121–122; E = 30–33; WT = 2–4 kg; males slightly larger than females.

Head, limbs, and tail glossy deep brown-black; back from neck to tail pale yellow to red-brown, palest on center of back between shoulders, darkening to brown on sides; **or back rich, dark brown. Face dark, crown hair growing** from a whorl forward and sideward **into two large tufts with sharp part on midline;** tufts much larger in males than females and supported by enlarged mounds of muscle on temples; **chin and cheeks with a long, thick, rounded beard,** small in females. **Tail nonprehensile, very bushy, often carried**

Map 103

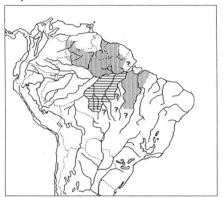

Brown bearded saki monkey, *Chiropotes satanus*
White-nosed bearded saki monkey, *C. albinasus*

in a high arc over back when walking, waved back and forth in this position, or sometimes waggled rapidly vertically below body in excitement. Underparts blackish except genital area naked, pure white. **Variation.** Animals from north of the Rio Amazonas have pale yellow or reddish backs; several colors are seen in a single troop. Those from south of the river have dark brown backs except close to the Rio Xingu, where they have pale brown backs. **Similar species.** Capuchins (*Cebus* spp.) have slender, prehensile tails carried in a coil; howler monkeys (*Alouatta* spp.) are either entirely red (N of Rio Amazonas) or black with brown hands (S of Rio Amazonas) and have slender, often coiled tails; spider monkeys (*Ateles* spp.) have long, slender limbs and tail and swing by their arms below branches. **Sounds.** High-pitched whistles rising in a crescendo and abruptly cut off; other whistles. **Natural history.** Diurnal; arboreal; groups of 2 to more than 30, usually more than 10. Feeds largely on the immature seeds of unripe fruits and on ripe fruits, flowers, and insects. Brown bearded sakis are most often seen in the canopy of tall forest. They often sit silently, but when they feed on unripe seeds a rain of discarded fruit parts patters to the ground, betraying their location. Sometimes they associate with other monkey species. They travel by walking on the tops of branches, tail held

high, and jump across gaps with powerful springs from the hind legs. They may hang by their hindfeet to reach down and pick fruits. Troops use large areas of several km², and each day they travel long distances. Found in mature lowland rainforest or, rarely, deciduous forest. **Geographic range.** South America: eastern Amazon Basin in Venezuela, Brazil, and the Guianas; north of the Amazon east of the Rios Negro and Orinoco, south of the Amazon east of the Rio Xingu. **Status.** CITES Appendix II, US-ESA endangered. Generally uncommon, hunted for food and rare or absent near settlements, but much of large geographic range is uninhabited. **Local names.** Cuxiú (Br); black saki, satan-aap, baard-aap, bisa (Su). **References.** Ayres, J. M. 1981. Observações sobre a ecologia e o comportamento dos cuxiús (*Chiropotes albinasus* e *Chiropotes satanus*, Cebidae: Primates). M.S. thesis, INPA, Manaus (CNPq, INPA, FUA).

Hershkovitz, P. 1985. A preliminary taxonomic review of the South American bearded saki monkeys genus *Chiropotes* (Cebidae, Platyrrhini), with the description of a new subspecies. *Fieldiana Zool.*, n.s., no. 27.

White-nosed Bearded Saki Monkey
Chiropotes albinasus
Plate 12, map 103
Identification. Measurements: HB = 360–460; T = 355–480; HF = 115–130; E = 25–30; WT = 2.2–3.6 kg; males slightly larger than females.
Upperparts glossy pitch black, back frosted with sheen of silvery olivaceous or silver; hair of back growing laterally from a part down midline. **Head** entirely black except muzzle from between eyes to upper lip a triangle of pure white; crown with long thick hair growing in two prominent tufts divided by sharp part; hair growing in a whorl from back of crown forward, sideward, and backward; tufts much larger in males than females; **beard of thick hair** much larger on males than females. **Tail nonprehensile, thickly furred and bushy; often carried in a high arc over back,** or waggled vertically below body. Underparts thinly haired, black; gen-

ital areas and scrotum naked, white. Infants have a prehensile tail. Locomotion similar to brown bearded saki.

Similar species. These are the only black monkeys with a white nose and bushy, nonprehensile tail; howlers (*Alouatta* spp.) are black with black faces, slender tails; spider monkeys (*Ateles* spp.) have long, slender limbs and tail; monk sakis (*Pithecia monachus*) are shaggy and grizzled with white.

Natural history. Diurnal; arboreal; groups of 2 to more than 26. Feeds on ripe fruits, many seeds of unripe fruits, and a few flowers and insects. White-nosed sakis most often use the middle and upper levels of the forest. A troop has a home range of about 1 km² and travels 2–5 km a day. Sometimes white-nosed bearded sakis associate with other monkeys, especially brown capuchins. Found only in undisturbed terra firme rainforest.

Geographic range. South America: Brazil, south of the Rio Amazonas between the Rios Madeira and Xingu.

Status. CITES Appendix I, US-ESA endangered. Apparently fairly common where not hunted, but rare from overhunting near settlements; threatened because of restriction to undisturbed forest.

Local names. Cuxiú, piroclucu, piroculú.

References. Van Roosemalen, M., R. A. Mittermeier, and K. Milton. 1981. The bearded sakis, genus *Chiropotes*. In A. F. Coimbra-Filho and R. A. Mittermeier, eds., *Ecology and behavior of Neotropical primates*, 419–41. Rio de Janeiro: Academia Brasileira de Ciências.

Red or White Uakari Monkey
Cacajao calvus
Plate 12, map 104

Identification. Measurements: HB = 360–570; T = 137–185; HF = 122–152; E = 26–33; WT = 2.3–3.5 kg; males larger than females.

There are two distinctive races, generally: **face and crown naked, bright red, forehead of males with large bulging muscles over temples,** giving head a square top; canines form large bulges under lips; **body hair long, shaggy, and coarse, forming a cape over shoulders; tail short, stumpy, one-third the length of head and body, thickly haired,** wagged in excitement.

Map 104

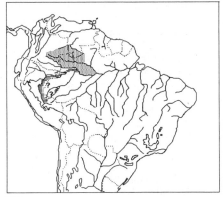

Black uakari monkey, *Cacajao melanocephalus*
Red or white uakari monkey, *C. calvus*

White uakari, *C. calvus calvus*. **Upperparts dirty white,** yellowish or pale gray frosted with blackish; crown frosted with black, with short cap of hairs growing forward; chin with fringe of blackish hairs; underparts and front of thighs buffy to red brown, or hindleg entirely washed with faint buff.

Red uakari, *C. calvus rubicundus*. **Upperparts entirely bright orange,** or cape paler, yellowish white; crown hair whitish yellow.

Variation. There are a number of local color variations of the forms above.

Similar species. These are the only monkeys with short tails and scarlet faces.

Sounds. Staccato series of sonorous, low-pitched, "ca-ca-ca-ca."

Natural history. Diurnal; arboreal; groups of 10 to at least 30, possibly up to 100. Large groups often split up into smaller groups of 1–10 that forage independently. Feeds on seeds of immature fruits, ripe fruits, leaves, nectar, and some insects, especially caterpillars. Husks of hard fruits are cracked open with the canines. Red and white uakaris seem restricted to permanently or seasonally flooded forests, especially near small streams (*várzea* forest). Usually they use the middle and upper levels of the forest, but in the dry season they will descend to the ground to feed on seedlings and fallen seeds. Uakaris are very active and spend more of the day moving, and they move farther per day than most other

New World monkeys. Sometimes they are seen in association with other monkeys. At night they climb into large trees and sleep on the highest thin branches. Apparently found only in primary forests.

Geographic range. South America: central Amazon Basin of Brazil, Colombia, and Peru. The white form is found only on or around the large delta island in the Rio Solimões below the mouth of the Rio Japurá; the red form is found south of the Solimões and west of the Juruá, and between the Ríos Ucayali and Putumayo in Peru. Range poorly known.

Status. CITES Appendix I; US-ESA endangered. Hunted for food in Peru and for bait in Brazil (where people apparently do not eat them because of their disconcertingly human-looking faces). A habitat specialist with limited, fragmented distribution in habitat that is particularly vulnerable to logging and hunting owing to accessibility by boat.

Local names. Uácari vermelho; macacoingles (Br); huapo colorado, mono ingles, puca huapo (Pe).

References. Ayres, J. M. 1986. Uakaris and Amazonian flooded forest. Ph.D. diss., University of Cambridge, Cambridge.

————. 1986. The conservation status of the white uakari. *Primate Conservation* 7:22–25.

Black Uakari Monkey
Cacajao melanocephalus
Plate 12, map 104

Identification. Measurements: HB = 300–500; T = 125–210; HF = 111–143; E = 19–36; WT = 2.4–4 kg; males slightly larger than females.

Head, forearms, hindfeet, and belly black; shoulders dark brown at top with cape of long hair, yellowish, or black, grading to dark red hindquarters and tail. Face and crown thinly haired, black with surrounding fringe of hair growing forward; male with large bulge of muscle at each temple. Tail short and stumpy, one-third the length of head and body, thickly haired. Chest with red skin showing below black hairs.

Variation. Animals from Venezuela have black capes; those from Brazil and Colombia have yellow capes.

Similar species. No other monkey in range has a short tail or similar color.

Natural history. Diurnal; arboreal; groups of 15 to 30 or more. Feeds on fruits and probably nuts. Black uakaris are apparently restricted to permanently or seasonally flooded forest, especially near blackwater streams, where they use the canopy layer. They have a patchy distribution in mature lowland rainforest.

Geographic range. South America: the upper Amazon Basin north of the river in SE Colombia, S Venezuela, and adjacent Brazil. Range poorly known.

Status. CITES Appendix I, US-ESA endangered. Said to be common in some remote parts of its range, but generally rare; hunted for food in Colombia.

Local names. Cauirí (Br); chucuto (Co).

References. Hershkovitz, P. 1987. Uacaries, New World monkeys of the genus *Cacajao* (Cebidae, Platyrrhini): A preliminary taxonomic review with the description of a new subspecies. *Amer. J. Primatol.* 12:1–58.

Mittermeier, R. A., and A. F. Coimbra-Filho. 1977. Primate conservation in Brazilian Amazonia. In Prince Rainier of Monaco and G. H. Bourne, eds., *Primate conservation,* 117–66. New York: Academic Press.

Red Howler Monkey
Alouatta seniculus
Plate 13, map 105

Identification. Measurements: HB = 439–690; T = 540–790; HF = 128–162; E = 30–48; WT = 3.6–11.1 kg; males larger than females.

Head, shoulders, limbs, tail, and usually underparts dark red to purplish red; back and sides paler, bright orange to gold. Head large, throat swollen; face naked, black; chin with forward-growing beard, longest in males. Shoulders robust, hindquarters disproportionately small and weak-looking; tail prehensile, often carried in a coil. Adult males often have blackish beard, limbs, and tail.

Variation. Some animals from Venezuela are almost entirely purplish red; those from Peru have head, limbs, and tail red-orange, back gold.

Similar species. These are the only large, red and gold prehensile-tailed monkeys. Some white-bellied spider monkeys (*Ateles*

Map 105

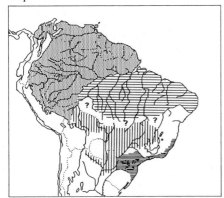

▦	Red howler monkey, *Alouatta seniculus*
▤	Red-handed howler monkey, *A. belzebul*
▥	Black howler monkey, *A. caraya*
▤	Brown howler monkey, *A. fusca*

belzebuth) are pale brown, with paler bellies; other spider monkeys in range have black backs.

Sounds. Loud choruses of roaring or howling, usually lasting many minutes, especially at dawn, late afternoon, and during rainstorms. Howling starts as an accelerating series of deep grunts by a male, which grades into long, deep, roars; the females join in with higher-pitched roars. In alarm, one monkey in the group gives a few soft grunts.

Natural history. Diurnal; arboreal; groups of three to nine, usually five to seven. Feeds on fruits and many leaves. Red howlers use the middle and upper levels of mature and disturbed forests. They are quiet, slow-moving monkeys that have small home ranges and move little during the day. Their inactivity makes them inconspicuous, but their recent presence is shown by copious droppings, with a strong, pungent stablelike smell. When alarmed they may position themselves to defecate and urinate on a person below. These are one of the monkeys most easily seen from a boat, as they sit high in riverside trees. In the dry season they come to the ground at the river's edge and drink or eat earth. They use salt licks. Because their home range needs are small, howlers can persist in small forest fragments or gallery forests. In habitats where there are few other mon-

key species, they can reach high densities of over 100/km². Mature and disturbed evergreen to deciduous forest and gallery forest.

Geographic range. South America: east of the Andes in Colombia, Venezuela, Trinidad, the Guianas, and Brazil north of the Amazon, and Ecuador, Peru, Bolivia, and Brazil west of the Purus. To 1,200 m elevation. SE limits of range poorly known.

Status. CITES Appendix II. Widespread and locally common but intensively hunted for meat in many areas. Rare or extinct from hunting near human settlements in the upper Amazon Basin, where its flesh is tasty; in French Guiana and some other areas, its meat is bitter and it is hunted only as a last resort.

Local names. Guariba (Br); mono colorado, cotudo, roncador, araguato, berreador (Co); coto mono (Ec, Pe); babouin (FG); brulaap, baboen (Su).

References. Crockett, C. M., and J. F. Eisenberg. 1987. Howlers: Variations in group size and demography. In B. B. Smuts, D. L. Cheney, R. M. Seyfarth, R. W. Wrangham, and T. T. Struhsaker, eds., *Primate societies,* 54–68. Chicago: University of Chicago Press.

Rudran, R. 1979. The demography and social mobility of a red howler (*Alouatta seniculus*) population in Venezuela. In J. F. Eisenberg, ed., *Vertebrate ecology in the northern Neotropics,* 107–26. Washington, D.C.: Smithsonian Institution Press.

Red-handed Howler Monkey
Alouatta belzebul
Plate 13, map 105

Identification. Measurements: HB = 374–560; T = 580–700; HF = 125–164; E = 30–45; WT = 4.8–8 kg; males larger than females.

Entirely black except hands, feet, tail tip, and sometimes forehead or back dull, rusty red. Scrotal area of males rust-red. Body shape like red howler.

Variation. At one locality near the Rio Madeira, all animals are completely black. Some males from between the Rios Tapajós and Madeira, and from the east bank of the Xingu are entirely black, but some have red feet and tail tip, like females. Males may tend to be entirely black in this region and females to have red extremities.

Similar species. Spider monkeys (*Ateles* spp.) in this region are entirely black, some with red faces; they have small pointed heads, no beard, and long, slender arms and legs. Bearded sakis (*Chiropotes* spp.) have bushy, nonprehensile tails often carried up over the back, never in a coil.
Sounds. Roars like red howler, but higher pitched.
Natural history. Diurnal; arboreal; groups of two to eight. Probably similar to red howler. Uses the upper levels of terra firme and flooded forests, including disturbed forests.
Geographic range. South America: Brazil south of the Amazon east of the Rio Purus to Alagoas.
Status. CITES Appendix II. Locally common, widespread.
Local names. Guariba.
References. Branch, L. C. 1983. Seasonal and habitat differences in the abundance of primates in the Amazon (Tapajós) National Park, Brazil. *Primates* 24:424–31.

Black Howler Monkey
Alouatta caraya
Plate 13, map 105
Identification. Measurements: HB = 420–550; T = 530–650; HF = 126–155; E = 30–38; WT = 3.5–7.3 kg; males larger than females.
Males entirely black or dark, blackish brown; scrotum rust-red. Females pale yellow-brown or dirty yellow; lower back sometimes brown darker than shoulders. Underparts sometimes tinged orange. Body shape like red howler. Young males to about 4.5 years colored like females.
Variation. Males from Bahia and Goiás are black; those from Matto Grosso and Paraná have brown back and hindparts, black head and forequarters. Those from São Paulo and Minas Gerais are brown-black with yellowish hands, feet, belly and tail tip; these parts vary from pale yellow to almost entirely brown-black.
Similar species. These are the only monkeys in their range with strikingly different blackish males and yellowish females; they are the only large monkeys in most of their range. See brown howler (*Alouatta fusca*).
Sounds. Like red howler.
Natural history. Diurnal; arboreal; groups of 3–19, usually 7–9. Probably similar to

red howler. Gallery forests and deciduous forests in Pantanal, Cerrado, and Chaco habitats. The southernmost howler monkey.
Geographic range. South America: southern Amazon Basin and Paraguay Basin from E Bolivia, W Paraguay, and N Argentina, to SE Brazil from Bahía to Rio Grande do Sul, in the latter region in the western higher elevations away from the coast.
Status. CITES Appendix II. Geographic range large, but suitable habitat highly fragmented in some regions. Locally common; little hunted in Paraguay because its meat is bitter.
Local names. Guariba, bugio, carajá (Br); mono aullador negro (Ar); carayá (TGua).
References. Thorington, R. W., Jr., J. C. Ruiz, and J. F. Eisenberg. 1984. A study of a black howling monkey (*Alouatta caraya*) population in northern Argentina. *Amer. J. Primatol.* 6:357–66.

Brown Howler Monkey
Alouatta fusca
Plate 13, map 105
Identification. Measurements: HB = 440–570; T = 510–610; HF = 129–140; E = 30–40; WT = 4.1–7.2 kg.
Males with upperparts brown, red-brown, or orange-red; back brown frosted with yellow or gold, rump red or orange. Beard dark red to blackish. Tail dark red-brown to dark red or orange. Underparts dark brown, red, or blackish. Females usually paler overall, yellowish brown to dark brown (but see below). Body shape like red howler.
Variation. Males from Minas Gerais are brown, with orange-brown rumps; females yellow-brown. Males from São Paulo are orange-red to red-brown, with red bellies. Males from Santa Catarina and Rio Grande do Sul are bright red-orange with dark brown feet; females dark brown or blackish or with back frosted with orange or yellow-brown. There is also much individual variation within populations; larger males seem reddest. The name *A. guariba* is also applied to this species.
Similar species. Woolly spider monkeys (*Brachyteles arachnoides*) are entirely beige, with prominent, furry ears; black howlers (*A. caraya*) have blackish males and yellow-brown females.

Sounds. Probably like other howlers.
Natural history. Diurnal; arboreal; groups
of 2–11. Feeds on leaves and fruits.
Brown howlers chiefly use the upper levels
of the forest. Like other howlers they are
slow-moving and sedentary. This species
has been associated with the endemic Paraná pine (*Araucaria angustifolia*). It favors
these trees for resting and sleeping, and its
brown color matches that of clusters of
dead leaves remaining on the tree. Found
in mature and secondary Atlantic coastal
forests.
Geographic range. South America: Brazil
coastal forests from Bahía to Rio Grande
do Sul; Misiones, Argentina; and possibly
a disjunct population in Beni, Bolivia.
Status. CITES Appendix II. Status unknown, but said to be drastically declining;
much of the original habitat has been deforested, and the species has been exterminated by hunting for meat in some of the
remaining forests.
Local names. Bugio, ruivo, guariba (Br);
mono aullador rufo (Ar).
References. Cordeiro da Silva, E., Jr.
1981. A preliminary survey of brown howler monkeys (*Alouatta fusca*) at the Cantar-
eira reserve (São Paulo, Brazil). *Rev. Brasil. Biol.* 41:897–909.

Mantled Howler Monkey
Alouatta palliata
Plate 13, map 106
Identification. Measurements: HB = 405–
555; T = 585–710; HF = 124–154; E =
25–39; WT = 4.8–7.7 kg; males larger
than females.
**Entirely black except sides with a fringe
(mantle) of long, pale hairs, yellow, gold,
pale brown, or buff; pale color sometimes
extending as a saddle across entire lower
back.** Scrotum of adult males white. Hair
of head, limbs, and tail relatively short.
Subadult males indistinguishable from females in the field. Body shape like red
howler.
Similar species. Mexican black howler
monkeys (*A. pigra*) are entirely black and
live in small groups; Central American spider monkeys (*Ateles geoffroyi*) are black,
brown, or red, with a paler belly and face
often pale around eyes; Colombian black
spider monkeys (*Ateles fusciceps*) are entirely black or have a brown head; spider

Map 106

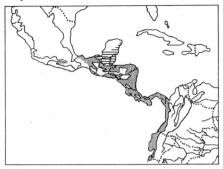

Mantled howler monkey, *Alouatta palliata*
Mexican black howler monkey, *A. pigra*

monkeys have small heads, no beard, long
arms and tail, and make whinnying calls.
Sounds. Roars as described for red howler
but higher pitched.
Natural history. Diurnal; arboreal; groups
of 2–45, usually 10–18. Feeds on fruit
and leaves. Mantled howlers use the upper
levels of the forest. They are slow-moving
and sedentary, often inconspicuous. Like
other howlers, their presence is shown by
strong-smelling droppings on the ground
below them, and they are prone to defecate
on the heads of people below. They have
small home ranges, so they can survive in
small fragments of forest. In protected
areas populations can be extremely high.
These are the monkeys most often seen in
parks and around ruins in archaeological
sites. In mature and old secondary evergreen, semideciduous, and riverine forests.
Geographic range. Central and South
America: E Mexico south through Panama,
except the Yucatán Peninsula (see following species); west of the Andes from Colombia to N Peru. To 1,500 m elevation.
Status. CITES Appendix I, US-ESA endangered. Threatened by deforestation in
many parts of its range, which is now
fragmented; hunted for food in some areas.
Local names. Mono negro (Co, Pn); mono
congo (CR); mono chongo, chongón (Co),
coto negro (Ec, Pe); saraguato, mono aullador (Me).
References. Carpenter, C. R. 1934. A
field study of the behavior and social rela-
tions of howling monkeys (*Alouatta palliata*). *Comp. Psychol. Monogr.* 10:1–168.

   Milton, K. 1980. *The foraging strategy*

of howler monkeys. New York: Colombia
University Press.

Mexican Black Howler Monkey
Alouatta pigra
Plate 13, map 106
Identification. Measurements: HB = 521–
639; T = 590–690; HF = 133–165.
Entirely black; hair long. Scrotum of
males over 3–4 months old white and
prominent. Body shape like red howler.
Similar species. Mantled howlers
(*A. palliata*) have a pale saddle or fringe
on sides; Central American spider monkeys
(*Ateles geoffroyi*) in same area are brown
or silvery with pale bellies and pale masks
around eyes.
Sounds. Like red howler. Usually howls in
late afternoon and early morning.
Natural history. Diurnal; arboreal; groups
of 2–10, usually 4–6. Feeds on fruits and
leaves. These monkeys can often be seen
in forests around archaeological sites. Their
behavior apparently is similar to that of
other howlers. Found in evergreen and riverine lowland forests.
Geographic range. Central America: Yucatán Peninsula of Mexico, Guatemala, and
Belize.
Status. CITES Appendix II, US-ESA endangered. Threatened by deforestation and
hunting for meat.
Local names. Mono aullador (Span); ba-
boon (Be); saraguate (Me); batz (May).
References. Horwich, R. H., and
C. Gebhard. 1983. Roaring rhythms in
black howler monkeys (*Alouatta pigra*) of
Belize. *Primates* 24:290–96.

   Schlichte, H.-J. 1978. A preliminary report on the habitat utilization of a group of
howler monkeys (*Alouatta villosa pigra*) in
the National Park of Tikal, Guatemala. In
G. G. Montgomery, ed., *The ecology of
arboreal folivores,* 551–59. Washington,
D.C.: Smithsonian Institution Press.

Common Woolly Monkey
Lagothrix lagothricha
Plate 14, map 107
Identification. Measurements: HB = 390–
580; T = 550–800; HF = 118–158; E =
25–37; WT = 3.6–10 kg.
**Upperparts dark brown, pale smoky
brown, dark smoky gray, pale gray, red-
brown, or olivaceous;** limbs and tail and/

Map 107

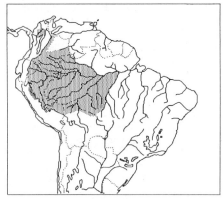

	Common woolly monkey, *Lagothrix lagothricha*
	Yellow-tailed woolly monkey, *L. flavicauda*

or head sometimes darker than back,
brown or reddish; but in general monkey
looks uniform-colored in the field; fur
soft and woolly, short and even, except old
animals have a fringe of longer hair on rear
of arms and legs and under belly. **Head
conspicuously round, face black;** crown
with short, even hair; **ears inconspicuous.
Tail strongly prehensile, thick and muscular at base, tapering to thinner tip.
Limbs and body robust and muscular;
potbelly rounded and protruding. Underparts dark.** Newborns straw-colored.
Variation. There is much individual and
geographic variation: in Ecuador and Colombia animals of different colors can be
seen in the same troop; gray or almost
black animals are found along the base of
the Andes in Colombia; olivaceous animals
with dark heads south of the Amazon in
Brazil and Bolivia; brown animals in Peru
and north of the Amazon in Brazil.
Similar species. Howler monkeys (*Alouatta* spp.) are red-orange or black, with
bearded chins; spider monkeys (*Ateles*
spp.) are black, or black or brown with
pale underparts, with small, pointed heads
and long, lanky arms, legs, and tail; capuchins (*Cebus* spp.) are smaller, yellow-
brown with pink or brown faces and black
caps.
Sounds. Loud, descending musical trills,
barks, screams, and other calls; alarm a
chorus of ''yoohk-yoohk.''
Natural history. Diurnal; arboreal; groups
of 2–60; where not hunted, groups are

large, 20–60. Common woolly monkeys feed on ripe fruits, seeds, palm nuts, a few leaves, and some arthropods. They use the upper and middle levels of the forest. Large groups are made up of collections of individual, permanent family units of 3–9 that may feed and travel independently of other such units in the group, or may travel and feed with them, or may only join them to sleep at night. Males may threaten humans by shaking branches and defecating. Common woolly monkeys seem to reproduce slowly, with females first giving birth at 6–8 years old and every other year thereafter. Found in mature terra firme and seasonally flooded rainforest; apparently not in secondary forests, but this may be because humans hunt them to extinction wherever they cut the forest.

Geographic range. South America: upper Amazon Basin of Colombia, Ecuador, Peru, Bolivia, and Brazil west of the Rios Negro and Tapajós, and isolated populations in Córdoba and Bolívar departments of Colombia. To at least 1,800 m elevation.

Status. CITES Appendix II. The most intensively hunted monkey, its meat is considered better than that of other large species; extinct or naturally absent from many areas. This species is unable to maintain its populations under much hunting pressure and is usually the first species to become locally extinct where subsistence hunting is high. Intolerance to disturbance of vegetation and low reproductive rate makes this species vulnerable to local extinction, but geographic range is large.

Local names. Macaco barrigudo (Br); choyo, churuco, chuluco (Co); choro, barrigudo (Ec, Pe).

References. Fooden, J. 1963. A revision of the woolly monkeys (genus *Lagothrix*). *J. Mammal.* 44:213–47.

Soini, P. 1986. A synecological study of a primate community in the Pacaya-Samiria National Reserve, Peru. *Primate Conservation* 7:63–71.

Yellow-tailed Woolly Monkey
Lagothrix flavicauda
Plate 14, map 107
Identification. Measurements: HB = 515–535; T = 560–610.

Upperparts glossy dark red-brown; hair long and thick, especially on legs, giving body a muscular, robust look. Face brown with pale yellow triangular patch covering mouth and nose, point of triangle between eyes. Tail prehensile, robust, dark brown above, yellow below for distal half. Underparts dark brown except genital area with a tuft of yellow or reddish hairs, prominent only in males.

Similar species. White-bellied spider monkeys (*Ateles belzebuth*) are black or brown, with distinctly pale underparts; red howler monkeys (*Alouatta seniculus*) are orange-red and have a beard; capuchins (*Cebus* spp.) are yellow-brown with black caps.

Sounds. A sharp, puppylike bark, like the bark of spider monkeys.

Natural history. Diurnal; arboreal; groups of 6–12. Feeds on fruits, flowers, roots of epiphytes, and leaf petioles. The largest male in a group may threaten an observer by shaking and dropping branches and by urinating and defecating. Yellow-tailed woolly monkeys are found only in steep, humid, premontane and montane forests. This species was discovered in 1802 by the explorer Von Humboldt, but it was not recorded again until 1925–26, when five specimens were taken. No more were recorded for 50 years, and the species was thought to be extinct, but it was rediscovered in 1974 by an expedition launched to search for it.

Geographic range. South America: NE Peru in a small region in the departments of Amazonas, San Martín, and Loreto, on the Andean slopes at 1,800–2,500 m elevation.

Status. CITES Appendix I, US-ESA endangered. Apparently rare; the geographic range is small and fragmented by valleys, although steep and inaccessible. The species is partly protected in Rio Abiseo National Park. Locally hunted for meat.

Local names. Paccorrunto, tupa, chú, choba, mono barroso, quillirunto.

References. Leo Luna, M. 1980. First field study of the yellow-tailed woolly monkey. *Oryx* 15:386–89.

Macedo-Ruiz, H. de, and R. A. Mittermeier. 1979. Redescubrimiento del primate peruano *Lagothrix flavicauda* (Humboldt 1812) y primeras observaciones sobre su biología. *Rev. Cienc. Univ. Nac. San Marcos* 71:78–92.

Map 108

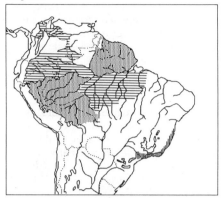

▨ Black spider monkey, *Ateles paniscus*
▤ White-bellied spider monkey, *A. belzebuth*
▥ Woolly spider monkey, *Brachyteles arachnoides*

Black Spider Monkey
Ateles paniscus
Plate 14, map 108

Identification. Measurements: HB = 430–620; T = 710–900; HF = 175–220; E = 30–40; WT = 7.5–13.5 kg.

Entirely black; face black, flesh-colored, or red. Head small, narrow at crown; pointed tufts of hair protrude sideways from in front of ear. Size large; arms, legs, and prehensile tail extremely long; hands with only four functional digits, thumb rudimentary. Often swings by arms below branches and hangs supported by tail.

Variation. Guianan black spider monkeys (*A. p. paniscus*), from north of the Amazon and east of the Negro in Brazil and the Guianas, have red faces and long hair, giving the body a robust, apelike appearance, with tail thickly haired for two-thirds of its length, tapering sharply at prehensile tip; they are the largest spider monkeys. Black-faced black spider monkeys (*A. p. chamek*), from the upper Amazon Basin south of the Amazon of Peru, Bolivia, and Brazil, have black faces, shorter hair, and are slightly smaller.

Similar species. North of the Amazon these are the only large, prehensile-tailed, completely black monkeys with a red face; south of the Amazon red-handed howlers (*Alouatta belzebul*) have completely black

faces with beards; bearded sakis (*Chiropotes* spp.) have bushy, nonprehensile tails, common woolly monkeys (*Lagothrix lagothricha*) are brown or gray, with shorter limbs and round heads. Spider monkeys are the only Neotropical monkeys that regularly travel by swinging by their arms from branch to branch.

Sounds. Loud, prolonged screams that can be heard for long distances; short quavering whinnies heard at close range; in alarm, barks like a pack of dogs. Noisy travelers that make long, crashing leaps through the canopy.

Natural history. Diurnal; arboreal; large social groups of about 20, but members of a group usually feed in small units of 1–5, and the whole social group is rarely seen together. Feeds on ripe fruits and some new leaves and flowers. Spider monkeys use the upper levels of the trees, usually in tall, open forest. They often sprawl quietly on a branch or hang motionless and soundless and may be difficult to spot. When disturbed they may threaten a human by standing on a branch and stamping and shaking the vegetation, breaking and dropping dead branches, and growling. They sometimes feed on bright moonlit nights (as do brown capuchins), and they have a distinct, moderately bright whitish eyeshine. They are nervous at night and sometimes bark for many minutes. Black spider monkeys have a low reproductive rate: females first give birth at 4–5 years old and have young only every three to four years thereafter. Found mainly in tall, undisturbed evergreen rainforest.

Geographic range. South America: Brazil and the Guianas east of the Rio Negro and north of the Amazon; the upper Amazon Basin of Peru, Bolivia, and Brazil from the Rio Madeira west; limits of range poorly known. To 1,800 m elevation.

Status. CITES Appendix II. Common where not hunted, locally extinct or threatened from overhunting for meat in inhabited areas, but widespread, and parts of range are uninhabited.

Local names. Coatá (Br); maquisapa (Pe); kwata (FG, Su); spinaap, slingeraap (Su); ca'í (Gua).

References. Mittermeier, R. A., and M. G. M. van Roosemalen. 1981. Preliminary observations on habitat utilization and diet in eight Surinam monkeys. *Folia Primatol.* 36:1–39.

Symington, M. 1987. Ecological and social determinants of party size in the black spider monkey, *Ateles paniscus chamek.* Ph.D. diss., Princeton University, Princeton.

White-bellied Spider Monkey
Ateles belzebuth
Plate 14, map 108
Identification. Measurements: HB = 416–582; T = 680–899; HF = 176–217; E = 31–46; WT = 5.9–10.4 kg.
Upperparts black, pale brown, dark brown, or reddish brown. Underparts sharply contrasting paler white, yellowish or pale brown, or black. Head with white or pale brow, triangular patch on brow, or white fringe completely surrounding face, face black or red. Hands, feet, and forearms like upperparts. Tail dark like upperparts, or dark above, pale below. Shape and posture like black spider monkey.
Variation. Animals from Amazonian Colombia and Venezuela are black, with yellow to gold underparts, legs and tail gold frosted with black, and little or no white on face. Those from N Colombia and Zulia, Venezuela (*A. b. hybridus*), are pale brown above with a darker brown head and white brow triangle, and whitish underparts; some have blue eyes.
A. b. marginatus in Pará is black, with a black belly and red mask; it probably belongs in the species *A. paniscus.* There is much individual variation within a group.
Similar species. These are the only large, pale-bellied, prehensile-tailed monkeys; in the range of the black-bellied form (*A. b. marginatus*) south of the Amazon, red-handed howlers (*Alouatta belzebuth*) have large, bearded heads and robust limbs; woolly monkeys (*Lagothrix lagothricha*) are uniform brown or gray; bearded sakis (*Chiropotes* spp.) have bushy, nonprehensile tails.
Sounds. Screams, barks, whinnies like black spider monkey.
Natural history. Diurnal; arboreal; groups of 1–20; members of larger groups often disperse into smaller units to feed. Feeds on ripe fruits and some dead wood and leaves. White-bellied spider monkeys use

the upper levels of tall forest. They spend much of the day resting and feed most intensively in early morning and late afternoon. Their postures and locomotion are like those of black spider monkeys. Found in mature rainforest and deciduous forest.
Geographic range. South America: east of the Andes from N Colombia and Venezuela to N Peru north of the Marañon; south of the Río Amazonas east of the Rio Purus. Range incompletely known. To 1,300 m elevation.
Status. CITES Appendix II. Threatened by deforestation in northern part of range; Amazonian populations are subject to similar pressures as black spider monkeys.
Local names. Marimonda (Co, Ec, Ve); choiba, coatá (Co); maquisapa (Pe).
References. Klein, L. L., and D. B. Klein. 1977. Feeding behaviour of the Colombian spider monkey. In T. H. Clutton-Brock, ed., *Primate ecology,* 153–81. London: Academic Press.

Kellogg, R., and E. A. Goldman. 1944. Review of the spider monkeys. *Proc. U.S. National Museum* 96:1–45.

Brown-headed or Colombian Black Spider Monkey
Ateles fusciceps
Map 109
Identification. Measurements: HB = 393–538; T = 710–855; HF = 157–190; E = 30–41.
Either entirely black, or head a contrasting brown. Shape and postures like black spider monkey.
Variation. Brown-headed animals are from west of the Andes in Ecuador. This may be a subspecies of the Central American spider monkey; there is evidence of hybridization where populations of the two species meet in Panama. Hybrids may be blackish with a reddish belly.
Similar species. In the region where they may occur together, Central American spider monkeys (*Ateles geoffroyi*) are red-brown; howler monkeys (*Alouatta palliata*) have large heads, bearded chins, and generally more robust build and do not often swing below branches by their arms.
Sounds. Similar to black spider monkey.
Natural history. Diurnal; arboreal. Probably similar to other spider monkeys. Found in lowland and lower montane rainforest.

Map 109

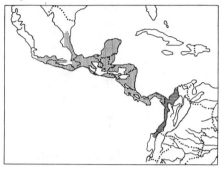

Central American spider monkey, *Ateles geoffroyi*
Brown-headed spider monkey, *A. fusciceps*

Geographic range. Central and South America: Darién, Panama, west of the Andes south to central Ecuador. To 1,600 m elevation.
Status. CITES Appendix II. Much of the geographic range is severely overhunted and deforested.
Local names. Mono negro, marimonda (Co); mona araña, marimonda, brasílargo (Ec); yerré (Pn).

Central American Spider Monkey
Ateles geoffroyi
Plate 14, map 109
Identification. Measurements: HB = 305–630; T = 635–840; HF = 135–196; E = 30–45; WT = 6.6–9 kg.
Upperparts black, brown, or reddish. Face often with "mask" of pale, unpigmented skin around eyes and muzzle. Forearms, lower legs, and feet usually black or dark. Underparts usually paler than back: white, pale brown, reddish, or buff. Body shape and posture like black spider monkey.
Variation. Highly variable both individually and geographically; animals from W Mexico, Guatemala, Honduras, and El Salvador are mostly blackish above; those from the Yucatán Peninsula, Nicaragua, and Costa Rica are brown to silvery above; those from Panama are red-brown with a reddish belly, except for one grayish population.
Similar species. Howler monkeys (*Alouatta* spp.) are black, with dark faces, large, bearded heads, and heavy shoulders; they

do not travel by swinging by their arms from branch to branch. There are no other large monkeys in Central America.
Sounds. Barks, whinnies, and screams like black spider monkey.
Natural history. Diurnal; arboreal; groups of 1–35; members of large groups frequently split into smaller groups to forage. Feeds on ripe fruits, with a few leaves and flowers. Spider monkeys use the upper levels of the forest. The locomotion, reproduction, and behavior of Central American spider monkeys are similar to those of black spider monkeys, except a female may have young every two to four years. Mature rainforest and deciduous forests.
Geographic range. Central America: both coasts of S Mexico south to the Colombian border.
Status. CITES Appendix I, US-ESA endangered (Pn, CR, Ni). Threatened by deforestation and fragmentation of range and by hunting for meat; common in a few parks, but locally extinct in many areas. Spider monkeys seem to need large areas of unbroken forest for survival.
Local names. Monkey (Be); mono araña, mico (Span); mono colorado (Pn, CR); maax (May).
References. Carpenter, C. R. 1935. Behavior of red spider monkeys in Panama. *J. Mammal.* 16:171–80.
  Richard, A. 1970. A comparative study of the activity patterns and behavior of *Alouatta villosa* and *Ateles geoffroyi*. *Folia Primatol.* 12:241–63.

Woolly Spider Monkey or Muriqui
Brachyteles arachnoides
Plate 14, map 108
Identification. Measurements: HB = 462–630; T = 650–800; WT = 12–15 kg.
Upperparts and underparts pale, smoky, grayish gold-beige; rump, basal half of tail, and sometimes rear of ankle pale orange; fur dense and woolly. Head round, face black in adults, fringed with pale ring; crown slightly darker grayish brown than body; ears thickly haired, standing out from head like teddy bear, dark brown at center. Arms, legs, and prehensile tail long and thin; belly large and protruding. Thumb vestigial. The largest Neotropical monkey. Swings below branches by

arms and tail. Young may have pale faces.
Similar species. Brown howler monkeys
(*Alouatta fusca*) are dark brown to dark
red, or yellow-brown with darker legs and
head, with large head, prominent beard,
and naked ears; brown capuchins (*Cebus
apella*) are smaller, dark brown, with black
limbs and tail; there are no other large
monkeys in range.
Sounds. Loud, piercing screams, whinnying or neighing sounds, barks in alarm,
staccato chutters.
Natural history. Diurnal; arboreal; groups
of one to eight, usually either all females
with young or all males. Feeds chiefly on
leaves, with some fruit and flowers. Woolly
spider monkeys use the upper levels of tall
forest. They are the most specialized New
World leaf-eating monkeys. They are generally inactive: two-thirds of the day is
spent resting, and they do not begin activity until well after dawn. Like spider monkeys, they may threaten an observer by
jumping and shaking the branches and defecating. They are not territorial and have
an unusual, promiscuous mating system.

Found in mature evergreen to deciduous
lowland Atlantic forest.
Geographic range. South America: Brazil, E coastal forests from Bahía to São
Paulo.
Status. CITES Appendix I, US-ESA endangered. Only 300–400 individuals are
thought to remain, and these are scattered
widely in small fragments of forest. Only
2–4% of the original forest habitat of this
species has not been deforested. Formerly
intensively hunted for meat; now close to
extinction owing to habitat loss. Considered by some to be the world's most endangered primate.
Local names. Muriqui.
References. Milton, K. 1984. Habitat,
diet, and activity patterns of free-ranging
woolly spider monkeys (*Brachyteles arachnoides* E. Geoffroy 1806). *Inter. J. Primatol.* 5:491–514.

Fonseca, G. A. B. da. 1985. Observations on the ecology of the muriqui
(*Brachyteles arachnoides* E. Geoffroy
1806): Implications for its conservation.
Primate Conservation. 5:48–52.

Carnivores (Carnivora)

Four of the five families of New World carnivores (Canidae, Procyonidae, Mustelidae, and Felidae) are found in the rainforest (bears are present in the Neotropics only at high elevations in the Andes). In general, carnivores are adapted to find, catch, and kill animal prey, although they also have other ecological roles. The skull, muscles, and teeth are fitted for giving a powerful puncturing, crushing, or slicing bite but relatively little grinding of the food. The canines are usually large, and the premolars are often developed into blades with a scissorlike shear. The range of diets is wide. While some are truly carnivorous or meat-eating species, others feed mainly on insects, fruits, or even leaves. Mammalian carnivores do not usually specialize on particular species of prey; most are opportunists that will eat whatever they encounter that can be caught and killed without danger. Perhaps for this reason, the meat-eating Carnivora usually have geographic ranges much larger than those of any single species of their prey (e.g., pumas eat different species of deer in different parts of their range). Purely meat-eating species generally live at low densities; those that also eat insects and fruit are often numerous in good habitat. Carnivores have an important and beneficial role in ecosystems.

Dog Family (Canidae)

Dental formula usually: i3/3, C1/1, P4/4, M2/3 = 42; the bush dog is unique, with M1/2 = 38. There are four weight-bearing toes on each foot; a few species have a fifth "dewclaw" that does not bear weight or appear in tracks. Canids are long-legged and digitigrade (walk on the toes, the heel not touching the ground), with long, pointed snouts, straight backs, bushy tails, and wet noses. Their senses of smell and hearing are acute, and most also see well. They can travel long distances and run economically, and many species track prey by following a scent trail. Canids are carnivores or scavengers that feed on vertebrate and/or invertebrate prey, but most species will eat fruit if prey is scarce. They do not have a "killing bite," and they kill their mammal prey by shaking it to break the back if it is small or by disabling it with bites to the legs or nose and tearing and worrying messily at vital organs if it is large. In a number of species, several individuals join in a pack to immobilize and kill prey larger than themselves. This requires complex social organization, cooperation, and strategy. Even species that forage solitarily for small prey often live in monagamous pairs or groups. They are in general highly intelligent (good at learning, remembering, and solving problems). Canids usually have large litters of pups that are born helpless and with eyes and ears closed (altricial). Pups are raised in a burrow and are usually fed prey by both parents (or all members of a pack) as they are weaned. The world's species are mostly occupants of savanna woodlands, grasslands, and woodlands and forests that are open-canopy or interspersed with open meadows. The only two deep tropical rainforest dogs in the world are the two described in this book, which are of particular interest because of their unique habitat. Unfortunately, both are exceedingly rare, and almost nothing is known about their behavior in the wild. Other canids occur on all the fringes of the entire rainforest zone, including coyotes *(Canis latrans)*, maned wolves *(Chrysocyon brachyurus)*, and several foxes *(Cerdocyon* spp., *Dusicyon* spp., *Urocyon* spp.). There are about 35 species in about 11 genera worldwide, and 2 genera each with 1 species in the Neotropical rainforest.

Short-eared Dog
Atelocynus microtis
Plate 15, map 110
Identification. Measurements: HB = 583–1,000; T = 260–350; HF = 140–150; E = 34–65; SH = 350–356; WT = 6.5–9 kg.

Upperparts grizzled blackish gray; hair short and stiff. **Head large,** grizzled more brownish than back; neck long and thick; **ears small but protruding well above crown,** rounded, brown; eyes hazel; eyeshine bright pale green; tips of upper canines visible when mouth is closed. **Tail**

Map 110

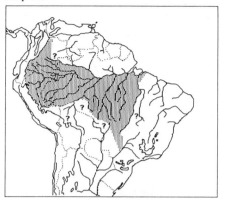

Short-eared dog, *Atelocynus microtis*

Map 111

Bush dog, *Speothos venaticus*

bushy, long enough to touch the ground, black except pale beneath base. Legs slender, dark brown or black, feet small. Underparts grizzled brown. Young like adults.
Variation. Placed by some in the genus *Dusicyon* or *Canis.*
Similar species. Bush dogs *(Speothos venaticus)* are smaller, with a pale tan head, short legs, and short tail; tayras *(Eira barbara)* are smaller, with inconspicuous ears and usually a contrasting gray head and yellow throat patch; jaguarundis *(Felis yagouaroundi)* have a small head and slender tail.
Sounds. Growls when threatened.
Natural history. Terrestrial; probably solitary. The habits of short-eared dogs are completely unknown; there is no firsthand published report of any sighting. All information comes from captive individuals. These dogs move with catlike grace. When excited, males emit a strong musky odor. Most records are from lowland tropical rainforest, but they may inhabit drier forests in S Pará, Brazil.
Geographic range. South America: east of the Andes in Colombia, Ecuador, and Peru; Brazil south of the Amazon from the Rio Toncantins to Matto Grosso, in the Rio Paraguay Basin. To 1,000 m elevation.
Status. Appears to be extremely rare throughout its range.
Local names. Cachorro do mato (Br); zorro negro, zorro de monte (Span).
References. Hershkovitz, P. 1961. On the South American small-eared zorro *Atelocynus microtis* Sclater (Canidae). *Fieldiana Zool.* 39:505–23.

Berta, A. 1986. *Atelocynus microtis.* Mammalian Species, no. 256.

Bush Dog
Speothos venaticus
Plate 15, map 111
Identification. Measurements: HB = 610–750; T = 110–130; HF = 110–120; E = 40–51; WT = 5–7 kg.
Head and neck to top of shoulders pale brown to tawny yellow, darkening gradually to black or dark brown hindquarters; fur long and soft. Muzzle short; ears short and rounded; eyes brown. **Tail short and stumpy, black,** thickly furred. **Legs very short, black** or dark brown; feet webbed. Underparts dark brown or black, sometimes chest with white spot. Back long, body cylindrical, thick. Young gray-black.
Variation. Some bush dogs from Brazil have completely pale brown backs.
Similar species. Tayras *(Eira barbara)* have longer tails, yellow throat patches, and small ears, and they climb trees; short-eared dogs *(Atelocynus microtis)* are grizzled blackish gray, with a bushy tail long enough to touch the ground, longer legs, and short fur.
Sounds. Groups communicate with high-pitched whining (so do coatis and several monkeys); they also bark.
Natural history. Apparently diurnal; terrestrial; usually in small groups of four to seven or more, but sometimes alone. The teeth of bush dogs are more specialized for meat-eating than those of other New World canids, and they may be strongly carnivorous. The only published firsthand account

of the prey describes a lone animal attacking a paca, which it could not kill. Bush dogs den in burrows. They swim well in captivity. They seem rare everywhere, and few forest Indians have ever seen one. Most information reported for wild bush dogs consists of undocumented anecdotes. Found in a range of habitats from rainforest to wooded savanna.

Geographic range. Central and South America: W Panama south to Bolivia, Paraguay, and N Argentina. To 1,500 m elevation.

Status. CITES Appendix I. Extremely rare in a large geographic range.

Local names. Cachorro-do-mata (Br); perro de monte (Ec, Pan, Pe); boshond, busdagoe (Su); zorro vinagre (Ar).

References. Deutch, L. A. 1983. An encounter between a bush dog *(Speothos venaticus)* and paca *(Agouti paca)*. *J. Mammal.* 64:532–33.

Raccoon Family (Procyonidae)

Dental formula: I3/3, C1/1, P4/4, M2/2 = 40; except kinkajou with P3/3. Five well-developed toes on all feet. Procyonids are of medium size; have pointed muzzles; broad, crushing molars; quite forward-set eyes; dense, dull fur; short legs; and plantigrade feet (their heels touch the ground when they are standing). Their hands are mobile, and they are able to dig and rummage for prey or hold and manipulate food items. They are omnivores that feed on invertebrates such as insects and crabs, small vertebrates such as frogs, snakes, and nestling birds, fruit, and flower nectar. Members of this family are all good tree-climbers, although some forage mainly on the ground. Several species are entirely arboreal, but even the most terrestrial species climb trees to escape danger when alarmed and to sleep during inactivity. All species raise their young in arboreal nests (although northern raccoons on the prairies may den on the ground). The young are tiny at birth. Kinkajous have a single young, the others a litter of three to seven. Weanlings follow the mother to forage. There are 6 genera and about 13–16 species, all in the New World, and all with ranges entering the Neotropics. (Pandas were once thought to belong to this family, but the giant pandas are now placed with the bears, and the lesser pandas are considered unrelated to procyonids, although it is not yet clear where they belong.) Five genera and 7–9 species are found in the rainforest region.

Crab-eating Raccoon
Procyon cancrivorous
Plate 15, map 112
Identification. Measurements: HB = 543–650; T = 252–380; HF = 130–150; E = 50–60; WT = 3.1–7.7 kg.
Upperparts blackish grizzled with gray or rusty brown; sides grizzled rusty brown or gray; fur short and stiff or medium length, usually with **no underfur; hair on back of neck "reversed," slanting forward.** Head broad, with pointed muzzle; **face grizzled whitish traversed by broad black "mask" around eyes, mask usually fading out just behind eyes; ears relatively short, white** inside and out. Tail about 50% as long as head and body, moderately bushy, **with prominent, wide, black and pale rings. Legs and feet dark brown.** Underparts deep rusty to whitish; chin and throat grizzled whitish like face; band across throat, if present at all, faint, dirty

Map 112

Crab-eating raccoon, *Procyon cancrivorus*

white or grizzled tan. Body thick; legs and feet slender, with long, dexterous fingers; rump much higher than forequarters, tail carried low; strong, distinctive odor.

Variation. Grizzled coloring is due to banded hairs; the tips of the hairs are black and the basal parts of back hair, belly hair, and paler rings on tail vary from dark, rust-red to whitish. Raccoons from Argentina and E Brazil have underfur and white bellies and more closely resemble northern raccoons.

Similar species. Northern raccoons *(P. lotor)* have whitish forelegs and usually whitish feet; normal, backward-slanting hair on neck; and prominent brown band across throat extending from mask that reaches well behind eyes. Coatis *(Nasua* spp.) have no mask and a long snout and carry their slender tails vertically; olingos *(Bassaricyon* spp.) are small, with long backs and tails and no mask, and tail rings are faint; cacomistles *(Bassariscus* spp.) are small and slender, with long tails. No other mammals in region except spotted cats have black and pale rings on tail.

Natural history. Nocturnal; terrestrial but climbs well; solitary. Feeds mainly on molluscs, fish, and crabs, with some amphibians and insects, and probably some fruits. Crabs can be found throughout the lowland rainforests in large and small, permanent and temporary bodies of water. Crab-eating raccoons seem to be restricted to waterside habitats such as swamps, rivers, streams, and beaches. During the day they den in hollow trees. This species is rarely seen deep in rainforest, although it is found throughout the region. It is also found in llanos and in deciduous as well as evergreen forests. In the zone of geographic overlap with the northern raccoon in Central America, the northern raccoon is found in mangrove swamps, while the crab-eating raccoon is found on inland rivers.

Geographic range. Central and South America: E Costa Rica and Panama south to Uruguay and NE Argentina.

Status. Unknown, but widespread and probably rarely hunted.

Local names. Mapache, osito lavador (Span); guaxinim, mao-pelada (Br); wasbeer, krabdagoe (Su); gato manglatero (Pn); ratón laveur (FG); aguará-popé, goáxiní (Gua); mayuato (Qui).

References. Bisbal, F. J. 1986. Food habits of some Neotropical carnivores in Venezuela (Mammalia, Carnivora). *Mammalia* 50:329–39.

Map 113

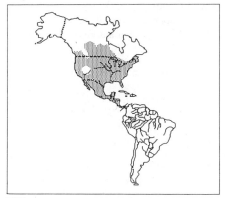

Northern raccoon, *Procyon lotor*

Northern Raccoon
Procyon lotor
Plate 15, map 113

Identification. Measurements (Central American animals only): HB = 440–625; T = 270–360; HF = 102–142; E = 60–75; WT = 2.7–6.4 kg.

Upperparts grizzled blackish gray; fur long and soft, on neck "normal," slanting backward. Head broad, with pointed muzzle; face white with broad black mask reaching across and well behind eyes to nearly below ear and extending below as sharply contrasting brown bar across throat; nose black; ears white inside and out, relatively large; eyes medium-sized, with bright greenish white eyeshine at night. Tail about 60% as long as head and body, moderately bushy, with broad, brightly contrasting black and gray or buff rings. Underparts whitish, covered with long, white hairs, soft brown underfur showing below. Forelegs whitish, hindlegs and feet usually whitish, sometimes brown. Shape and postures like crab-eating raccoon.

Variation. Five Neotropical island species of raccoons have been described: *P. gloveralleni* (Barbados), *P. insularis* (Maria Madre, Maria Magdalena, Mexico), *P. maynardi* (Bahamas), *P. minor* (Guadeloupe), and *P. pygmaeus* (Cozumel). All are probably forms of *P. lotor.*

Similar species. See crab-eating raccoon.

Sounds. A variety of whines, yips, and growls; not usually heard in field.

Natural history. Nocturnal; terrestrial and arboreal; solitary except females with young or congregations at food sources. Feeds on fruits and small animals, especially aquatic ones such as crayfish, crabs, and fish; also raids garbage bins and dumps and is a pest when it eats young ears of corn. Northern raccoons are better tree-climbers than are crab-eating raccoons, but nevertheless they do most of their foraging on the ground. When alarmed, they will often escape by running up the nearest tree into the canopy. Before eating, they rub and manipulate their food with their hands. During the day they usually den in hollow trees. Northern raccoons are common on the beaches of both coasts of Central America and are also found inland in mangrove swamps and near rivers. They adapt well to humans and thrive in towns and cities.

Geographic range. North and Central America; S Canada south to Chiriquí, Panama.

Status. Widespread and common.

Local names. Mapache, osito lavador (Span); tzil (May).

References. Sanderson, G. C. 1983. *Procyon lotor.* In D. N. Janzen, ed., *Costa Rican natural history,* 485–88. Chicago: University of Chicago Press.

Lotze, J.-H., and S. Anderson. 1979. *Procyon lotor.* Mammalian Species, no. 119.

South American Coati
Nasua nasua
Plate 15, map 114

Identification. Measurements: HB = 470–580; T = 420–550; HF = 84–100; E = 38–45; WT = 3–7.2 kg; males larger than females.

Upperparts dark brown, dark or bright rust red-orange, or gray; fur thick and dull. Head narrow, **snout long, narrow, mobile;** nose slightly upturned, wet, black; cheeks often grizzled gray; muzzle brown; **prominent pale spots usually above, below, and behind eye; ear small,** rounded, **fringed with white** on inner rim, dark brown on back and base; canine teeth large, especially lowers. **Tail long,** well-haired, **tapering** to point at tip, **blackish brown with bright or faint pale yellow, brown, or buff rings; often carried vertically upright.** Chin and throat grizzled

Map 114

▦ South American coati, *Nasua nasua*
▤ White-nosed coati, *N. narica*

white, underparts whitish to rich buff. Feet dark brown except in orange individuals; **forefeet with long claws,** hindfeet with strong, curved claws. Small young darker than adults, with faint markings.

Variation. Color highly variable, from almost black, with gray face without white markings (Peru), to entirely bright red-orange (Pará, Brazil) or gray (Minas Gerais, Brazil). Facial markings and pale rings on tail range from sharply contrasting to faint or absent.

Similar species. Raccoons *(Procyon cancrivorous)* have black masks, short tails, and normal snouts; olingos *(Bassaricyon* spp.) are much smaller, with no facial markings; tayras *(Eira barbara)* have short snouts and no rings on the tail. At high elevations in the Andes (> 2,000 m), Andean coatis *(Nasuella olivacea)* are externally similar to South American coatis but smaller, with small canine teeth. The tails of spotted cats may be held vertically and resemble those of coatis in the undergrowth: cat tails are not tapered and are irregularly spotted with pure black and white.

Sounds. Members of active groups emit constant soft whining sounds; alarm calls consist of explosive woofs and clicks.

Natural history. Diurnal; terrestrial and arboreal; solitary and in groups of up to 30. Omnivorous, feeds on fruit, invertebrates, and other small animals. Coatis can be seen feeding on fruit high in the canopy of a tree, or searching the forest floor for animal prey by poking their long noses into

crevices, turning over rocks, or ripping apart dead logs with their claws. Males are often solitary, but females and young travel in conspicuous troops; they are often seen as a group of vertical tail tips waving among the shrubbery. When one coati spots danger and gives the alarm, all run partway up trees to look. After a few moments they drop to the ground and disperse rapidly through the undergrowth. At night coatis sleep in a treetop. Females leave the group when they give birth to young, which are raised in a nest in a tree. Found in forested habitats ranging from tropical rainforest and gallery forest to chaco, cerrado, and dry scrub.
Geographic range. South America: east of the Andes in all countries from Colombia and Venezuela south to Argentina and Uruguay.
Status. CITES Appendix III (Ur). Generally uncommon, but ubiquitous in rainforest; widespread. Hunted moderately for meat and the live animal trade.
Local names. Coatí (Span, Gua); achuni (Pe); tejón (Ec); cuzumbo (Co); quati, quatimundé (Br); ncusbeer, kwaskwasi (Su).
References. See white-nosed coati.

White-nosed Coati
Nasua narica
Plate 15, map 114
Identification. Measurements: HB = 463–690; T = 490–620; HF = 93–146; E = 34–42; WT = 4.5 kg.
Upperparts dark brown, shoulders or entire forequarters grizzled gray. Head sometimes rusty brown; **muzzle and chin white, extending as a white stripe above eye; other white spots above and behind eye;** ear short, lined with whitish fur. **Tail** 75–100% as long as head and body, thickly furred but not bushy, tapering at the tip; **dark brown with paler rings usually inconspicuous to invisible.** Feet dark brown, forefeet often contrasting with gray shoulder. Throat and chest frosted white, belly dark brown or frosted white. **Body shape and postures like South American coati.** Young dark brown with pale facial markings like adult.
Variation. The above description applies to coatis from the rainforest region; animals from more arid habitats may be paler, tawny brown. Coatis from Cozumel Island, Quintana Roo, Mexico, are sometimes

considered a distinct species *(N. nelsoni).* They are smaller and have silkier fur than the mainland form. White-nosed coatis are often considered a subspecies of the South American coati *(N. nasua).*
Similar species. See South American coati.
Sounds and natural history. Like South American coati.
Geographic range. North, Central, and South America: Arizona and Texas south through all of Central America to the west coasts of Colombia, Ecuador, and Peru.
Status. CITES Appendix III (Ho). Locally common to rare.
Local names. Quash (Be); pizote, pizote solo (CR, Ho); tejón (Ec, Me); cuzumbo (Co); gato solo (Pa); chic, sis (May).
References. Kaufmann, J. H. 1962. Ecology and social behavior of the coati, *Nasua narica,* on Barro Colorado Island, Panama. *Univ. Calif. Pub. Zool. 60:95–222.*
_____. 1983. *Nasua narica.* In D. H. Janzen, ed., *Costa Rican natural history,* 478–80. Chicago: University of Chicago Press.

Olingo
Bassaricyon gabbii
Plate 15, map 115
Identification. Measurements: HB = 360–411; T = 370–520; HF = 62–92; E = 30–44; WT = 1.1–1.4 kg.
Upperparts drab **brown,** sometimes darker on midline; fur soft, dense, quite long (1.5 cm at midback). Crown often dark brown; **face grizzled gray; muzzle pointed;** ears rounded, set low on head, brown with fringe of white hairs on rim; eyes large, round, brown; eyeshine bright. **Tail slightly longer than head and body; nonprehensile; well-haired, tip almost tufted** with longer hair than base (2–4 cm); **brown with indistinct dark rings,** tip usually darker than base, or dirty white; carried straight out behind during travel. Underparts cream or buff. Tiny young graybrown, sides with faint dark stripes. A slender-bodied, short-legged, monkeylike animal.
Variation. Olingos from higher elevations in Panama are frosted with gray; many from Panama have white tail tips. Six species of olingos from different regions are sometimes recognized: *B. gabbii,* from Central America; *B. alleni,* from the west-

Map 115

Olingo, *Bassaricyon gabbii*
Cacomistle, *Bassariscus sumichrasti*

ern Amazon Basin; *B. beddardi,* from northern South America; *B. pauli,* from one locality in Panama; and *B. lasius,* from one locality in Costa Rica. The differences between these forms are minor, and they probably are all conspecific *(B. gabbii)*.
Similar species. Kinkajous *(Potos flavus)* are similar and often confused with olingos: they are about twice the weight, usually reddish, and have short fur and a tapered, prehensile tail that is often curled under branches during travel; cacomistles *Bassariscus sumichrasti)* have the tail prominently ringed; night monkeys *(Aotus* spp.) have a white-rimmed face with dark stripes on crown, large, close-set eyes, flat face, and inconspicuous ears; opossums have small eyes and prehensile tails either naked at the tip or strongly tapered.
Sounds. Usual alarm call a two-toned "wake-up," less often a sneezing. Both calls similar to those of kınkajou. A noisy traveler that jumps from tree to tree in the canopy, but more lightly than kinkajou.
Natural history. Nocturnal; arboreal; solitary. Feeds on fruits and probably some invertebrates, in the dry season drinks the nectar of flowers such as balsa. Olingos are agile and active and usually travel high in the canopy, lightly leaping and running through the branches, feeding quickly and running on. They often feed on fruits, especially figs, in the same trees with kinkajous and night monkeys, and in the dry season feed on nectar in the same trees as these and also opossums. They are restricted to humid forests, and are generally less common than kinkajous and do not seem to

adapt as readily to disturbed or secondary forests, or plantations and gardens. Found in mature and disturbed rainforest.
Geographic range. Central and South America: Nicaragua south, west of the Andes to N Ecuador; east of the Andes from Venezuela to Bolivia, in the western half of the Amazon Basin only. To 1,600 m elevation.
Status. CITES Appendix III (CR). Locally common, not hunted, and widespread; in Central America may be locally threatened by destruction of mature forests.
Local names. Usually the same as kinkajou: jupará (Br); ocate, chosna pericote (Pe); olingo (Pn).

Kinkajou
Potos flavus
Plate 15, map 116
Identification. Measurements: HB = 390–547; T = 400–570; HF = 70–108; E = 30–55; WT = 2–3.2 kg.
Upperparts reddish brown to smoky gray-brown, often with a dark brown stripe on midback; fur dense, soft, and short (about 1 cm at midback). Head and face red-brown to blackish; **head round, muzzle short, pointed;** nose brown; **eyes brown, large and round, set wide apart;** eyeshine bright orange; **ears** brown, thinly haired, **set low on sides of head;** tongue long and highly extensible. **Tail** slightly longer than head and body, **prehensile, tapered toward tip, brown or darkening to black at tip.** Feet colored like back, all with five long, curved claws. Underparts yellow to orange-buff. Juveniles gray with dark stripes behind shoulders. An agile, muscular, short-legged, long-backed, monkeylike animal.
Variation. Occasional animals have a small white tail tip; some animals from Central America are gray-brown.
Similar species. Olingos *(Bassaricyon* spp.) are similar but smaller, with a non-prehensile tail, slightly bushy at the tip; cacomistles *(Bassariscus sumichrasti)* have a brightly ringed tail and white markings on face; night monkeys *(Aotus* spp.) have a flat face, white brow, three black stripes on crown, close-set eyes, and long, nonprehensile, black-tipped tail with no taper.
Sounds. Highly vocal: when alarmed emits sneezing calls most frequently, less often a two-syllable "wake-up" (calls of olingo

Map 116

Kinkajou, *Potos flavus*

are similar); jumps noisily from tree to tree at night.

Natural history. Nocturnal; arboreal; solitary, in pairs, or several may congregate in a fruit tree. Feeds on fruit (80% of diet), primarily figs (*Ficus* spp.), and insects (20%), and in the dry season drinks flower nectar. Kinkajous are animals of the forest canopy, and they are the most commonly seen larger, nocturnal, arboreal mammal. They are agile and can travel quickly, running and jumping noisily from tree to tree. When seen at night in the canopy, they can usually be distinguished from olingos and night monkeys (which may be feeding in the same tree) by the tapered, prehensile tail, which is frequently curled partially under or around a branch during travel or feeding. During the day they den in tree hollows. Found in mature, disturbed and secondary rainforest, gardens and plantations, gallery forest, and occasionally in deciduous forest.

Georgraphic range. Central and South America: S Mexico south to Matto Grosso, Brazil. Rarely above 500 m, occasionally to 1,750 m elevation.

Status. CITES Appendix III (Ho). Widespread and often common; hunted for meat and for the pet trade.

Local names. Jupará, macaco-de-noite (Br); night walker (Be); perro de monte, oso mielero (Co); marta, martucha (Co, Me); martilla (CR); martica, tutamono, chuche, cuchicuchi (Ec); chosna (Pe); meti-keskesi (Su); cusumbí (Pn). Sometimes not locally distinguished by name from night monkey.

References. Bisbal, F. J. 1986. Food hab-

its of some Neotropical carnivores in Venezuela (Mammalia, Carnivora). *Mammalia* 50:329–39.

Cacomistle
Bassariscus sumichrasti
Plate 15, map 115

Identification. Measurements: HB = 389–468; T = 400–545; HF = 77–94; E = 42–55; WT = 900 g.

Upperparts tawny brown; midline of back blackish. Crown blackish; **eyes narrowly ringed with black, bordered by prominent pale, dirty yellow rings encircling outside of eye;** eyeshine bright yellow-white; **ears large and rounded,** standing above crown, **rims white** inside and out, inside of pinnae pale. **Tail longer than head and body, thickly furred; strongly banded with black and whitish or tawny rings,** pale rings disappearing distally, **tip entirely black** for about terminal quarter. **Feet and toes dark** brown or blackish. Underparts yellowish white, grayish around midsection. A small, slender, long-bodied animal with a relatively small head and long neck.

Similar species. Ringtails (*B. astutus,* in dry habitats) are paler, with pale feet, and pale rings to near tip of tail; olingos (*Bassaricyon* spp.) are similar in shape, with no facial markings and indistinct rings on tail.

Sounds. Frequently calls with loud barks.

Natural history. Nocturnal; arboreal; solitary. Feeds on fruits, insects, and probably small vertebrates. Cacomistles use the middle and upper levels of the forest. Found in mature and secondary lowland and montane rainforest.

Geographic range. Central America: S Mexico to W Panama. To about 2,000 m elevation.

Status. CITES Appendix III (CR). Common in remnant forests of Veracruz; rare in Panama.

Local names. Olingo, cacomistle (CR, Pn); guía de león, guayanoche (Gu); uayuc (Ho); cacomixtle (Me).

References. Estrada, A., and R. Coates-Estrada. 1985. A preliminary study of resource overlap between howling monkeys (*Alouatta palliata*) and other arboreal mammals in the tropical rainforest of Los Tuxtlas, Mexico. *Amer. J. Primatol.* 9:27–37.

Weasel Family (Mustelidae)

Dental formula: I3/3, C1/1, P3 or 4/3 or 4; M1/1 or 2 = 32–40. Five toes on all feet. Mustelids have shearing premolars and crushing molars. They are plantigrade, with broad heads, small ears and eyes, tails shorter than the head and body, and usually long slinky bodies and short legs, convenient for using burrows and holes. The back is humped when the animal is standing. Most have soft, dense, glossy fur, valuable in the fur trade. Although some are good climbers, the New World rainforest species all forage terrestrially or in the water. Mustelids have an extremely powerful bite for their size (legendary in the case of the wolverine), and some of them are among the few carnivores in the world that can single-handedly kill prey much larger than themselves. They do this with a strong killing bite to the head or neck. The family includes pure carnivores, some that feed on specific prey types (e.g., black-footed ferrets); worm eaters (some badgers); omnivores (tayras, skunks); and fish eaters (otters). The senses of smell and hearing are acute, but most species do not seem to see particularly well, and a motionless observer can often watch them undetected. Most species have large anal glands that produce strong-smelling musk: this has been carried to an extreme in the skunks, zorillas, and stink badgers, which spray vile-smelling fluid at their enemies. Litter size is 2–18 altricially born young. Most species sleep and raise their young in burrows or hollow logs. The family includes about 26 genera and 67 species worldwide; 7 genera and 8 species occur in Neotropical rainforest.

Amazon Weasel
Mustela africana
Plate 16, map 117
Identification. Measurements: HB = 260–333; T = 170–234; HF = 50–56; E = 22.

Upperparts entirely glossy chestnut-brown. Head broad, narrowing sharply to blunt snout; ear short, broad at base; eyes small; upper lip, chin, and lower cheek to below ear yellowish white. Tail shorter than head and body, well furred. **Throat and belly pale buff or yellowish with a sharp, dark brown stripe down midline** from throat or chest to belly. A small, slender mammal with long body and neck, strongly humped back and very short legs.
Similar species. These are the only true weasels in their range and the smallest lowland carnivores in the region; no other mammals are of similar color and shape; long-tailed weasels and Don Felipe's weasels (*M. frenata* and *M. felipei*) have no dark ventral stripe.
Natural history. Unknown. Amazon weasels probably feed on rodents and other small mammals, like other members of the genus. Several were found denning in a hollow tree stump. Found in lowland rainforest.
Geographic range. South America: east of the Andes in the lowland Amazon basins of Peru, Ecuador, and Brazil.
Status. This is one of the rarest carnivores

Map 117

▨ Amazon weasel, *Mustela africana*

in South America; known from fewer than 50 specimens from widely scattered localities.
Local names. Comadreja (all weasels, Span); furão (Br).
References. Izor, R. J., and L. de la Torre. 1978. A new species of weasel (*Mustela*) from the highlands of Colombia, with comments on the evolution and distribution of South American weasels. *J. Mammal.* 59:92–102.

Long-tailed Weasel
Mustela frenata
Plate 16, map 118
Identification. Measurements: HB = 215–

Map 118

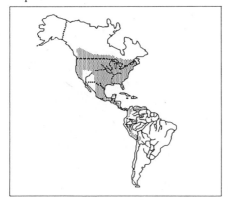

Long-tailed weasel, *Mustela frenata*

320; T = 115–207; HF = 35–55; E =
10–20; WT = 85–340 g; males usually
much larger than females.
**Upperparts entirely rich to dark, glossy
chocolate-brown. Head with or without
following facial markings: broad white
stripe from nose over eye and down neck
in front of ear, or a white spot or spots
between eyes, and a diagonal white
stripe in front of ear. Tail short,** about
60% as long as head and body, thickly
furred, with slight tuft, **black at tip. Underparts entirely creamy white, or chin
and throat creamy white, grading to pale
orange on belly.** Shape like Amazon
weasel.
Variation. Much geographic variation in
size and color: the white facial stripe is
most highly developed in Mexico and
gradually disappears southward, from a
broad, solid stripe in Mexico, to striking
white spots in Nicaragua, faint traces in
Colombia, head entirely brown above, with
no white spots, in Peru; body color varies
from almost black (Colombia) to paler
brown (Central America).
Similar species. See Amazon weasel.
Natural history. Diurnal and nocturnal;
terrestrial; solitary. Feeds chiefly on small
mammals such as rabbits and rodents but
will also take birds and reptiles. These
weasels have been called snakes with legs:
their short legs and long neck and back,
fitted for running down narrow burrows of
their prey, give them a graceful, sinuous
bounding gait. They are active and nervous, and it is rare to get more than a brief

glimpse of one. When alert they briefly
stand on their hindlegs and crane their
necks. They den in burrows or holes under
roots or among rocks. Long-tailed weasels
are highly adaptable, but they are not a
rainforest species. In general they are
found either in drier cleared or open country, agricultural land, or montane forest at
high elevations, but they are occasionally
found below 1,000 m in rainforest in Central America. In South America they are
chiefly found at high elevations in the Andes or at low elevations on the west coasts
of Colombia, Ecuador, and Peru.
Geographic range. North, Central, and
South America: Canada to Bolivia. To
4,000 m elevation.
Status. Widespread and locally common.
Local names. Comadreja (Span); oncilla
(Me); lince (Pa); tolompeo (Pe).
References. Hall, E. R. 1951. American
weasels. *Univ. Kansas Pub. Mus. Nat.
Hist.* 4:1–466.

Grison or Huron
Galictis vittata
Plate 16, map 119
Identification. Measurements: HB = 465–
552; T = 135–195; HF = 75–97; E =
20–30; WT = 1.5–2 kg.
Upperparts grizzled pale gray mixed with
dark brown. **Head tricolor: crown gray,
forehead with broad white band above
eyes, across ears and down side of neck,
muzzle to eyes black, continuous with
black chin, throat, and chest.** Ears small
and inconspicuous, white; eyes small,
black. **Tail short and stumpy,** one-third as
long as head and body, well haired with
fur grizzled gray like back. Legs and feet
black. Underparts black anteriorly, grizzled
gray posteriorly. **A big weasel with very
short legs and tail, long neck and back.**
Young like adults.
Similar species. No other animals in region are gray with a broad white band
across the face and black muzzle and
throat. Similar grisons *(G. cuja)* live in
south temperate latitudes and at higher elevations; hog-nosed skunks *(Conepatus semistriatus)* are black with a pure white cape
over crown, neck, and back; tayras *(Eira
barbara)* have black bodies and long legs
and tails.
Sounds. Snorts in alarm; growls in
defense.

Map 119

Map 120

[grid] Grison or huron, *Galictis vittata*

[grid] Tayra, *Eira barbara*

Natural history. Probably diurnal; terrestrial; solitary and pairs that travel together. Carnivorous; feeds on small vertebrates and probably some fruit. Found in rainforest and savanna.

Geographic range. Central and South America: S Mexico south to Peru and Bolivia.

Status. CITES Appendix III (CR). Widespread but apparently uncommon or rare everywhere.

Local names. Bushdog (Be); furão, furax (Br); hurón (Co, Ec); grisón (CR, Me); lobo gallinero, tigrillo rosillo (Pn); wetiaira (Su); yaguagumbé, yuguapé (Gua); zabin (May).

Tayra
Eira barbara
Plate 16, map 120

Identification. Measurements: HB = 559–712; T = 365–460; HF = 90–123; E = 30–42; WT = 2.7–7 kg.

Upperparts glossy dark brown to black, including legs, feet, and tail; fur of head short and stiff; fur of body and tail long and glossy, underfur brown. **Head and neck sharply contrasting grizzled tan, gray-brown, or yellowish,** rarely dark brown slightly paler than back, or same color as back. **Ears small and round,** same color as head, **not protruding above crown.** Underparts completely black or dark brown, except for a **bright, pale yellow to orange spot, often triangular, on chest and throat.** Tail quite bushy, two-thirds as long as head and body. A large, muscular weasel, **much like a small dog** with a long, slightly humped back and long

tail. Young entirely black, sometimes with white throat patch and/or white head.

Variation. Some tayras from Costa Rica and Panama are completely blackish, with dark brown head and no throat patch; a pale phase, almost pure pale yellow, occurs in Guyana, in mixed populations with typical individuals; occasional animals from northern South America have a large bright yellow spot on the shoulders, or even a complete yellow collar joining with the throat patch.

Similar species. Bush dogs *(Speothos venaticus)* are similar in size and color, but have pale brown shoulders and back, stumpy tail, ears that project above crown, straight back, and thick, cylindrical body; short-eared dogs *(Atelocynus microtis)* are larger, uniformly grizzled dark gray, with ears that protrude above crown; grisons *(Galictis vittata)* are gray, with white band across head and very short legs and a stumpy tail; jaguarundis *(Felis yagouaroundi)* are uniformly brown or reddish, with a slender, feline tail and small head.

Sounds. Snorts in alarm; growls and squeals when cornered.

Natural history. Diurnal except near human habitations, when also crepuscular; terrestrial and arboreal; solitary or in pairs that travel together. Feeds on small mammals, other small vertebrates, insects, and fruit. Tayras are generally seen as they travel quickly through the forest on the ground during the day. They often travel along the tops of fallen logs, where they also leave seed-filled scats. When alarmed at close quarters they may climb a tree and growl and spit in threat or run away through the branches. If disturbed from a distance, they gallop away noisily on the ground. Tayras den in hollow trees or holes in the ground. They are wary and not often seen but are much commoner than any of the similar species. They inhabit mature and secondary rainforests, gallery forests and drier areas, gardens and plantations, and cloud forest.

Geographic range. Central and South America: S Mexico south to N Argentina. To elevations of 2,400 m.

Status. CITES Appendix II (Ho). One of the most widespread and commonest carnivores, tayras can live in disturbed habitats near man.

Local names. Bushdog (Be); irara (Br); tayra (Co); tolomuco (CR); tejón, manco, perro de monte (Pe, Ec); lepasil (Ho); gato eirá (Pa); comadreja grande, gato negro, gato cutarra (Pn); cabeza de viejo (Me); guache, guanico (Ve); eirá (Gua); sacol (May).

Striped Hog-nosed Skunk
Conepatus semistriatus
Plate 16, map 121

Identification. Measurements: HB = 332– 500; T = 166–317; HF = 70–102; E = 26–35; WT = 1.4–3.4 kg.

Entirely dark brown or black with a broad white band from crown of head over neck which splits at shoulder into two parallel stripes down top of back with a narrow black stripe between, stripes ending short of rump; fur coarse and thick. **Head conical, snout long and naked; ears** short, not protruding above crown, black with a few white hairs at base; eyes black, eyeshine bright green. **Tail bushy, entirely white distally, black near base, often carried high.** Forefeet with very long, black claws. Underparts dark brown or blackish. In defense squirts vile-smelling fluid from its anal glands.

Similar species. Grisons (*Galictis* spp.) have a grizzled gray body and tail. Several other skunks are found in open habitats fringing the rainforest: other species of hog-nosed skunks in dryer areas of Mexico have a solid white back and tail; hooded skunks (*Mephitis* spp.) have black tails; spotted skunks *(Spilogale putorius)* are small, with short stripes and spots all over the body.

Natural history. Nocturnal; terrestrial; soli-

Map 121

▨ Striped hog-nosed skunk, *Conepatus semistriatus*

tary. Feeds mainly on insects and other invertebrates, and probably small vertebrates and occasional fruit. When alarmed at close range, this and other skunks aim their rear at the enemy, raise the tail, and spray. Well protected, skunks move slowly, ambling along snuffling the ground, scratching and rooting for their prey. By day they rest in or under logs or in burrows. Skunks are not truly rainforest animals, but this species occasionally is found in Central American rainforest, especially in secondary or disturbed forests near the clearings and gardens that are its more usual habitat. Occurs in pastures, clearings, roadsides, and other cultivated areas, in rainforest, dry forest, and scrub. Around its fringes but not in Amazonian rainforest.

Geographic range. Central and South America: Veracruz south through Central America, absent from eastern Panama; in South America in coastal Peru and Ecuador, N Colombia and Venezuela, and at high elevations in the Andes, and in E central Brazil in caatinga and cerrado. To at least 4,100 m elevation.

Status. Widespread, adaptable, and rarely hunted.

Local names. Polecat (Be); jaritataca, zorilho (Br); mofeta, mapuro, mapurito (Co, Ec); zorrillo, zorrillo pijón (Me, Co); gambá, zorrino, anas (Pe); gato cañero (Pn); yaguaré (Gua).

References. Kipp, H. 1965. Beitrag zur Kenntnis der Gattung *Conepatus* Molina, 1782. *Z. Saügetierk.* 30:193–232.

In all the plates that illustrate text accounts to the level of genus only (e.g. bats, murid rodents), distinguishing features listed are those useful for identifying the genus, not the species if there is more than one. Likewise, for those examples only, distributions refer to the genus in the rainforest region.

Plate 1

Large Opossums (Didelphidae)
All have prehensile tails; five toes on all feet.

1. Black-shouldered opossum
Caluromysiops irrupta (p. 13)
Black shoulders; tail furred on top to tip. SA.

2. Bare-tailed woolly opossum
Caluromys philander (p. 12)
Dark stripe down center of face; tail naked except at base. SA.

3. Western woolly opossum
Caluromys lanatus (p. 11)
Dark stripe down center of face; tail furred above for half its length. SA.

4. Central American woolly opossum
Caluromys derbianus (p. 13)
As above; forefeet white; gray patch on midback. C, SA.

5. Bushy-tailed opossum
Glironia venusta (p. 14)
Broad black stripes through eye from crown to nose; tail furred to tip. SA.

6. Common gray four-eyed opossum
Philander opossum (p. 17)
Back uniform gray; cream spots above eyes; tail furred at base for 5–8 cm. C, SA.

7. Anderson's gray four-eyed opossum
Philander andersoni (p. 18)
As above but midback black with glossy guard hairs, or entirely black. SA.

8. Water opossum
Chironectes minimus (p. 18)
Marbled with broad black bands across back; hindfeet webbed. C, SA.

9. Brown four-eyed opossum
Metachirus nudicaudatus (p. 19)
Back brown; pale yellow spots above eyes; tail naked from near base. C, SA.

10. White-eared opossum
Didelphis albiventris (p. 16)
Like common opossum, except: ears or ear tips white; sharp facial markings. SA.

11. Common opossum
Didelphis marsupialis (p. 14)
Fur in two layers, pale underfur showing through black or gray guard hairs; ears black; white part of tail usually longer than black; cheek below ear not white. C, SA.

12. Virginia opossum
Didelphis virginiana (p. 16)
As above but cheek below ear white; white part of tail equal to or shorter than black. CA.

Plate 2

Mouse Opossums (Didelphidae)

All are small; have prehensile tails; no pouch; large ears. (Several species not illustrated.)

1. Brazilian gracile mouse opossum
Gracilinanus microtarsus (p. 26)
Small; fur long and soft; underparts cream. Southern SA.

2. Delicate slender mouse opossum
Marmosops parvidens (p. 22)
Tiny; legs thin; eye rings indistinct; underparts white with gray patches. SA.

3. Gray slender mouse opossum
Marmosops incanus (p. 22)
Large; underparts creamy white. Southeastern SA.

4. White-bellied slender mouse opossum
Marmosops noctivagus (p. 21)
Large; underparts white. SA, western Amazon Basin.

5. Red mouse opossum
Marmosa rubra (p. 24)
Medium; upperparts red; underparts orange. SA, western Amazon Basin.

6. Little rufous mouse opossum
Marmosa lepida (p. 24)
Tiny; red with pinkish underparts; tail very long. SA western Amazon Basin.

7. Murine mouse opossum
Marmosa murina (p. 23)
Medium; warm brown with glowing salmon underparts. SA.

8. Woolly mouse opossum
Micoureus cinereus (p. 20)
Large; dense woolly fur; tail base densely furred for 3–4 cm. C, SA.

9. Robinson's mouse opossum
Marmosa robinsoni (p. 24)
Large; fur short; underparts yellow-orange. C, SA.

Plate 3

Short-tailed Opossums
(Didelphidae)

All are small; have short, prehensile tails; short ears; no pouch. (Several species not illustrated.)

1. Sepia short-tailed opossum
Monodelphis adusta (p. 28)
Tiny; fur extremely short. Andean region only. SA.

2. Three-striped short-tailed opossum
Monodelphis americana (p. 30)
Three black stripes on back. Eastern SA.

3. Emilia's short-tailed opossum
Monodelphis emiliae (p. 27)
Rump and head forward of ears rufous. Amazon Basin. SA.

4. Red-legged short-tailed opossum
Monodelphis brevicaudata (p. 27)
Head and rump gray. Amazon Basin and Guianas. SA.

5. Gray short-tailed opossum
Monodelphis domestica (p. 29)
Large; pale gray. Atlantic coast and Paraguay Basin. SA.

Plate 4

Sloths (Bradypodidae, Megalonychidae)

All are large; have long shaggy hair; hands and feet modified to hooks.

1. Maned three-toed sloth
Bradypus torquatus (p. 37)
Long black plumes of hair on neck and shoulder. E Brazil, SA.

2. Pale-throated three-toed sloth (female)
Bradypus tridactylus (p. 37)
Three claws on forefoot; short, stumpy tail; male with orange-and-black spot on back; pale throat. SA.

3. Brown-throated three-toed sloth (male)
Bradypus variegatus (p. 36)
As above, but throat brown. C, SA.

4. Southern two-toed sloth
Choloepus didactylus (p. 38)
Two claws on forefoot; no tail; throat the same color as chest. SA.

5. Hoffmann's two-toed sloth
Choloepus hoffmanni (p. 37)
As above, but throat paler than chest. C, SA.

Anteaters (Myrmecophagidae)

All have long narrow heads; long tongues; no teeth; greatly enlarged foreclaws.

6. Collared anteater or **southern tamandua**
Tamandua tetradactyla (p. 32)
Large; blond with black vest, or pure blond, or black; spotted prehensile tail. SA.

7. Silky or **pygmy anteater**
Cyclopes didactylus (p. 34)
Small; dense woolly fur with iridescence; prehensile tail; small eyes and ears. C, SA.

8. Giant anteater
Myrmecophaga tridactyla (p. 31)
Very large; large bushy tail; black stripe from throat to shoulder. C, SA.

Plate 5

Sheath-tailed Bats
(Emballonuridae)

1. Long-nosed bat
Rhynchonycteris naso (p. 45)
Elongated muzzle; white tufts of hair on
forearm; wavy white lines on back. C,
SA.

2. White-lined sac-winged bat
Saccopteryx bilineata (p. 45)
Wavy white lines on back; wing sac close
to forearm bone beyond elbow. C, SA.

3. Ghost bat
Diclidurus albus (p. 49)
White color; short, round ear; glands in
membrane around tail. C, SA.

Bulldog Bats (Noctilionidae)

4. Greater bulldog or fishing bat
Noctilio leporinus (p. 50)
Lip split from below nose; forward-
pointing nose and ears. C, SA.

Mustached Bat (Mormoopidae)

5. Mustached bat
Pteronotus parnellii (p. 52)
Lips flared forward; tail short; eye small.
C, SA.

Spear-nosed Bats
(Phyllostominae)

6. Fringe-lipped bat
Trachops cirrhosus (p. 60)
Lips and chin studded with long
tubercules. C, SA.

7. Hairy-nosed bat
Mimon crenulatum (p. 57)
Long, narrow noseleaf; large pointed ears;
tail half as long as membrane. C, SA.

Plate 6

Long-tongued Bats
(Glossophaginae)

1. Common long-tongued bat
Glossophaga soricina (p. 62)
Lower jaw about the same length as
upper; chin with V bordered by tiny
tubercules; fur bicolored pale at base. C,
SA.

Short-tailed Fruit Bat
(Carolliinae)

2. Short-tailed fruit bat
Carollia perspicillata (p. 67)
Short tail present; short narrow muzzle;
chin with V of tiny warts flanking large
central wart; fur banded. C, SA.

Neotropical Fruit Bats
(Stenodermatinae)

3. Tent-making bat
Uroderma bilobatum (p. 70)
No tail; strong dorsal and facial stripes;
tongue blackish; tail membrane without
hairy fringe. C, SA.

4. Honduran white bat
Ectophylla alba (p. 75)
No tail; tiny size; white. CA.

5. Large fruit-eating bat
Artibeus jamaicensis (p. 75)
No tail; no back stripe; U-shaped jaws;
blackish tongue. C, SA.

Vampire Bats (Desmodontinae)

6. Common vampire
Desmodus rotundus (p. 79)
Short M-shaped fold behind nostrils
backed by another fold; large sharp
triangular incisors larger than canines; no
tail; tail membrane a complete short band
between legs. C, SA.

Plate 7

Funnel-eared Bats (Natalidae)

1. Funnel-eared bat
Natalus stramineus (p. 81)
Tiny; mouth and ears funnel-shaped; tail
and legs very long; color pale. C, SA.

Sucker-footed Bats
(Thyropteridae)

2. Sucker-footed bats
Thyroptera tricolor (p. 83)
Foot and thumb base with tiny sucker
disks; tail longer than tail membrane. C,
SA.

Vespertilionid Bats
(Vespertilionidae)

3. Hoary bat
Lasiurus ega (p. 87)
Ears short and wide; tail membrane
thickly furred above for more than half its
length. N, C, SA.

4. Big brown bat
Eptesicus furinalis (p. 85)
Fur dark at base with pale tips; tail
membrane comes to long point at tip of
tail; no gap between canine and first large
tooth behind it. Genus N, C, SA.

Free-tailed Bats (Molossidae)

5. Brazilian free-tailed bat
Tadarida brasiliensis (p. 90)
Long, naked tail free beyond edge of
membrane; upper lip with vertical
wrinkles; ears meet but do not join on
center of crown. N, C, SA.

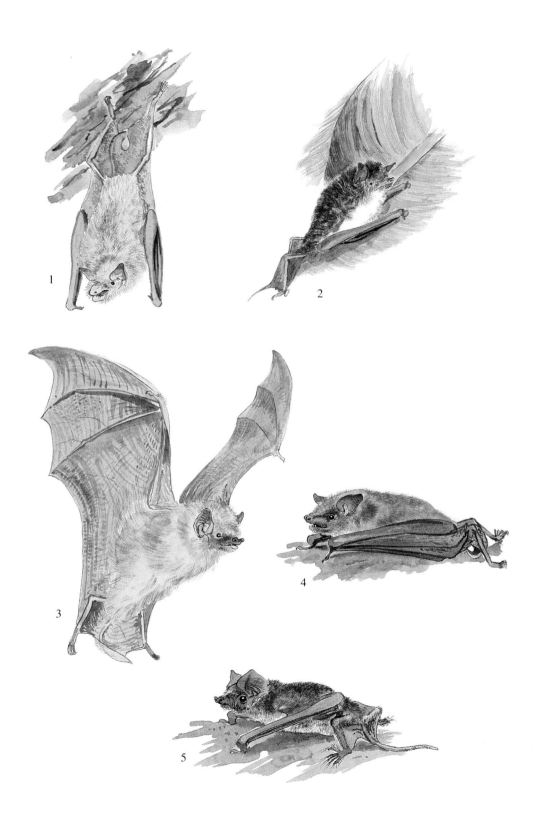

1

2

3

4

5

Plate 8

Marmosets, Tamarins, Goeldi's Monkey
(Callitrichidae, Callimiconidae)

1. Silvery marmoset
Callithrix argentata (p. 97)
Ears naked; white or pale brown; tail
black or brown. SA.

2. Tassel-ear marmoset
Callithrix humeralifer (p. 97)
Ear hidden by long tufts of pale hair; tail
banded. SA.

3. Tufted-ear marmoset
Callithrix jacchus jacchus (p. 98)
Lower back finely banded; tail banded;
white tufts surrounding ear. SA.

4. Gold-and-black lion tamarin
Leontopithecus rosalia chrysomelas (p.
107)
Black with gold crown and forelimbs.
SA, Bahía, Brazil.

5. Golden lion tamarin
Leontopithecus rosalia rosalia (p. 107)
Entirely golden. SA, Rio de Janeiro,
Brazil.

6. Goeldi's monkey
Callimico goeldii (p. 108)
Entirely black; head hair in two tiers, a
short, even cap fringed by a long ruff.
SA.

Plate 9

Marmosets and Tamarins
(Callitrichidae)

1. Pygmy marmoset
Cebuella pygmaea (p. 96)
Tiny; tawny yellow-gray with fine
striations; tail faintly banded. SA, western
Amazon Basin.

2. Saddleback tamarins (two forms)
Saguinus fuscicollis (p. 100)
Back variegated black and yellowish;
hindquarters reddish; forequarters reddish
(2a north of Río Marañon) or black (2b
south of Río Marañon). SA.

3. Golden-mantle tamarin
Saguinus tripartitus (p. 101)
Black head; white muzzle; sharply
contrasting golden shoulders. SA.

4. Black-mantle tamarin
Saguinus nigricollis (p. 102)
Forequarters black or olivaceous;
hindquarters dark red or blackish. SA.

5. Red-chested mustached tamarin
Saguinus labiatus (p. 103)
Sharp white mustache and spot on nape;
underparts bright red. SA.

6. Black-chested mustached tamarin
Saguinus mystax (p. 103)
Sharp white mustache; dark brown
underparts. SA.

7. Golden-handed or **midas tamarin**
Saguinus midas (p. 101)
Blackish with bright gold hands and feet,
or black hands and feet. SA.

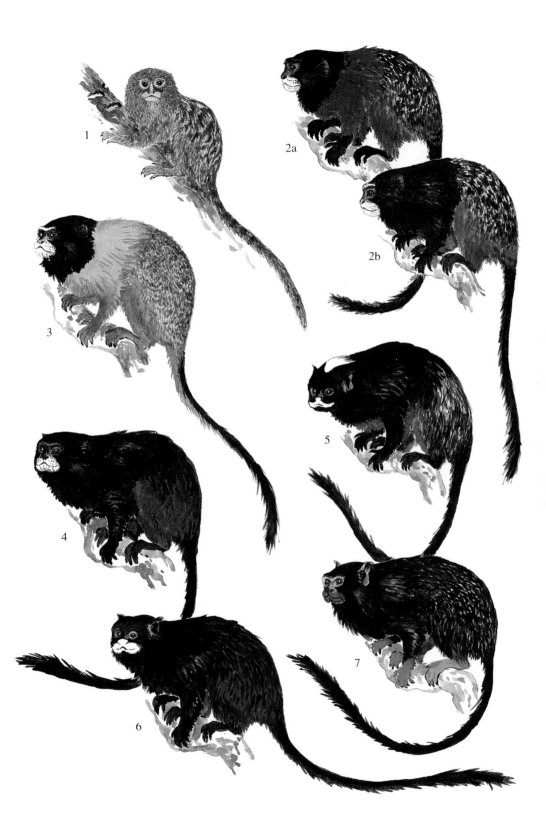

Plate 10

Tamarins (Callitrichidae)

1. Emperor tamarin
Saguinus imperator (p. 104)
Long downcurling white mustaches to
below chin; upperparts pale grayish; tail
orange or orange below. SA.

2. Mottled-face tamarin
Saguinus inustus (p. 106)
Face naked, mottled white and black. SA.

3. Brazilian bare-face tamarin
Saguinus bicolor bicolor (p. 104)
Face and head to ears naked; tail brown
above, orange below. SA.

4. Silvery-brown bare-face tamarin
Saguinus leucopus (p. 105)
Face naked; ruff dark brown; back silvery
brown; underparts orange. SA.

5. Cotton-top tamarin
Saguinus oedipus (p. 106)
Face naked; crest of long white hair; back
brown; underparts and feet cream. SA.

6. Red-naped or Geoffroy's tamarin
Saguinus geoffroyi (p. 107)
Back variegated black and yellow; face
naked; crown with short white hair; nape
dark red; underparts pale yellow. C, SA.

Plate 11

Small and Medium-Sized Monkeys (Cebidae)

1. Night monkey (red-necked form, Peru)
Aotus sp. (p. 110)
Eyes very large; three black stripes on crown; nonprehensile tail. C, SA.

2. Dusky titi monkeys (three forms)
(a) *Callicebus moloch moloch;* (b) *C. m. brunneus;* (c) *C. m. ornatus* (p. 111)
Face small; fur long and dense; tail thickly furred, nonprehensile. SA.

3. Yellow-handed titi monkey
Callicebus torquatus (p. 112)
As above; blackish with yellow collar and hands. SA.

4. Masked titi monkey
Callicebus personatus (p. 113)
As (2) above; wide black fringe encircling face. SA, E Brazil.

5. Common squirrel monkey (two forms)
(a) *Saimiri sciureus sciureus;* (b) *S. s. boliviensis* (p. 113)
Small; slender; short hair; white mask around eyes; dark muzzle; nonprehensile tail. SA.

6. Central American squirrel monkey
Saimiri oerstedii (p. 115)
As above. CA.

7. Brown capuchin monkey
Cebus apella (p. 116)
Dark cap, limbs and tail; dark bar in front of ear; prehensile tail. SA.

8. White-fronted capuchin monkey
Cebus albifrons (p. 118)
Dark cap; pale limbs and tail; pale cheeks to ear; prehensile tail. SA.

9. White-throated capuchin
Cebus capuchinus (p. 118)
Black with whitish face, chest, and shoulders; prehensile tail. C, SA.

10. Wedge-capped or **weeping capuchin**
Cebus olivaceus (p. 117)
Dark cap and line down center of face; no dark bar in front of ear; hands, feet, and tail dark; prehensile tail. SA.

Plate 12

Saki, Bearded Saki, and Uakari Monkeys (Cebidae)

All have nonprehensile tails; long fur; mouth pulled into a frown by large bulging canine teeth under lips; South America only.

1. Guianan saki (male and female)
Pithecia pithecia (p. 119)
Male black with white-rimmed face; female gray-brown. SA.

2. Equatorial saki
Pithecia aequatorialis (p. 120)
Face ringed with white; crown of male with mat of short white hair; underparts orange or brown. SA.

3. Monk saki
Pithecia monachus (p. 120)
Crown almost bald, covered by mop of grizzled hair; chest whitish or buffy. SA.

4. Buffy saki
Pithecia albicans (p. 121)
Ruff, arms, legs, sides, and underparts buffy to reddish; back and tail blackish. SA.

5. Brown bearded saki (male)
Chiropotes satanus (p. 121)
Brown; large tufts over temples; bearded chin; face brown. SA.

6. White-nosed bearded saki (female)
Chiropotes albinasus (p. 122)
Black; white triangle covering nose and mouth. SA.

7. Red uakari
Cacajao calvus rubicundus (p. 123)
Tail short; face naked, red. SA.

8. Black uakari
Cacajao melanocephalus (p. 124)
Tail short; face naked, black. SA.

Plate 13

Howler Monkeys (Cebidae)

All are large; bearded; with strongly
prehensile tails naked below the tip.

1. Red howler monkey
Alouatta seniculus (p. 124)
Red to orange. SA.

2. Red-handed howler monkey
Alouatta belzebul (p. 125)
Entirely black or with hands, feet, tail tip,
or sometimes forehead or back dull red.
SA.

3. Black howler (male and female)
Alouatta caraya (p. 126)
Male black or dark brown; female pale
yellow or brown. SA.

4. Brown howler monkey
Alouatta fusca (p. 126)
Brown, red-brown, or yellowish brown.
SA.

5. Mantled howler monkey
Alouatta palliata (p. 127)
Black with pale-frosted fringe on sides or
cape over back. C, SA.

6. Mexican black howler monkey
Alouatta pigra (p. 128)
Entirely black. CA.

1 ♂

2 ♂

♀

3 ♂

4 ♂

5 ♂

6 ♂

Plate 14

Spider, Woolly, and Woolly Spider Monkeys (Cebidae)

All are large; have strongly prehensile tails naked under the tip.

1. Black spider monkey (two forms)
(a) *Ateles paniscus chamek* (Peru);
(b) *A. p. paniscus* (Guianas) (p. 130)
Black; underparts black. SA.

2. White-bellied spider monkey (two forms)
(a) *Ateles belzebuth belzebuth* (Colombia);
(b) *A. b. hybridus* (N Venezuela) (p. 131)
Black or brown with pale belly. SA.

3. Central American spider monkey (two forms)
(a) *Ateles geoffroyi panamensis* (Panama);
(b) *A. g. geoffroyi* (Nicaragua) (p. 132)
Highly variable; pale mask around eyes; any spider monkey north of Darién, Panama. CA.

4. Woolly spider monkey
Brachyteles arachnoides (p. 132)
Entirely pale; ears hairy and protruding prominently; belly large. E Brazil, SA.

5. Common woolly monkey
Lagothrix lagothricha (p. 128)
Brown, gray, or olivaceous above and below; robust limbs and tail; round head. SA.

6. Yellow-tailed woolly monkey
Lagothrix flavicauda (p. 129)
Dark red; yellow under tail tip and on muzzle. Peruvian Andes only, SA.

Plate 15

Raccoon Family (Procyonidae)

1. Cacomistle
Bassariscus sumichrasti (p. 141)
Tail bushy, brightly banded; white mask
around eyes; dark feet. CA.

2. Olingo
Bassaricyon gabbii (p. 139)
Tail faintly banded, nonprehensile. C,
SA.

3. Kinkajou
Potos flavus (p. 140)
Tail prehensile, slightly tapered, not
banded. C, SA.

4. South American coati
Nasua nasua (p. 138)
Muzzle and foreclaws elongated; muzzle
dark; tail banded, tapered. SA.

5. White-nosed coati
Nasua narica (p. 139)
As above but muzzle white, shoulders
gray. CA.

6. Northern raccoon
Procyon lotor (p. 137)
Tail shorter than head and body, brightly
banded; black mask over eyes continues
as dark bar across throat; forelegs whitish.
N, CA.

7. Crab-eating raccoon
Procyon cancrivorous (p. 136)
As above but legs and feet dark; mask
ends behind eyes, no brown band on
throat. C, SA.

Dogs (Canidae)

8. Bush dog
Speothos venaticus (p. 135)
Tail and legs short; head and neck pale;
fur long and soft. C, SA.

9. Short-eared dog
Atelocynus microtis (p. 134)
Tail long enough to touch ground; entirely
dark; fur short and stiff. SA.

Plate 16

Weasel Family (Mustelidae)

1. Long-tailed weasel
Mustela frenata (p. 142)
Small; dark brown with pale belly; with
or without white facial markings. C, SA.

2. Amazon weasel
Mustela africana (p. 142)
As above, but dark brown stripe down
underparts; no facial markings. SA.

3. Grison or huron
Galictis vittata (p. 143)
Head tricolor; gray crown, white brow
and side of neck, black muzzle and
throat. C, SA.

4. Striped hog-nosed skunk
Conepatus semistriatus (p. 145)
Bicolored black and white; tail bushy,
entirely white distally. C, SA.

5. Tayra
Eira barbara (p. 144)
Head usually pale (may be dark); chest
usually with pale spot. C, SA.

6. Southern river otter
Lutra longicaudis (p. 147)
Underparts pale; feet webbed. C, SA.

7. Giant otter
Pteronura brasiliensis (p. 147)
Very large; throat usually spotted; belly
dark. SA.

Plate 17

Cats (Felidae)

1. Jaguarundi
Felis yagouaroundi (p. 151)
Small; uniform color, red, brown, gray or tawny. C, SA.

2. Oncilla
Felis tigrina (p. 150)
Small; spotted; built like house cat, small head and feet; hair on neck not reversed. C, SA.

3. Margay
Felis wiedii (p. 150)
Small; spotted; large head and feet; tail longer than hindleg; neck striped, hair reversed. C, SA.

4. Ocelot
Felis pardalis (p. 149)
As above but larger; tail shorter than hindleg. C, SA.

5. Puma or **mountain lion**
Felis concolor (p. 152)
Large; uniform color.

6. Jaguar
Panthera onca (p. 153)
Large; spotted or black; spots, not stripes on neck; head and legs large. C, SA.

Plate 18

Deer (Cervidae)

1. Gray brocket deer (male)
Mazama gouazoubira (p. 161)
Gray-brown; underparts whitish; males
with short, straight, unbranched antlers;
no facial markings; back slightly humped.
C, SA.

2. Red brocket deer (female and young)
Mazama americana (p. 160)
As above but body and belly red. C, SA.

3. White-tailed deer (male)
Odocoileus virginianus (p. 162)
Gray to red; belly white; back straight;
markings on muzzle and around eyes;
males with branched antlers. C, SA.

Capybara (Hydrochaeridae)

4. Capybara
Hydrochaeris hydrochaeris (p. 204)
Large square muzzle; no tail; webbed
feet. C, SA.

Peccaries (Tayassuidae)

5. White-lipped peccary
Tayassu pecari (p. 159)
Large; narrow, piglike snout; no tail;
white on chin. C, SA.

6. Collared peccary
Tayassu tajacu (p. 158)
As above but smaller; faint pale collar on
neck; chin not white. C, SA.

Tapirs (Tapiridae)

7. Brazilian tapir (and young)
Tapirus terrestris (p. 156)
Very large; upper lip elongated in
proboscis; crown between ears raised in
humped crest. SA.

8. Baird's tapir
Tapirus bairdii (p. 157)
As above but crown flat. C, SA.

1 ♂

2 ♀

3 ♂

4

5

6

7

7

8

Plate 19

Larger Central and South American Squirrels (Sciuridae)

All have long ears that protrude above crown.

1. Red-tailed squirrel (three forms)
Sciurus granatensis (a) Parts of N
Colombia; (b) parts of Venezuela and
Central America; (c) N coast of Colombia
and Ecuador (p. 170)
Medium size; tail reddish. C, SA.

2. Guayaquil squirrel
Sciurus stramineus (p. 171)
Ears black; tail black frosted white. SA.

3. Southern Amazon red squirrel
Sciurus spadiceus (p. 168)
Large; red or rarely black; feet with
mixture of black and red hairs (not visible
at distance). SA.

4. Northern Amazon red squirrel
Sciurus igniventris (p. 167)
Large; red or rarely black; feet pure red;
side sometimes with black line. SA.

5. Junín red squirrel
Sciurus pyrrhinus (p. 168)
Medium; back pure red unmixed with
black. Andean Peru, SA.

Plate 20

Small Squirrels (Sciuridae)

1. Sanborn's squirrel
Sciurus sanborni (p. 170)
Large ears; lower belly pure orange or
whitish; feet may be paler than back. SA.

2. Guianan squirrel
Sciurus aestuans (p. 169)
Large ears; lower belly mixed grayish;
females with 4 pairs mammae. SA.

3. Bolivian squirrel
Sciurus ignitus (p. 169)
Large ears; lower belly grayish; feet the
same color as back; females with 3 pairs
mammae. SA.

4. Amazon dwarf squirrel
Microsciurus flaviventer (p. 174)
Small ears; tail shorter than head and
body. SA, Amazon Basin.

5. Western dwarf squirrel
Microsciurus mimulus (p. 175)
As above; midback may have black stripe;
underparts orange. C, SA west of Andes.

6. Central American dwarf squirrel
Microsciurus alfari (p. 175)
As above but underparts grayish. C, SA
to NW Colombia.

7. Neotropical pygmy squirrel
Sciurillus pusillus (p. 176)
Tiny size; large white patches behind
ears; grayish. SA.

Plate 21

Larger Central American Squirrels (Sciuridae)

Black individuals can occur in most species.

1. Deppe's squirrel
Sciurus deppei (p. 173)
Finely grizzled brownish; forelegs usually gray; underparts gray to white; tail fairly slender. CA.

2. Yucatan squirrel
Sciurus yucatanensis (p. 173)
Coarsely grizzled blackish or brownish; underparts pale gray or white. CA.

3. Red-bellied squirrel (two forms from Mexico)
Sciurus aureogaster (p. 174)
Finely grizzled gray to brown; nape, shoulders, or rump may have orange patches; underparts red to white. CA.

4. Variegated squirrel (three forms)
Sciurus variegatoides (p. 172)
(a) Nicaragua, El Salvador; (b) Nicaragua, Costa Rica; (c) Costa Rica. Highly variable; large; tail very bushy; fur often coarsely variegated; pale patch behind ear; tail black frosted white above; underparts red to white. CA.

Plate 22

Small Terrestrial Rats
(Muridae, Heteromyidae)
All in the range of about 30–120 g.

1. Grass mouse
Akodon urichi (p. 190)
Tail short; whiskers short. SA.

2. Brazilian shrew mouse
Blarinomys breviceps (p. 187)
Tail very short; eyes and ears tiny. SA.

3. Atlantic forest rat
Delomys dorsalis (p. 185)
Dark midback stripe (not present in all members of genus); whiskers fine, short. SA.

4. Spiny pocket mouse
Heteromys desmarestianus (p. 177)
Spiny fur; externally opening cheek pouches; sharply demarcated white underparts. C, SA.

5. Long-nosed mouse
Oxymycterus cf. *inca* (p. 191)
Long narrow muzzle and strongly undershot jaw; long claws on all feet. SA.

6. Macconnell's rice rat
Oryzomys macconnelli (p. 178)
Long muzzle; orange sides; long slender tail and feet. SA.

7. Common rice rat
Oryzomys capito (p. 178)
Tawny brown; fur fine and soft; underparts grayish white. SA.

8. Mexican deer mouse
Peromyscus mexicanus (p. 192)
Ears large; dark ring around eye; whiskers long; tail long. CA.

9. Isthmus rat
Isthmomys (p. 193)
Cinnamon or orange; ears large; whiskers long; tail very long. CA.

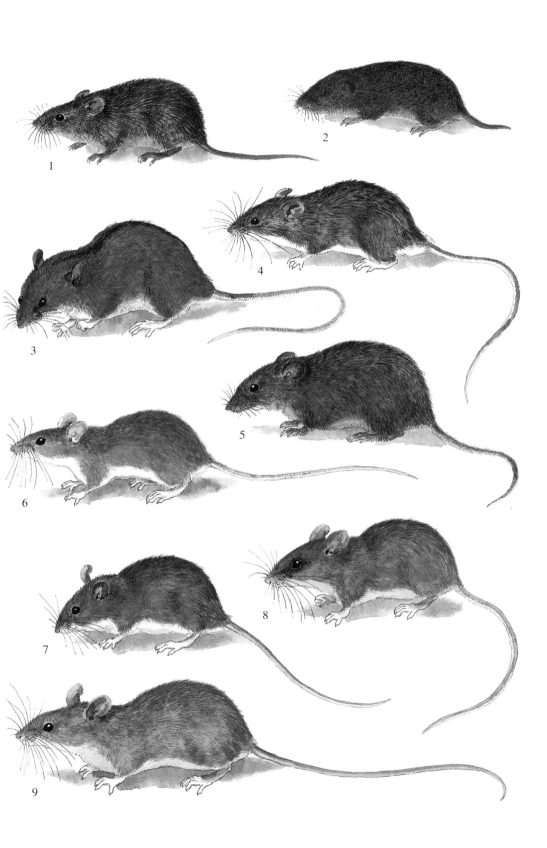

Plate 23

Small or Tiny Mice (Muridae)
All in the range of about 8–30 g.

1. Spiny mouse
Neacomys (p. 181)
Fur spiny; tail as long as head and body;
underparts white or orange. C, SA.

2. Gray spiny mouse
Scolomys melanops (p. 182)
Fur spiny; tail much shorter than head
and body; entirely gray. SA.

3. Pygmy rice rat
Oligoryzomys fulvescens (p. 180)
Tail much longer than head and body;
underparts gray; head and neck grayish.
SA.

4. Bicolored arboreal rice rat
Oecomys bicolor (p. 181)
Short, broad, pink-soled hindfeet; reddish;
underparts pure white; tail slightly hairy.
C, SA.

5. Harvest mouse
Reithrodontomys sumichrasti (p. 193)
Tail longer than head and body; sides of
head and neck tinged orange; whiskers
long; upper incisors grooved.

6. House mouse
Mus musculus (p. 195)
Tail about same length as head and body;
whiskers short; underparts not sharply
contrasting with sides.

Plate 24

Large Terrestrial and Semiaquatic Rats (Muridae, Echimyidae)

All in the range of 70–550 g.

1. Water rat
Nectomys squamipes (p. 182)
Hindfoot partly webbed, fringed with hair, heel with scales on sole; ear tips naked; underparts not sharply demarcated from sides; fur soft and dense. SA.

2. Marsh rat
Holochilus brasiliensis (p. 191)
Hindfoot partly webbed, fringed with hair, heel without scales; ear hairy to tip; underparts not sharply contrasting with sides; fur soft and dense. SA.

3. South American water mouse
Neusticomys sp. (p. 187)
Tail completely covered with flat dark brown hair; eyes and ears tiny; hindfoot with fringe of hair but not strongly paddle-shaped; underparts not sharply contrasting with sides. SA.

4. Central American water mouse
Rheomys mexicanus (p. 186)
Tail completely covered with hair; eyes and ears tiny; hindfoot paddle-shaped, fringed with hair; underparts silvery; forefoot with four large pads on palm; CA only.

5. Crab-eating rat
Ichthyomys pittieri (p. 185)
As above but forefoot with five pads. SA, CA.

6. Black rat
Rattus rattus (p. 194)
Black or tawny; fur coarse and sparse; tail longer than head and body; whiskers long.

7. Norway rat
Rattus norvegicus (p. 195)
Tawny brown; fur coarse and sparse; tail shorter than head and body; whiskers medium.

8. Spiny rat
Proechimys steerei (p. 212)
Fur bristly or spiny, spines usually lie flat; head long and narrow; underparts sharply demarcated, often snow white; tail shorter than head and body. Genus C, SA.

9. Armored rat
Hoplomys gymnurus (p. 214)
As above but spines on lower back strong, upstanding, geometrically spaced; underparts may be dark. C, SA.

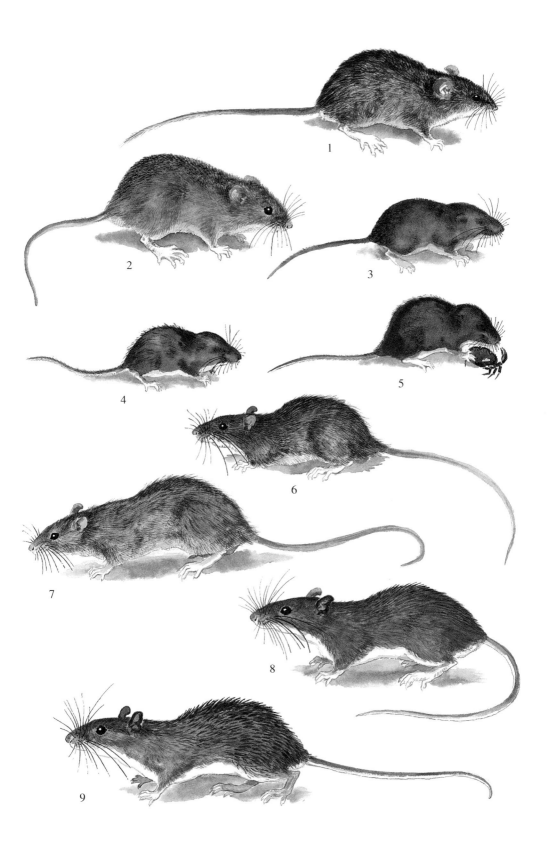

Plate 25

Small and Medium-Sized Arboreal Rats (Muridae, Echimyidae)

All in the range of 40–300 g; with large eyes and broad, strong feet.

1. Arboreal rice rat
Oecomys trinitatus (p. 181)
Whiskers and hair on tail fine; ears quite short. Genus C, SA.

2. Climbing rat
Rhipidomys mastacalis (p. 184)
As above but whiskers coarse and hair on tail prominent. SA.

3. Big-eared climbing rat
Ototylomys phyllotis (p. 188)
Large ears; tail naked, shiny, dark. CA.

4. Naked-tailed climbing rat
Tylomys watsoni (p. 188)
Large; tail naked, shiny, white at tip. C, SA.

5. Vesper rat
Nyctomys sumichrasti (p. 189)
Eyes very large; whiskers very long; tail very hairy. CA.

6. Spiny tree rat
Mesomys hispidus (p. 223)
Heavily spined; ears short; tail sparsely but prominently haired, with slight tuft. SA.

7. Tuft-tailed spiny rat
Lonchothrix emiliae (p. 224)
Heavily spined; ears short; tail scaly proximally, tip with large tuft of long, coarse hair. SA.

Plate 26

Larger Spiny Tree Rats
(Echimyidae)

All in range of 220–900 g; have bristly or spiny fur; short ears; strong, broad feet; South America only.

1. Giant tree rat
Echimys grandis (p. 216)
Large; black and gold; fur bristly, not spiny; tail thickly haired, pitch black.

2. Peruvian tree rat
Echimys rhipidurus (p. 216)
Tail sparsely but prominently haired throughout, or with hairs decreasing gradually down tail above; fur bristly, not spiny; sides of muzzle rusty.

3. Dark tree rat
Echimys saturnus (p. 215)
Large; black and chestnut; tail thickly haired, black with white tip.

4. Speckled tree rat
Echimys semivillosus (p. 218)
Heavily spined; tawny grayish speckled with white tips on spines; tail looks naked.

5. Red-nosed tree rat
Echimys armatus (p. 217)
Moderately spined; muzzle to between eyes rusty; tail appears naked but slightly hairy, haired at base for 3–4 cm; whiskers coarse; belly usually grayish.

6. Bare-tailed tree rat
Echimys occasius (p. 218)
Heavily spined; tail conspicuously naked, haired at base for 1–2 cm; underparts pure orange with white patches; muzzle rusty.

7. White-faced tree rat
Echimys chrysurus (p. 214)
Heavily spined; white blaze or spot on face or crown; white or chestnut tail tip.

8. Black-spined Atlantic tree rat
Nelomys nigrispinis (p. 219)
Red-brown streaked with black; tail sparsely haired.

9. Golden Atlantic tree rat
Nelomys blainvillei (p. 220)
Yellow or red-gold; tail thickly haired, with tuft.

10. Pallid Atlantic tree rat
Nelomys lamarum (p. 220)
Heavily spined; drab, yellow ochraceous; tail sparsely haired.

Plate 27

Large Soft-furred Tree Rats
(Echimyidae)
All in range of 300–700 g; have soft fur
and no spines; small ears; strong, broad
feet; large eyes.

1. Amazon bamboo rat
Dactylomys dactylinus (p. 224)
Large, square muzzle; naked tail; fingers
long, separated by central gap, with nails.
SA.

2. Southern bamboo rat
Kannabateomys amblyonyx (p. 225)
As above, but tail well haired throughout.
SA.

3. Red crested tree rat
Diplomys rufodorsalis (p. 223)
Rust-red; tail completely haired, tip
white. SA.

4. Rufous tree rat
Diplomys labilis (p. 222)
Reddish brown; face grayish with pale
spots at base of whiskers and over eye;
tail completely haired. Genus C, SA.

5. Yellow-crowned brush-tailed rat
Isothrix bistriata (p. 221)
Crown with pale yellow patch bordered
on each side by a black stripe; tail fully
haired, dusky and/or orange. SA.

6. Plain brush-tailed rat
Isothrix pagurus (p. 221)
Crown without markings; tail fully haired,
dusky. SA.

7. Painted tree rat
Echimys pictus (p. 218)
Sharply patterned black and white; tail
fully haired, tip white. SA.

Plate 28

Porcupines (Erethizontidae)

All have stout, sharp spines; prehensile tails; feet with long claws; inconspicuous ears.

1. Bicolor-spined porcupine
Coendou bicolor (p. 198)
Hairless; large; black with variable number of white tips on spines; long tail. SA.

2. Brazilian porcupine
Coendou prehensilis (p. 197)
Hairless; large; gray or yellowish. SA.

3. Rothschild's porcupine
Coendou rothschildi (p. 198)
Like (1), but smaller. CA.

4. Mexican hairy porcupine
Coendou mexicanus (p. 202)
Black with pale head; body covered with long hair. CA.

5. Black dwarf porcupine
Coendou sp. nov. (p. 199)
Small; hairless; blackish. SA.

6. Brown hairy dwarf porcupine
Coendou vestitus (p. 201)
Small; hairy; warm smoky brown; tail short. SA.

7. Black-tailed hairy dwarf porcupine
Coendou melanurus (p. 199)
Small; hairy; black sprinkled with yellow bristles; tail long, black. SA.

8. Frosted hairy dwarf porcupine
Coendou pruinosus (p. 201)
Small; hairy; black frosted whitish; spines very short; tail short. SA.

9. Orange-spined hairy dwarf porcupine
Coendou villosus (p. 200)
Small; hairy; hair dark at base; some spines orange-tipped; tail usually long. SA.

10. Bahía hairy dwarf porcupine
Coendou insidiosus (p. 200)
Small; hairy; smoky brown to gray; hair usually pale at base; tail usually short. SA.

11. Paraguay hairy dwarf porcupine
Coendou paragayensis (p. 200)
Small to tiny; hairy; neck and shoulders with cape of longer yellow spines; tail long. SA.

12. Bristle-spined porcupine
Chaetomys subspinosus (p. 202)
No hair; nose not bulbous; spines on lower back kinky and flexible like broom straws; short, sharp spines on head and shoulders. SA.

Plate 29

Large Rodents
Paca (Agoutidae)

1. Paca
Agouti paca (p. 204)
Large; red or brown with rows of spots;
no visible tail. C, SA.

Pacarana (Dinomyidae)

2. Pacarana
Dinomys branickii (p. 205)
Large; black, with spots from shoulder to
rump; prominent tail. SA.

Agoutis and Acouchys
(Dasyproctidae)

3. Green acouchy
Myoprocta pratti (p. 211)
Olivaceous; small, thin, white-tipped tail.
SA.

4. Red acouchy
Myoprocta acouchy (p. 210)
Red with black rump; tail as above. SA.

5. Brown agouti
Dasyprocta variegata (p. 207)
Uniform brown to orange; tail a short,
naked black stub. SA.

6. Black-rumped agouti
Dasyprocta prymnolopha (p. 208)
Small; reddish with black rump; tail as
above. SA.

7. Mexican black agouti
Dasyprocta mexicana (p. 210)
Grizzled black; rump black; tail as in (5).
CA.

8. Central American agouti (two forms)
Dasyprocta punctata (p. 209)
Either (a) uniform reddish brown; or (b)
blackish with midbody yellowish and
rump with long, black, white-tipped hairs;
tail as in (5). S, CA.

9. Azara's agouti
Dasyprocta azarae (p. 208)
Tawny gray, midbody washed with
orange; tail as in (5). SA.

10. Black agouti
Dasyprocta fuliginosa (p. 207)
Grizzled black; tail as in (5). SA.

11. Red-rumped agouti
Dasyprocta agouti (p. 206)
Brownish with red to orange rump; tail as
in (5). SA.

Rabbits (Leporidae)

12. Brazilian rabbit or **tapiti**
Sylvilagus brasiliensis (p. 227)
Long oblong ears; short furry tail; reddish
nape and feet. C, SA.

Plate A

Armadillos (Dasypodidae)

All have body covered above with bony armor plates.

1. Yellow armadillo
Euphractus sexcinctus (p. 39)
Many long hairs over armor plates; short, widely set ears; front claws not enlarged. SA.

2. Southern naked-tailed armadillo
Cabassous unicinctus (p. 40)
Large, widely set ears; naked tail; greatly enlarged front claws. SA.

3. Nine-banded long-nosed armadillo
Dasypus novemcinctus (p. 42)
Closely set ears; long, narrow muzzle; 8–10 bands around midbody. N, C, SA.

4. Great long-nosed armadillo
Dasypus kappleri (p. 43)
Closely set ears; long, narrow muzzle; large size; projecting scutes on hind knees; 7–8 bands. SA.

5. Seven-banded long-nosed armadillo
Dasypus septemcinctus (p. 43)
Closely set ears, long, narrow muzzle; small size; 6–7 bands. SA.

6. Giant armadillo
Priodontes maximus (p. 41)
Enormous size; small, widely set ears; greatly enlarged front claws and limbs. SA.

Plate B

Bats (Chiroptera)

Sheath-tailed Bats
(Emballonuridae)
All are small, with no noseleaf; tail shorter than tail membrane with tip emerging free above membrane about halfway down its length.

1. Long-nosed bat
Rhynchonycteris naso (p. 45)
Elongated muzzle; white tufts of hair on forearm; wavy white lines on back. C, SA.

2, 2a. White-lined sac-winged bat
Saccopteryx bilineata (p. 45)
Wavy white lines on back; wing sac close to forearm bone beyond elbow. C, SA.

3. Shaggy bat
Centronycteris maximiliani (p. 46)
Long woolly hair; hairy face; no wing sac. C, SA.

4. Doglike sac-winged bat
Peropteryx macrotis (p. 47)
Tuft on crown ends abruptly on naked face; stiff mustache; wing sac on forward edge of membrane (propatagium). C, SA.

5. Chestnut sac-winged bat
Cormura brevirostris (p. 48)
Wide, heavily ridged, rounded ear; wing membrane attaches near base of toe. C, SA.

6. Least sac-winged bat
Balantiopteryx plicata (p. 48)
Tuft on crown ends abruptly on naked face; wing sac in center of membrane. C, SA.

7. Ghost bat
Diclidurus albus (p. 49)
White color; short round ear; glands in membrane around tail. C, SA.

Bulldog Bats (Noctilionidae)
No noseleaf, tail shorter than tail membrane.

8. Greater bulldog or fishing bat
Noctilio leporinus (p. 50)
Lip split from below nose; forward-pointing nose and ears. C, SA.

Leaf-chinned and Mustached Bats (Mormoopidae)
No noseleaf; tail shorter than tail membrane.

9. Mustached bat
Pteronotus gymnonotus (p. 52)
Lips flared forward; eye small. C, SA.

10. Leaf-chinned bat
Mormoops megalophylla (p. 53)
Platelike folds on chin; short rounded ear encircling eye. C, SA.

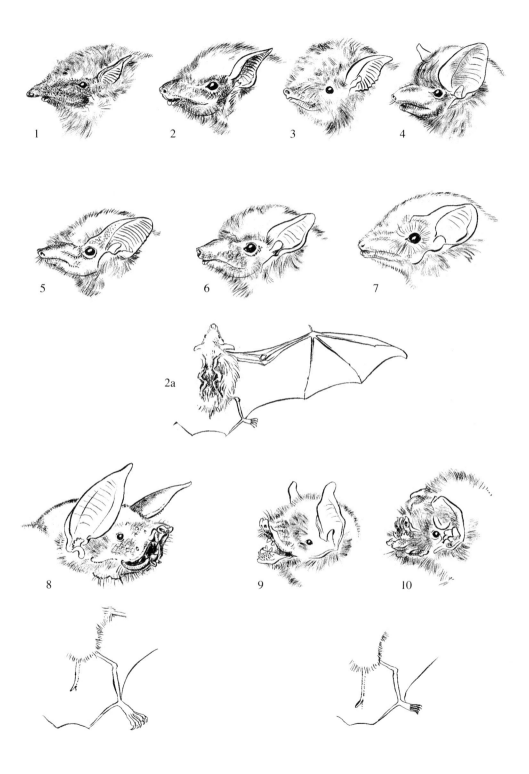

Plate C

Spear-nosed Bats
(Phyllostominae)
All have large noseleaves, narrow muzzles, large ears; tails except where noted.

1. Little big-eared bat
Micronycteris daviesi (p. 54)
Chin with single, smooth pads in V. C, SA.

2. Sword-nosed bat
Lonchorhina aurita (p. 55)
Extremely long, sword-shaped noseleaf; tail to point at edge of membrane. C, SA.

3. Long-legged bat
Macrophyllum macrophyllum (p. 56)
Tail to edge of membrane; membrane square at edge with parallel rows of dots. C, SA.

4. Round-eared bat
Tonatia bidens (p. 56)
Chin with tiny warts in U; large rounded ears. C, SA.

5. Hairy-nosed bat
Mimon crenulatum (p. 57)
Long, narrow noseleaf; large pointed ears; tail half as long as membrane. C, SA.

6. Fringe-lipped bat
Trachops cirrhosus (p. 60)
Lips and chin studded with long tubercules. C, SA.

7. Spear-nosed bat
Phyllostomus hastatus (p. 58)
Short, broad noseleaf and ears; horseshoe below nostrils free from lip. C, SA.

8. Pale-faced bat
Phylloderma stenops (p. 59)
Horseshoe below nostrils continuous with lip at center; skin of face pale and mottled; large size. C, SA.

9. Woolly false vampire bat
Chrotopterus auritus (p. 60)
Noseleaf and horseshoe form continuous hollow cup around nostrils; fur long and woolly; tail tiny. C, SA.

10. False vampire bat
Vampyrum spectrum (p. 61)
Enormous size; noseleaf and horseshoe form hollow cup around nostrils; short fur; midback stripe; no tail. C, SA.

Long-tongued Bats
(Glossophaginae)
All are small; have elongated, narrow
muzzles; highly extensible long tongues;
short noseleaves and ears; no free
horseshoe at center of lip under nostrils.

1. Common long-tongued bat
Glossophaga soricina (p. 62)
Lower jaw about the same length as
upper; chin with V bordered by tiny
tubercules; fur bicolored pale at base. C,
SA.

2. Spear-nosed long-tongued bat
Lonchophylla robusta (p. 63)
Lower jaw longer than upper; fur bicolor
pale at base; chin pads slightly rippled on
upper edge. C, SA.

3. Chestnut long-tongued bat
Lionycteris spurrelli (p. 63)
Lower jaw about same length as upper;
fur unicolored; chin pads smooth on
sides. C, SA.

4. Hairy-legged long-tongued bat
Anoura geoffroyi (p. 64)
No tail or tiny tail; tail membrane reduced
to hairy band down side of leg. C, SA.

5. Dark long-tongued bat
Lichonycteris obscura (p. 65)
Fur tricolored dark at base and tip; elbows
thinly furred above; muzzle robust. C,
SA.

6. Long-nosed long-tongued bat
Choeroniscus cf. *intermedius* (p. 66)
Muzzle narrow and tubelike; fur slightly
bicolored pale at base; first and second
thumb joints about equal in length. C,
SA.

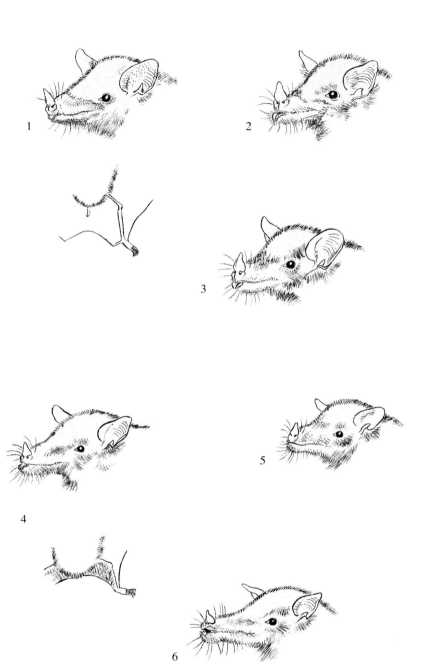

Plate E

Short-tailed and Little Fruit Bats (Carolliinae)

1. Short-tailed fruit bat
Carollia perspicillata (p. 67)
Short tail present; short narrow muzzle; chin with V of tiny warts flanking large central wart; fur banded. C, SA.

2. Little fruit bat
Rhinophylla fischerae (p. 68)
Muzzle narrow; no tail; fur unicolored; chin with three or four large pads flanking a large central wart. C, SA.

Neotropical Fruit Bats (Stenodermatinae)
All have broad muzzles; no tails; often white stripes.

3. Yellow-shouldered fruit bat
Sturnira lilium (p. 70)
No tail membrane; no stripes; yellow-stained shoulders; fur tricolor dark at base and tip. C, SA.

4. Tent-making bat
Uroderma bilobatum (p. 70)
Bright facial and midback stripes; blackish tongue. C, SA.

5. White-lined fruit bat
Vampyrops (tail membrane only) (p. 71)
Facial and midback stripes present; tail membrane with hairy edge; tongue pink. C, SA.

6. Big-eyed bat
Chiroderma villosum (p. 72)
Fur with dark band at base; muzzle short; upper middle incisor teeth much longer than outer incisors, with simple points; facial stripes if present do not extend past middle of ear. C, SA.

7. Large fruit-eating bat
Artibeus jamaicensis (p. 75)
No midback stripe; tongue blackish; central upper incisors short, bilobed; upper tooth row horseshoe-shaped. C, SA.

8. Great stripe-faced bat
Vampyrodes major (p. 72)
Sharp facial and midback stripes interdigitate on head; large size; yellowish ear and noseleaf rims. C, SA.

9. Double-lipped bat
Pygoderma bilabiatum (p. 77)
White spot on shoulder; fold from horseshoe under nose to corner of lip. SA.

10. Visored bat (male and female)
Sphaeronycteris toxophyllum (p. 78)
White spot on shoulder; horizontal fold across brow. SA.

11. Wrinkle-faced bat (male)
Centurio senex (p. 78)
Grotesquely wrinkled face; ladderlike pattern in wing. C, SA.

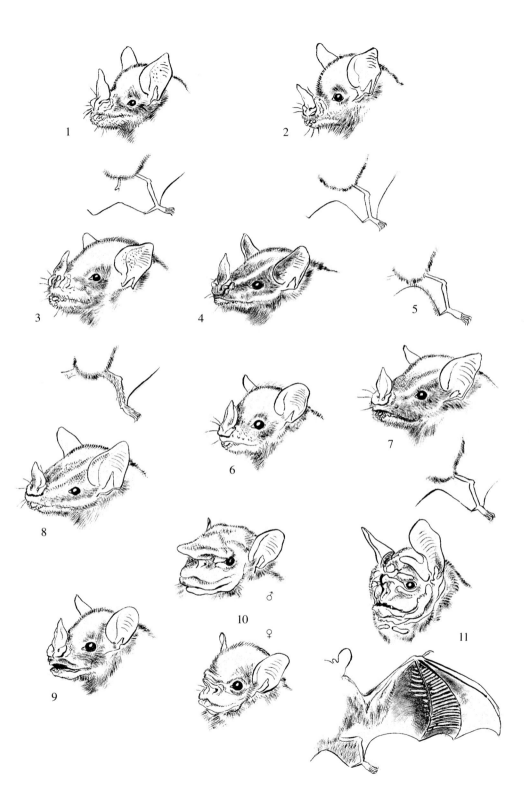

Vampire Bats (Desmodontinae)

1. Common vampire
Desmodus rotundus (p. 79)
Short M-shaped fold behind nostrils
backed by another fold; large sharp
triangular incisors larger than canines; no
tail; tail membrane a complete short band
between legs. C, SA.

2. Hairy-legged vampire
Diphylla ecaudata (p. 80)
Fold behind nostrils ∩-shaped; incisors as
above; tail membrane almost absent; legs
very hairy. C, SA.

Funnel-eared Bats (Natalidae)

3. Funnel-eared bat
Natalus stramineus (p. 81)
Tiny; mouth and ears funnel-shaped; tail
and legs very long; color pale. C, SA.

Thumbless Bats (Furipteridae)

4. Thumbless bat
Furipterus horrens (p. 82)
Miniscule (3 g); tail half as long as
membrane and entirely within it; thumb a
stub with no claw. C, SA.

Sucker-footed Bats
(Thyropteridae)

5. Sucker-footed bat
Thyroptera tricolor (p. 83)
Foot and thumb base with tiny sucker
disks; tail longer than membrane. C, SA.

Vespertilionid Bats
(Vespertilionidae)
All have no noseleaf; long tail with
membrane coming to a point at tip.

6. Little brown bat
Myotis nigricans (p. 84)
Tiny size (4–7 g); color usually dark; fur
unicolor dark or hair tips frosted; calcar
shorter than foot; gap between canine and
first large tooth behind. C, SA.

7. Big brown bat
Eptesicus furinalis (p. 85)
Larger size (6–16 g); fur dark at base
with paler tips imparting rich sheen;
calcar longer than foot; no gap between
canine and first large tooth behind. C,
SA.

8. Big-eared brown bat
Histiotis montanus (p. 86)
Ears enormous. SA.

9. Black-winged little yellow bat
Rhogeessa tumida (p. 85)
Tiny size; fur pale yellow frosted brown;
wings blackish. C, SA.

10. Hoary bat
Lasiurus cinereus (p. 87)
Ears short and wide; tail membrane
thickly furred above for more than half its
length. C, SA.

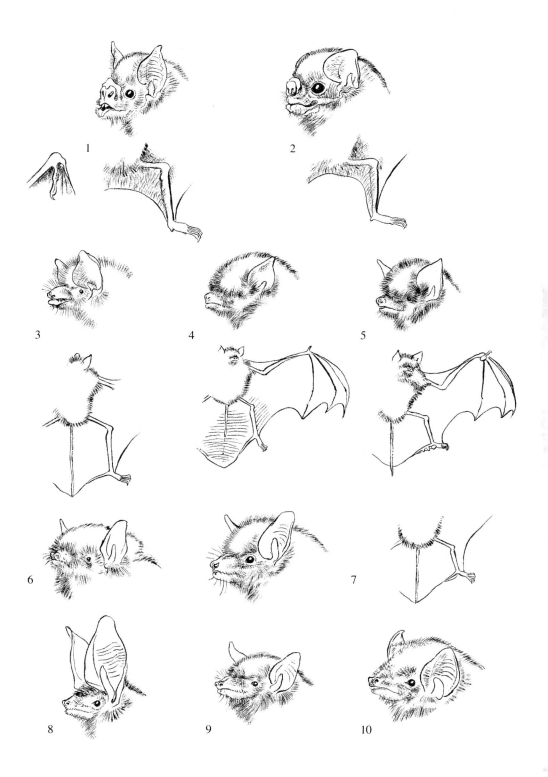

1

2

3

4

5

6

7

8

9

10

Plate G

Free-tailed or Mastiff Bats
(Molossidae)
All have naked tail extending one-third or more of its length beyond membrane; no noseleaf; flattened body; hairy feet.

1. Brazilian free-tailed bat
Tadarida brasiliensis (p. 90)
Upper lip with vertical wrinkles; ears meet but do not join on center of crown. C, SA.

2. Broad-eared free-tailed bat
Nyctinomops laticaudatus (p. 91)
Upper lip with vertical wrinkles; ears join on stalk at center of crown. C, SA.

3. Flat-headed bat
Neoplatymops mattogrossensis (p. 90)
Head and body extremely flattened; forearm skin above sprinkled with tiny bumps. SA.

4. Dog-faced bat
Molossops temminckii (p. 88)
Ears simple, triangular, pointed, widely separated on crown; fur diminishes gradually down face. C, SA.

5. Doglike bat
Cynomops greenhalli (p. 89)
Ears do not meet on crown; fur ends abruptly at naked face; muzzle and chin broad. C, SA.

6. Bonneted bat
Eumops perotis (p. 91)
Large ears flattened forward like hat over brow, join on midcrown; muzzle pointed, slanting sharply back from nose to chin. C, SA.

7. Crested mastiff bat
Promops nasutus (p. 92)
Top of muzzle raised in central ridge; ears just meet with horizontal fold on crown; chin broad and rounded; rear lower edge of ear narrow; four lower incisors; palate inside mouth deeply concave. C, SA.

8. Mastiff bat
Molossus molossus (p. 93)
Top of muzzle raised in ridge; ears just meet with horizontal fold on crown; rear lower edge of ear with broad lateral fold; two lower incisors; palate inside mouth flat. C, SA.

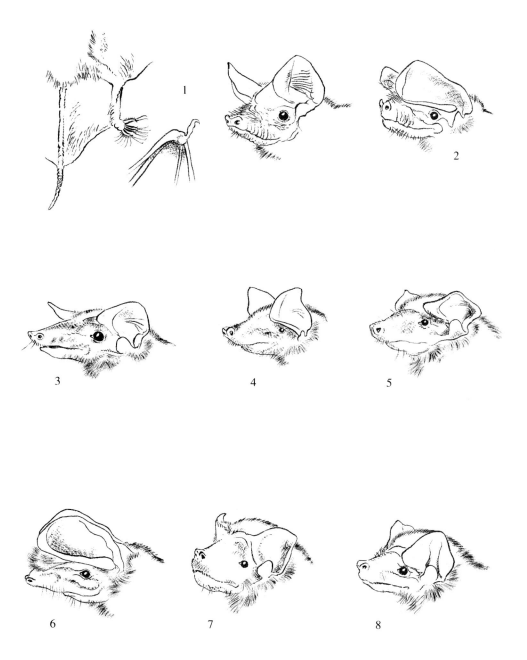

Map 122

░ Southern river otter, *Lutra longicaudis*

Map 123

░ Giant otter, *Pteronura brasiliensis*

Southern River Otter
Lutra longicaudis
Plate 16, map 122
Identification. Measurements: HB = 532–809; T = 360–500; HF = 80–130; E = 19–23; WT = 5–14.75 kg; males larger than females.
Upperparts entirely glossy dark brown; fur short and dense. **Upper lip, lower cheek, throat, and belly silvery whitish to yellowish. Head small, flat; muzzle broad;** neck thicker than head; eyes small; **ears short and rounded. Tail long, thick at the base and tapering.** Legs short and stout, **feet webbed.** On land head and tail are carried low and back humped high. Young like adult.
Similar species. Giant otters *(Pteronura brasiliensis)* are much larger, with dark belly and throat spotted brown on pale.
Sounds. Probably whistles, hums, and screeches like other river otters.
Natural history. Proabably diurnal and nocturnal; semiaquatic; solitary or in pairs. Feeds chiefly on fish and crustaceans, supplemented by other aquatic animals. River otters are always in or near water and are graceful swimmers and divers. On land they are awkward and move with a humping gallop or waddling walk. These otters appear to favor clear, fast-flowing rivers and streams, and they may be rare in or absent from sluggish, silt-laden lowland rivers. Found in riverine habitats in both deciduous and evergreen forests, and in warm and cool climates.
Geographic range. Central and South America: N Mexico south to Uruguay. To 3,000 m elevation.
Status. CITES Appendix I, US-ESA endangered. Widespread but rarely seen; formerly intensively hunted for the fur trade, but amount of current hunting and population status are unknown.
Local names. Nutria, lobito del río (Span); guaiao (Ar); water dog (Be); lontra, cachorro-d'agua, nútria (Br); perro de agua (Me); gato de agua (Pn); watradagoe (Su).
References. Van Zyll de Jong, C. G. 1972. A systematic review of the Nearctic and Neotropical river otters (genus *Lutra,* Mustelidae, Carnivora). Roy. Ontario Mus. Life Sci. Contrib., no. 80.

Giant Otter
Pteronura brasiliensis
Plate 16, map 123
Identification. Measurements: HB = 1,000–1,200; T = 530–700; HF = 176–207; WT = 24–34 kg.
Upperparts rich brown, almost black when wet; fur short, dense, velvety. Head round, muzzle blunt, whiskers stout, ears small, set low on sides of head. Upper lip and **throat with irregular splotches of cream and brown. Underparts apart from throat** about **the same color as back. Tail thick at base tapering to dorsoventrally flattened tip.** Legs short and thick; feet large, toes completely webbed to tips of digits. Young like adults.
Variation. Each individual has a different pattern of spots on throat.
Similar species. Southern river otters *(Lu-*

tra longicaudis) are smaller, with pale bellies and no spots on throat.

Sounds. Alarm call a loud explosive snort; group members constantly interact with loud, high-pitched hums and whining squeals and screeches.

Natural history. Diurnal; semiaquatic; usually in groups of five to nine, rarely solitary. Feeds primarily on large fish, but will kill and eat other vertebrates such as snakes and small caiman. Groups are territorial and consist of an adult pair and their offspring from several years. Giant otters are always found in or near large bodies of water. When hunting fish, they often travel swimming in a phalanx underwater together, alternately surfacing to breathe. When alarmed, all members of the group surface and crane their necks high out of water with a chorus of loud snorts. An otter will climb out onto the bank or a half-submerged log to eat a large fish, but fish are often eaten while the otter swims on its back holding the prey in its forepaws. Group members constantly squabble over food, growling and squealing. Their territories are marked by large trampled places on the waterside, usually around a log projecting into the water. These have a strong, unpleasant fishy odor and much residue of scats and fish scales. At night the group sleeps in a large burrow in a bank, with the entrance above a worn slide into the water. Giant otters fiercely defend their young, attacking in a group, and can even drive a jaguar away. They occupy lowland forest rivers and lakes of many types, from large, silt-laden, fish-rich waterways to clearwaters, blackwaters, and flooded forest in the rainy season.

Geographic range. South America: east of the Andes from S Venezuela and Colombia south to N Argentina. There is a historical record from Uruguay, but probably none remain there.

Status. CITES Appendix I, US-ESA endangered. Quite common where undisturbed, but rare or extinct in much of its former range and can be seen only in a few remote and protected areas. Decimated by overhunting for the skin trade and locally killed because of competition with man for fish. Giant otters are much more endangered than spotted cats because their waterside habitat is limited and accessible and their behavior is highly conspicuous.

Local names. Ariranha (Br); lobo del rio, arirai (Pe, Ec); grote waterhond, watradagoe (Su); ariraí (Gua).

References. Duplaix, N. 1980. Observations on the ecology and behavior of the giant river otter *Pteronura brasiliensis* in Suriname. *Rev. Ecol.* 34:496–620.

Munn, M. B., and C. A. Munn. 1988. The Amazon's gregarious giant otters. *Animal Kingdom*, Sept.–Oct., 34–41.

Cat Family (Felidae)

Dental formula: I3/3, C1/1, P2 or 3/2, M1/1 = 28–30. Four weight-bearing toes on all feet and a fifth, non-weight-bearing claw on the forefoot (the first digit); forefeet with retractile claws (except in cheetahs). Cats have teeth that are highly specialized for killing and meat-eating, and their sharp, retractile claws and strong shoulders allow them to single-handedly grasp and drag down large prey, which they kill either by a powerful bite to the head or neck or by throttling with a bite to the throat or muzzle, but most species eat small prey. Cats see and hear well; their vision is binocular, and they see colors. They are purely carnivorous. Cats are major predators in tropical rainforests worldwide. Most species are solitary hunters that capture their prey by surprise, either by stealthy approach or by patient waiting in ambush, followed by a short charge or pounce. They prey on almost anything they encounter that is not too large, including mammals, birds, snakes, turtles, caiman, fish, and even large insects. Some are territorial; others appear to share their home ranges but avoid an area where another cat is hunting. They communicate their presence to each other with marking behavior that includes spraying urine, scratching the ground and trees, and leaving their scats in prominent places. The Neotropical cats scratch fallen logs, and logs with scratch marks can be been wherever cats are common. The litter size is about one to four. The kittens are sheltered in a den and raised by the mother alone, who brings them prey as they are weaned. Mothers must leave their young and hunt for many hours to find food; healthy kittens found alone have not been abandoned and should not be touched. Cats can be active at any time of day or night, but rainforest species are most active nocturnally; the best time to see them is after nightfall, when they walk on trails and their bright eyeshine makes them visible. There are about 35 species in 4 genera worldwide, with 2 genera and 6 species in Neotropical rainforest. There is entrenched disagreement about the classification of cat genera: the 28 species of *Felis* are sometimes divided into as many as 8 genera. All cat species are listed under CITES Appendix II.

Ocelot
Felis pardalis
Plate 17, map 124

Identification. Measurements: HB = 710–875; T = 320–410; HF = 140–170; E = 50–65; WT = 8–14.5 kg; adult males are usually 11–12 kg, females 8–9 kg.
Upperparts tawny yellowish to buff with black spots and lines in longitudinal rows, many spots in open rosettes; **fur usually short, smooth,** and slightly stiff, rarely soft and woolly. **Neck with heavy black stripes dorsally, fur "reversed," slanting forward.** Eyes large, eyeshine very bright pale yellow; muzzle profile slightly convex. **Tail distinctly shorter than hind leg, banded and spotted with black.** Underparts white with black spots. Feet large; forefeet broader than hindfeet. Kittens spotted. A long-legged gracile cat the size of a medium dog.
Variation. Each individual has a different spot pattern; ground color varies from bright buff to gray (in arid regions of Mexico).
Similar species. Jaguar *(Panthera onca)* are much larger, with spots, not stripes, on neck; margays *(F. wiedii)* and oncillas *(F. tigrina)* are smaller, with tails longer than hindleg.
Sounds. None usually heard in field.
Natural history. Nocturnal and diurnal; terrestrial; solitary. Entirely carnivorous, feeds chiefly on rodents, supplemented by birds, snakes, lizards, and other small mammals and vertebrates. Ocelots hunt and capture their prey on the ground; they rarely climb trees, but will climb to cross over a stream on a branch or sometimes to rest. They are mainly active at night, when they spend many hours walking, often on man-made trails. When active by day they tend to keep hidden in dense brush. At rest they shelter under treefalls or between the closed buttress roots of large trees. Ocelots are the most commonly seen spotted cats,

Map 124

Ocelot, *Felis pardalis*

and their tracks can be found on muddy spots in trails and on riverbanks. They leave fine scratch marks on horizontal fallen logs. Found in habitats with good cover from rainforest to riverine scrub in deserts. Where they are not hunted for their skins, ocelots adapt well to disturbed habitats around villages, where they sometimes kill poultry.
Geographic range. North, Central, and South America: S Texas south to N Argentina. To 1,000 m elevation.
Status. CITES Appendix II, Appendix I (CR, southern Br), US-ESA endangered (two subspecies). Formerly intensively hunted for the skin trade, much less so currently. Widespread, common in some areas, naturally or artifically rare in others.
Local names. Tigrillo, ocelote (Span); gato-maracajá (Br, Pa); maracajú-açu, gato mourisco (Br); tiger cat (Be); manigordo (CR, Pn); gato tigre, tigre chico (Pn); onsa, yaguareté-í (Gua); tigri-kati (Sar).
References. Emmons, L. H. 1988. A field study of ocelots in Peru. *Rev. Ecol.* 42:133–57.

Map 125

Margay, *Felis wiedii*

Margay
Felis wiedii
Plate 17, map 125
Identification. Measurements: HB =
501–720 (few > 600); T = 351–490; HF
= 107–137; E = 45–60; WT = 3–9 kg;
males larger than females.
**Upperparts tawny yellowish to grayish
brown, with rows of black spots and
lines in longitudinal rows, some spots in
open rosettes; neck with heavy black
stripes, hair "reversed," slanting forward; fur often soft, parted here and
there into clumps,** more rarely stiff and
smooth. **Eyes very large,** eyeshine bright;
**whiskers long, muzzle area at base of
whiskers large and bulging. Tail longer
than hindleg, spotted and banded with
black.** Feet large, fore- and hindfeet about
the same width. This is the only New
World cat with ankle joints that can rotate
sufficiently for it to climb headfirst down
vertical trees, with the hindfeet turned facing the trunk (like a squirrel). Kittens spotted. Somewhat larger than a large house
cat.
Variation. Much individual variation in
ground color and spot pattern.
Similar species. Ocelots *(F. pardalis)* have
a shorter tail and are usually much larger,
but the largest margay males can be as
large as the smallest female ocelots; oncillas *(F. tigrina)* have the nape hair growing
in normal slant and are usually smaller,
with smaller, solid dotlike spots and body
and head shaped like a house cat's, but
they may not be distinguishable from margays with certainty in the field.

Natural history. Probably mostly nocturnal; arboreal and terrestrial; solitary. Feeds
on arboreal and terrestrial small mammals,
birds, and reptiles. Margays may not adapt
well to human disturbance of the habitat.
Found in evergreen and deciduous forests.
Geographic range. Central and South
America: Mexico south to Uruguay and
Argentina. To 900 m elevation.
Status. CITES Appendix I (two subspecies), US-ESA endangered. Widespread,
but always seems rarer than ocelot, status
unknown. Fur has less value, but margays
are caught in traps set for ocelots. Threat-
ened by deforestation in Central America.
Local names. Tigrillo, gato tigre (Span);
gato pintado, gato brasilieño (Ar); maraca-
já-peludo, gato-do-mato (Br); tiger cat
(Be); caucel (CR, Ho); pichigueta
(Cent Am); burricón (Ec); mbaracayá
(Gua); chulul (May); kuichua (Gy).
References. Tewes, M. E., and D. J.
Schmidly. 1987. The Neotropical felids:
Jaguar, ocelot, margay and jaguarundi. In
M. Novak, ed., *Wild furbearer manage-
ment and conservation in North America*.
Ontario Trappers Association, Ontario.

Oncilla
Felis tigrina
Plate 17, map 126
Identification. Measurements: HB = 452–
648; T = 255–330; HF = 96–145; E =
39–78; WT = 1.5–3 kg; males larger
than females.
**Similar to margay, but spots often solid,
small dots, sometimes rosettes; or body
entirely black; neck hair slants backward
in "normal" direction;** fur soft or coarse.
**Head shaped like house cat's; muzzle at
base of whiskers not bulging. Feet small,
like house cat's.** Kittens spotted. **Body
size and shape of slender house cat.**
Variation. Color highly variable, from
margaylike to quite distinct tawny brown,
gray, or blackish with rows of tiny solid
spots; black individuals not uncommon.
Similar species. See margay *(F. wiedii);*
other species of small spotted cats, *F. geoffroyi* and *F. colocolo,* occur in parts of
range outside the rainforest region; the
small spotted cats are difficult to tell apart
in the field and are often confused. Found
at higher elevations than margay or ocelot.
Natural history. Feeds on small mammals

Map 126

Oncilla, *Felis tigrina*

Map 127

Jaguarundi, *Felis yagouaroundi*

and birds. Found in evergreen and deciduous forest and bush; lowlands in some regions and in cloud forest and high elevation thickets in others.

Geographic range. Central and South America: Costa Rica south to S Brazil and N Argentina; Atlantic coastal forests of SE Brazil. Range poorly known. To 3,200 m elevation.

Status. CITES Appendix I (Central America), US-ESA endangered. Apparently always rare.

Local names. Tigrillo (Span); gato tigre, chivi (Ar); gato-do-mato (Br); caucel (CR); tigrillo chico (Ec); chat-tigre (FG); ocelot-cat, tigricati (Su).

References. Gardner, A. L. 1971. Notes on the little spotted cat, *Felis tigrina oncilla* Thomas, in Costa Rica. *J. Mammal.* 52:464–65.

Jaguarundi

Felis yagouaroundi
Plate 17, map 127

Identification. Measurements: HB = 505–645; T = 330–609; HF = 120–152; E = 25–40; SH ∼ 350; WT = 4.5–9 kg.
Upperparts usually uniform grizzled gray or brown, or red, black, or tawny yellow; no spots. Head small; ears small, rounded. Underparts the same as back, or slightly paler. **Feet small and doglike. A long-backed, slender-bodied cat with a long slender tail, long neck, and short legs.** At least some kittens are unspotted; some are reported to be dark-spotted.

Variation. Probably the most variable in color of all wild cats; red and gray animals are found in the same populations; rainforest animals tend to be dark brown or blackish; sometimes grizzled gray on head with black bodies; specimens from dry habitats are usually paler, gray, yellowish brown, or red; some from N Venezuela are tawny yellow like puma.

Similar species. Tayras (*Eira barbara*) usually have a pale spot on throat, inconspicuous ears, slightly bushy tail; bush dogs (*Speothos venaticus*) have thick bodies and short stumpy tails; puma (*F. concolor*) are much larger, with dark tail tips and pale muzzles; in dry habitats could be mistaken for a fox.

Sounds. Probably none usually heard in field; makes birdlike chirps.

Natural history. Diurnal and nocturnal; terrestrial but can climb trees; solitary and in pairs. Feeds on birds, reptiles, and small mammals. Jaguarundis travel widely in a huge home range. They den in hollow logs, treefalls, and thickets. Found in many habitats from rainforest to dense thickets in scrub and can live in secondary vegetation near villages, where they may raid poultry. They may be more common in deciduous or secondary forest than in rainforest.

Geographic range. North, Central, and South America: Texas south to S Brazil and Argentina. To 2,200 m elevation.

Status. CITES Appendix I, US-ESA endangered (Central and North America only). Widespread and not hunted for the fur trade but apparently uncommon or rare everywhere.

Local names. Yaguarundi (Span); halari
(Be); maracajá-preto, gato-preto, gato
mourisco (Br); león breñero (CR); gato
pardo, gato moro (Co); gato cerban (Ho);
tigrillo congo, tigrillo negro (Pn); anushi-
puma (Pe); weasel-cat, boesikati (Su);
mbaracaya-eirá, yaguarundí (Gua); ekmuch
(May).
References. See margay *(F. wiedii)*.

Puma
Felis concolor
Plate 17, map 128
Identification. Measurements: HB = 860–
1,540; T = 630–960; HF = 230–290; E
= 83–102; SH = 530–790; WT = 29–
120 kg; males larger than females.
**Upperparts uniform tawny yellow-brown
to dark reddish.** Head relatively small;
facial markings prominent: muzzle
around mouth white, patch at base of
whiskers blackish; eyeshine bright pale
yellow; throat whitish. **Tail darkening to
blackish tip.** Underparts paler than back.
Young spotted with dark brown. Large,
long-legged, rangy cats, slightly sway-
backed when standing.
Variation. Pumas from dry habitats are
pale gray or yellow; those from rainforest
range from yellow-brown to dark red-
brown.
Similar species. Puma are the only large,
uniformly colored cats in the region; jaguarundi *(F. yagouaroundi)* are much smaller,
with no contrasting dark tail tip.
Sounds. None usually heard in field.
Natural history. Nocturnal and diurnal;
terrestrial; solitary. Feed chiefly on medium
and large mammals such as deer, agoutis,
and pacas, but also eat smaller prey such
as snakes and rats. Puma are found
throughout the rainforests, where they
seem to keep mostly to dry ground. They
are lovers of wilderness, shy and wary of
man, and are rarely seen even where common. They can sometimes be encountered
walking down a rainforest trail at night.
They mark their paths by scraping up a
patch of dirt with their hind feet (but so do
jaguar and collared peccaries), and like
other cats they scratch fallen logs. Puma
will attack livestock where their habitat has
been converted to pasture and natural prey
reduced, and many populations have been
hunted to complete extinction by ranchers.

Map 128

Puma, *Felis concolor*

In centuries of historical association with
man, only a few attacks on humans have
ever been recorded, although puma sometimes follow humans discreetly from apparent curiosity. Puma are the most adaptable
of the world's cats and are found in many
climates from boreal to tropical, desert to
rainforest, and lowland to montane, in forest, woodland, and scrub habitats with
abundant game.
Geographic range. North, Central, and
South America: Canada to S Argentina and
Chile to at least 51° S. To 4,500 m
elevation.
Status. CITES Appendix I, US-ESA endangered (CR and SE U.S. subspecies
only). Widespread but always uncommon
or rare; locally hunted as a predator of
livestock, extinct over much of former
range, and locally threatened by deforestation and overhunting of its prey.
Local names. León, león colorado, león
de montaña, puma (Span); red tiger (Be);
onça-vermelha (Br); poema, redi-tigri (Su);
guasura, yaguá-pytá (Gua); cabcoh (May).
References. Anderson, A. E. 1983. A
critical review of literature on puma *(Felis
concolor)*. Colorado Div. Wildl. Special
Rep., no. 54.
   Seidensticker, J. C., IV, M. G. Hor-
nocker, W. V. Wiles, and J. P. Messick.
1973. *Mountain lion social organization in
the Idaho Primitive Area*. Wildl. Monogr.,
no. 35.

Map 129

Jaguar, *Panthera onca*

Jaguar
Panthera onca
Plate 17, map 129
Identification. Measurements: HB = 1,100–1,850; T = 440–560; HF = 220–250; E = 64–88; SH = 640–760; WT = 31–158 kg; males larger than females. **Upperparts tawny yellow with black spots, many on back and sides in open circles or rosettes; neck spotted above and below, not striped;** fur short and smooth. **Head very large, canine teeth long and stout; eyeshine bright greenish yellow; ears rounded, white inside, black behind tips.** Tail long, spotted or banded with black. **Underparts white with black spots.** Kittens spotted. **Very large and heavy, built for power not speed, with a short back, thick body, robust, short legs, and large feet.**
Variation. There is much size variation; jaguars from open habitats seem on average larger than those of rainforest. Dark brown or black jaguars (with spots always faintly visible at certain angles to light) occur rarely in all populations; Indians give these a separate name and often consider them a separate species, claiming different habits than tawny jaguars.
Similar species. Jaguars are the only very large spotted cats in the New World; ocelots *(Felis pardalis)* are much smaller, with stripes on the neck.
Sounds. Occasionally roar, day or night, with a pulsed series of single, deep, hoarse grunts that can be heard for several hundred meters.

Natural history. Nocturnal and diurnal; terrestrial; solitary. Feeds chiefly on the larger mammals such as capybara, peccary, and deer; also on turtles, tortoises, caiman, birds, fish, and smaller mammals such as sloths and agoutis. Jaguars hunt at any time of the day or night. They are fond of walking on man-made trails at night (as are other cats). Jaguars often use wet or waterside habitats, where they hunt capybara, turtles, caiman, and fish. Large cat tracks on river beaches are usually those of jaguars. In remote areas free from hunting, jaguars are most often seen when they lie stretched out on a log over the water in the morning sun. They rest, and carry their kills to feed, in dense vegetation. Jaguars may kill livestock where their natural habitat has been destroyed and their prey replaced by cattle. Although most jaguars flee quickly from man and attacks are very rare, these cats are potentially dangerous. Never run away from a big cat, since that may cause it to give chase; face it or walk toward it making loud noises (e.g., shouting and clapping hands). Found in a wide range of habitats from rainforest to wet grasslands and arid scrub.
Geographic range. North, Central, and South America: Mexico to Argentina. Formerly in SW United States and Uruguay, where now extinct. To 2,000 m elevation.
Status. CITES Appendix I, US-ESA endangered. It is difficult to estimate populations, but jaguars are rare or extinct in many parts of their former range—from overhunting for the fur trade, loss of habitat by deforestation, persecution by ranchers, and probably loss of their prey. Only a few hundred are thought to remain in all of Mesoamerica, but they are still widespread and can be locally common in Amazonia.
Local names. Tigre, jaguar (Span); tiger (Be); onça, onça pintada (Br); otorongo (Pe); penitigri (Su); yaguareté (Gua); zacbolay (May).
References. Rabinowitz, A. R., and B. G. Nottingham, Jr. 1986. Ecology and behaviour of the jaguar *(Panthera onca)* in Belize, Central America. *J. Zool., Lond.* 210:149–59.

Emmons, L. H. 1987. Comparative feeding ecology of felids in a Neotropical rainforest. *Behav. Ecol. Sociobiol.* 20:271–83.

Dolphins (Cetacea)

River Dolphins and Dolphins (Platanistidae and Delphinidae)

The dolphins are toothed whales (Odontoceti), with many conical teeth. They have long fusiform bodies, long, narrow beaks, pectoral flippers, a dorsal "fin," and a horizontally flattened tail. They breathe through a blowhole on top of the head. Dolphins are fast and agile predators that feed mainly on fish. They find their prey by echolocation (sonar) and communicate with each other by sound. The river dolphins (Platanistidae) are a primitive group of four genera and five species worldwide, which mostly live in muddy river systems; one genus and species is endemic to the Amazon Basin. They have degenerate eyes, highly developed echolocation, and flexible necks. Little is known of their behavior; they are apparently often solitary. The single young is apparently weaned before 8–9 months of age. The dolphins (Delphinidae) are an advanced group of largely marine species that have large brains, good vision, acrobatic movements, and complex social behavior. The single young may nurse for up to 1.5 years. There are 17 genera and 32 species worldwide, with 1 in Neotropical rainforest rivers.

Boutu or Pink River Dolphin
Inia geoffrensis
Figure 6, map 130
Identification. Measurements: total length to 1.8–2.6 m, weight to 160 kg.
Dorsal "fin" a long, low hump with inconspicuous point. Head with long, narrow beak and large bulge on forehead (the melon); eyes small. **Flippers broad at base, long.** Color variable, often pale, sometimes with a pinkish tinge, especially on paler underparts. **When they surface to breathe, only a small portion of back emerges from water.** Young usually dark gray.
Variation. Animals from the Orinoco are somewhat smaller than those from the Amazon; those from the upper Rio Madeira in Bolivia are considered by some to be a distinct species (*I. boliviensis*).
Similar species. Tucuxis (*Sotalia fluviatilis*) are smaller, with pointed, triangular dorsal fins; their backs rise well out of the water when they surface.
Sounds. Snorts or sighs when breathing; a variety of underwater calls can be heard when one is swimming near them.
Natural history. Diurnal and nocturnal; aquatic; often solitary, more rarely in groups of two to four. Feeds on fish, including many bottom-feeding species such as catfish, and sometimes other animals such as crabs and turtles. Boutus rarely leap above the water surface. They are curious and will approach swimmers but will not attack. They are easy to see almost anywhere along the main Amazon and in

Map 130

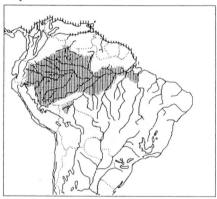

Boutu, *Inia geoffrensis*
Tucuxi, *Sotalia fluviatilis*

many tributaries, especially near the mouths of rivers and streams. Found in fresh water only.
Geographic range. South America: rivers of the Amazon and Orinoco drainages, below significant waterfalls except for an isolated population in the upper Madeira drainage in Beni, Bolivia.
Status. Often common; widespread. River dolphins are the object of many superstitions and myths, and they rarely are deliberately killed by local fishermen or hunters. However, where large-scale mechanized commercial fishing is carried out, or poisons or dynamite are intensively used, their food supply may be threatened and accidental deaths may occur.
Local names. Bufeo colorado, boto

(Span); boto vermelho (Br); uyara (Ec); tonina (Ve); uynia (TGua).

References. Best, R. C. 1984. The aquatic mammals and reptiles of the Amazon Basin. In H. Sioli, ed., *The Amazon,* 371–412. Dordrecht: W. Junk.

Tucuxi or Gray Dolphin
Sotalia fluviatilis
Figure 6, map 130
Identification. Measurements: Total length = 1.3–1.5 m; weight to 53 kg (for riverine animals).
Dorsal fin triangular and prominent. Head with quite short beak, hump on forehead small. Bicolored gray, brown, or bluish above, paler gray, whitish, or pinkish below. Flippers quite short, narrow at junction with body. **When surfacing, back and often top of head emerge high out of water.**
Variation. Coastal forms are sometimes considered a distinct species.
Similar species. See boutu *(Inia geoffrensis).*
Sounds. Snorts or puffs when surfacing to breathe; many underwater calls.

Natural history. Apparently diurnal; aquatic; usually in groups of two to nine, sometimes solitary. Feeds on fish, chiefly pelagic, schooling species in the Amazon, armored catfish in Surinam. These little dolphins are more acrobatic than the larger boutus and will more often leap clear of the water. They are easy to see in many areas, especially near the mouths of rivers and streams. Found in both fresh and salt water, in rivers below significant falls, and in estuaries and nearshore coastal waters.
Geographic range. South America: rivers draining into the Atlantic and Caribbean, and coastal waters to Panama.
Status. CITES Appendix I. Widespread and often common; status like boutu.
Local names. Bufeo negro, bufeo plomo, boto (Span); tucuxi (Br); profossoe (Su).
References. Magnusson, W. E., R. C. Best, and V. M. F. da Silva. 1980. Numbers and behaviour of Amazonian dolphins, *Inia geoffrensis* and *Sotalia fluviatilis fluviatilis,* in the Rio Solimões, Brasil. *Aquatic Mammals* 8:27–32.

a

b

Figure 6. Dolphins: (*a*) **boutu,** *Inia geoffrensis;* (*b*) **tucuxi,** *Sotalia fluviatilis.*

Tapirs (Perissodactyla)

Tapiridae

Dental formula: I3/3, C1/1, P4/3, M3/3 = 42. Forefeet with four toes; hindfeet with three. Most of the body weight is borne on the third toe, which is the largest and centrally located, but the toes on each side are also functional; the small, outer digit of the forefoot (actually the fifth digit) touches the ground only when the animal sinks into a soft substrate. There are hooves on all the toes. Tapirs are the only extant native New World odd-toed ungulates (Perissodactyla). They are the largest terrestrial mammals in the region. They have large, rotund bodies; thick, muscular necks; upper lip elongated into a proboscis; and short stumpy tails. Members of this order, which includes horses and rhinoceroses, are large herbivores. Mammals do not themselves possess the enzymes with which to digest the cellulose that makes up the structural material of plants. They have thus evolved a variety of systems to house and maintain microorganisms that can digest (''ferment'') the plant material for them. In the Perissodactyla the stomach is simple, and the cecum is enlarged to form a chamber in which microorganisms live and digest plant cellulose. This system is not extremely efficient, so these animals must eat large amounts of forage each day, and spend most of the day eating, in order to obtain enough energy from the leaves of plants they eat. They produce corresponding amounts of droppings, which are full of undigested, chopped fibers and seeds of fruits. The large teeth of tapirs are fitted for grinding up plants. The prehensile proboscis is used to reach out and sweep browse into the mouth. Tapirs have single, precocial young. There is one genus, with four species of tapirs worldwide—one in Asia and three in the New World.

Brazilian Tapir
Tapirus terrestris
Plate 18, map 131
Identification. Measurements: HB = 1,700–2,010; T = 46–100; HF = 330–350; E = 120–137; SH = 800–1,100; WT = 227–250 kg.
Body, head, and legs uniform blackish brown; hair short and smooth, skin gray, sometimes not completely covered with hair; short, stiff mane or crest of longer, dark hair in narrow strip from forehead to shoulder. **Upper lip elongated in a downcurved proboscis; forehead rises in distinct convex hump under base of mane in front of ears; ears brown, tips white,** base with small white spot; eyes small, eyeshine moderate. **Tail a short stump. Hindfeet with three toes, forefeet with three large toes** that appear in tracks **and a fourth, smaller toe** that appears as a small crescent in tracks on soft ground. Throat and sometimes chest grayish or white; belly brown, paler than back. Young with bright yellow or white longitudinal stripes alternating with lines of spots, legs and cheeks heavily spotted, belly white. **A robust animal the size of a small pony, with a cylindrical body, thick neck, and convex back highest above kidneys.**

Map 131

Brazilian tapir, *Tapirus terrestris*
Baird's tapir, *T. bairdii*

Variation. Color varies from tan to blackish or reddish; undersides more or less white. There seems to be much size variation, but few measurements or weights of wild adult specimens are available.
Similar species. Baird's tapirs *(T. bairdii)* are only found with Brazilian tapirs in N Colombia; they have a flat crown between the ears and a poorly developed crest; capybaras *(Hydrochaeris hydrochaeris)* are much smaller, with a blunt muzzle and no proboscis.

156

Sounds. Usually silent but may snort in alarm; communicates with a whistling call. (They answer imitations of this call by hunters, who then easily find and shoot them.)

Natural history. Mostly nocturnal, partly diurnal; terrestrial; solitary, but several use the same area. Feeds on terrestrial and swamp browse, grass, and fruits. Tapirs favor waterside habitats with some herbaceous vegetation, such as river edges, swamps, and lush stream bottoms, but they are all-terrain animals that travel widely over the forest, including terra firme far from water, and they negotiate almost vertical slippery hillsides. By day they rest in thick vegetation, especially in swamps. Their tracks are frequent where they occur, but they are shy, silent, and rarely seen. The dung is a large heap of fibers, seeds, and leaf fragments often (but not always) deposited in water. When strongly alarmed they run for the nearest water, plunge in, and swim beneath the surface; they can flatten everything in their path in a desperate rush. Tapirs use salt licks, often the same ones used by peccaries. Found in rainforest, gallery forest, and more open grassy habitats with water and thick vegetation for refuge.

Geographic range. South America: east of the Andes from N Colombia to S Brazil and N Argentina and Paraguay. To 1,700 m elevation.

Status. CITES Appendix II, US-ESA endangered. Locally common, but scarce or absent in overhunted regions; tapir meat is much prized, and tapirs are easy to locate with dogs or calls and thus vunerable to local extinction.

Local names. Tapir, anta (Span); anta, marebis (Br); danta, gran bestia (Co, Ec); sacha vaca, huagra (Pe); tapií, mborebí (Gua); boskoe, bosfroe (Su); maypouri (Qui).

References. Hershkovitz, P. 1954. Mammals of northern Colombia, preliminary report no. 7: Tapirs (genus *Tapirus*), with a systematic review of American species. *Proc. U.S. Nat. Mus.* 103:465–96.

Baird's Tapir
Tapirus bairdii
Plate 18, map 131

Identification. Measurements: HB = 1,930–2,000; T = 70–100; HF = 372–380; E = 130–140; WT = 150–300 kg.

Similar to Brazilian tapir, except: head with a **flat crown; mane** often **poorly developed** or absent; proboscis slightly longer; lower cheeks and chest usually more white or gray; body more often reddish. Small young chestnut-red with white stripes and spots, older young browner.

Similar species. See Brazilian tapir.

Sounds. When surprised may stamp foot loudly; whistles like Brazilian tapir.

Natural history. Nocturnal and diurnal; terrestrial; usually solitary, family groups seem to use a small area. The habits of Baird's tapirs seem similar to those of Brazilian tapirs. They spend almost 90% of their waking time feeding on browse, grass, and fruits. Sometimes they sleep in the water. Found in rainforest and montane forest, swamps, and flooded grasslands.

Geographic range. Central and South America: S Mexico south to Panama; west of the Andes from N Colombia to the Gulf of Guayaquil, Ecuador.

Status. CITES Appendix I, US-ESA endangered. Threatened by hunting and deforestation.

Local names. Tapir, anta, danta (Span); mountain cow (Be); macho de monte (CR) antiburro (Me); tzimin (May).

References. Terwilliger, V. J. 1978. Natural history of Baird's tapir on Barro Colorado Island, Panama Canal Zone. *Biotropica* 10:211–20.

Peccaries and Deer (Artiodactyla)

There are two families of Artiodactyla, or even-toed ungulates, in Neotropical rainforest—the peccaries (Tayassuidae) and deer (Cervidae). In Artiodactyla the body weight is borne evenly on the third and fourth toes, with the center of gravity falling between them. The hooves on the tips of these toes form the familiar "cloven hoof," but there are often one or a pair of smaller toes behind them that normally do not touch the ground.

Peccaries (Tayassuidae)

Dental formula: I2/3, C1/1, P3/3, M3/3 = 38. Forefeet with two large weight-bearing toes and two small rear toes that do not touch the ground except on soft substrates, hindfeet with two large toes and one small toe; all toes with hooves. Peccaries are large, chunky animals with slender legs, large heads, thick necks, and virtually no tails. They have large, sharp canines: the upper ones point downward (instead of out or up as in pigs); their snouts are piglike, with nostrils opening in the center of a mobile disk that is used for rooting. They have simple digestive tracts and are omnivores that feed on fruit, seeds, some browse, and animal matter. The peccaries' nearest relatives are the pigs, but they are in a separate family and differ from true pigs in a number of features. Unlike pigs (which have large litters of altricial young that are raised in a nest), peccaries have small litters of one or two precocial young that can walk and follow their mother shortly after birth. There are two genera and three species, all in the New World.

Collared Peccary
Tayassu tajacu
Plate 18, map 132
Identification. Measurements: HB = 800–980; T = 25–45; HF = 170–200; E = 70–90; SH = 300–500; WT = 17–30 kg.
Upperparts uniformly grizzled gray-black, faint but distinct collar or stripe of pale yellow hairs from top of shoulder forward to lower cheek; midback from head to rump with crest of long hairs, raised in excitement; large scent gland along the spine; hair sparse, coarse bristles, banded black and white or yellow. **Head large, sharply tapering from large jowls to narrow nose; nostrils in a naked, pink, mobile disk; canines large, form distinct lumps under lip,** do not protrude; eyes small, eyeshine weak, reddish; ears small. Tail tiny, not visible. Forefeet with two large toes and two smaller rear toes that do not touch the ground (do not appear in tracks); hindfeet with two large toes and one smaller toe. Young grizzled reddish brown. **Piglike, with a stout body, thick neck, and thin, delicate legs.**
Variation. Collared peccaries from dry, open habitats are small and gray; rainforest animals are large and blackish.

Map 132

Collared peccary, *Tayassu tajacu*

Similar species. White-lipped peccaries *(T. pecari)* are black with a white chin and no collar; capybaras *(Hydrochaeris hydrochaeris)* are tan or dark red-brown, with large, square muzzles, robust legs, and no markings.
Sounds. Usually quiet; if surprised at close range may give very loud doglike barks, "whoof," while in flight; sharp clacks of the teeth and grunts (perhaps threats) are heard rarely; chews nuts and snails with loud cracking.

Natural history. Diurnal in rainforest; terrestrial; groups of 1–20; usually 6–9; loose groups often change composition, males sometimes solitary. In rainforest feeds on fruit, palm nuts, browse, snails, and probably other small animals. Collared peccaries travel in single file along small paths and disperse in the undergrowth to feed. They release a strong odor like cheese or chicken soup, especially when alarmed, and they frequently mark trails by scraping the ground in front of a pole with their hooves, defecating in the scrape, and rubbing their back gland on the pole. They regularly use mud wallows or salt licks in favored, traditional spots. Collared peccaries are wary and surprisingly quiet and difficult to approach: they stampede in panic when they detect humans, and they are not dangerous (but those raised as pets may be very aggressive and can inflict serious wounds with their canines). In Amazonian rainforest, herds sleep at night in deep burrows, often under the roots of trees. These are the most commonly seen peccaries, and they are resident throughout the rainforest. Found in habitats from dry, shrubby, Sonoran desert and chaco to deep rainforest. Note that the behavior is different in hot, dry habitats, where peccaries are active at night and do not use burrows.

Geographic range. North, Central, and South America: SW United States to Argentina.

Status. CITES Appendix II. Widespread and locally common, but hunted intensively for meat, sport, and hides; rare or absent near many settlements but not generally threatened.

Local names. Sajino, saíno (Span); chancho de monte (Ar); peccari (Be); caitetu (Br); jabalí (Co); pakira (FG, Su); báquiro de collar, chácharo (Ve).

References. Robinson, J. G., and J. F. Eisenberg. 1985. Group size and foraging habits of the collared peccary *Tayassu tajacu*. *J. Mammal.* 66:153–55.

Castellanos, H. G. 1983. Aspectos de la organización social del báquiro de collar, *Tayassu tajacu* L., en el Estado Guarico—Venezuela. *Acta Biol. Venez.* 11:127–43.

Map 133

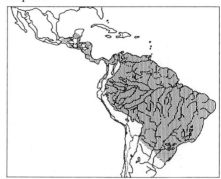

White-lipped peccary, *Tayassu pecari*

White-lipped Peccary
Tayassu pecari
Plate 18, map 133

Identification. Measurements: HB = 950–1,100; T = 28–56; HF = 210–230; E = 80–90; SH = 500–600; WT = 25–40 kg.

Upperparts uniform black, sometimes brownish; **hair long and coarse,** few or no bands on individual hairs. **Chin and area around corner of mouth and lower cheek white,** cheek below eye sometimes with pale band. Chest, lower legs, and belly sometimes white. Young grizzled gray, reddish, or dark tan with white throat and sometimes dark midback stripe. Similar in shape to collared peccary, but larger.

Variation. Usually black in rainforest, sometimes brownish or reddish in sunny habitats (e.g., captives); amount of white on jaw and body varies from faint sprinkling of white hairs to striking snow white.

Similar species. Collared peccaries *(T. tajacu)* are smaller, grizzled gray-black with pale collar; capybaras *(Hydrochaeris hydrochaeris)* are tan with a square muzzle.

Sounds. Large active herds produce a continuous racket of bellowing, screaming, and loud tooth-clacking, which can be heard for several hundred meters. Small groups seem to be quiet.

Natural history. Mostly diurnal; terrestrial; large herds of 50–300 or more. Feeds mostly on fruits, palm nuts, and browse. White-lipped peccaries travel long distances, periodically visiting each area for a few hours or a day or two and then moving on. Their presence in a given area is episodic and unpredictable. They forage with much rooting; when they have passed, the ground is pocked and churned. Wide bands of densely packed tracks on beaches mark their river crossings. They have a strong odor distinct from that of collared peccaries. Their reputation for aggressiveness has been greatly exaggerated and may be partly due to impressive appearance, numbers, and acoustic performance. A herd generally runs away from a detected human, but their vision seems poor; they will walk close past a stationary person without noticing. Much less commonly seen in rainforest than collared peccaries. Found with extremely patchy distribution in rainforest, dry forest, and chaco scrub; in Central America only in humid tropical forests, but at the southern end of its range occurs in dry scrub.

Geographic range. Central and South America: Mexico south to N Argentina.
Status. CITES Appendix II. Poorly known; probably threatened. Extensively hunted for meat and hides; many animals from a herd are often killed at a time. Extinct from hunting near many human settlements and naturally absent or extinct from many rainforest areas within its theoretical range. Anecdotes suggest it is most common on the northern and southern central edges of the Amazon forest (Rio Branco and Matto Grosso in Brazil and Chaco of Paraguay, where it outnumbers collared peccary). In Central America threatened by habitat destruction. In areas where they seem to be disappearing, small herds of fewer than 10 are now seen.

Local names. Wari (Be); queixada (Br); huangana (Ec, Pe); pingo (FG, Su); pecari labiado (Co, Ar); puerco de monte, chancho de monte (Pn, CR); báquiro (Ve).
References. Kiltie, R. A., and J. Terborgh. 1983. Observations on the behavior of rain forest peccaries in Peru: Why do white-lipped peccaries form herds? *Z. Tierpsychol.* 62:241–55.

Sowls, L. K. 1984. *The peccaries.* Tucson: University of Arizona Press.

Deer (Cervidae)

Dental formula: I0/3, C0 or 1/1, P3/3, M3/3 = 32–34. All feet with two large, weight-bearing toes and two small rear toes that do not touch the ground. These are large graceful animals with long slender legs, long necks, and short tails. Males have antlers that are shed and regrown each year; females have antlers only in caribou and reindeer. Deer are true ruminants with complex stomachs in which plant material is fermented. Before digestion they must take time to rest and regurgitate and finely ruminate (chew cuds) forage that has been stored temporarily in part of the stomach while they feed. Members of the family are mostly browsers and grazers that feed on leaves, twigs, and grasses, with occasional fruits, but the rainforest brockets are largely frugivorous. Brocket deer tend not to use trails but to wander throughout the forest, so that a fusiform, forest-adapted shape for pushing through thickets is strongly developed. Their droppings consist of small, smooth, oval pellets deposited in groups. Deer give birth to one or two well-developed young that for the first weeks of life lie in hiding, well camouflaged by their spotted coats, while the mother feeds elsewhere. Fawns found lying alone have not been abandoned; they should be left undisturbed. Little is known of the behavior of rainforest species in the wild. There are about 36 species of deer in 16 genera worldwide; with 2 genera and 3 species in the Neotropical rainforest region. Several other genera and species occupy montane, scrub, and grassland habitats around the fringes of the rainforest.

Red Brocket Deer
Mazama americana
Plate 18, map 134
Identification. Measurements: HB = 1,050–1,440; T = 120–150; HF = 313–340; E = 84–106; SH = 670–760; WT = 24–48 kg.

Body and legs chestnut-red; head and neck gray-brown. Face without prominent markings; forehead with tuft of long, dark-tipped hairs; ear thinly haired, rim at inner base with thin fringe of white hair, inner surface naked, pinkish; eyes large, eyeshine brilliant yellow-white; **antlers on**

Map 134

Red brocket deer, *Mazama americana*

males only, short, straight, unbranched, directed backward. Legs long and slender, as dark as midback or darker. Tail red above, white below; raised vertically in alarm to show undersurface. Throat at angle of jaw whitish; **belly chestnut, slightly paler but not demarcated from sides;** hind thighs white in inner surface under tail. **Rump large, higher than shoulders, back slightly humped in profile, while walking head carried low, level with back.** Young red with prominent or faint white spots, belly reddish.
Variation. Animals from Central America are deep, rich red with dark brown or black faces; some from northern Colombia are grayish.
Similar species. Gray brockets *(M. gouazoubira)* are smaller, gray-brown with white bellies; white-tailed deer *(Odocoileus virginianus)* are red or gray-brown with white bellies, prominent facial markings, branched antlers, and a straight back.
Sounds. Alarm a soft to very loud explosive, whistling snort.
Natural history. Diurnal and nocturnal; terrestrial; solitary. Feeds on fruits, fungi, browse, and fallen flowers; browse is mainly eaten when fruits are scarce in the dry season. These deer favor the dense vegetation with abundant herbaceous understory of *platanillos,* swampy areas, riversides, and old plantations, but they forage throughout the forest. They are adapted for forest life; with their low forequarters and simple antlers they can slip easily through dense vegetation. When inactive they lie down in a fairly sheltered spot.

Found in mature and secondary rainforest, gallery forest, forest edges, gardens and plantations, and savannas near the forest.
Geographic range. Central and South America: S Mexico south to N Argentina. To 2,000 m elevation.
Status. CITES Appendix III (Gu). Widespread and often common. Much hunted for meat and scarce in some areas, but seems to persist where other large mammals are exterminated, probably because of solitary habits and liking for dense vegetation.
Local names. Venado colorado (or rojo; Span); corzuela roja (Ar); antelope (Be); veado pardo, veado mateiro (Br); cabro de monte (CR); biche (FG); güitsizil (Gu); tilopo, antilope (Ho); corzo (Pn); temazate (Me); locho (Ve); guazú-pytá (Gua); grootboshert, prasara-dia, redi-dia (Su); chan yuc (May).
References. Branan, W. V., M. C. M. Werkhoven, and R. L. Marchinton. 1985. Food habits of white-tailed deer and red brocket deer in Suriname. *J. Wildl. Mgmt.* 49:972–76.

Gray or Brown Brocket Deer
Mazama gouazoubira
Plate 18, map 135
Identification. Measurements: HB = 870–1,250; T = 100–150; HF = 244–293; E = 100; SH = 350–610; WT = 11–18 kg.
Head and body gray-brown, darkest at midback, paling to sides. Face without prominent markings; forehead with tuft of dark hairs; ears thinly haired, rim at inner base with small tuft of white hairs; eyes large, eyeshine brilliant yellow-white. Throat whitish; neck gray, paler below; **belly whitish.** Legs dark as center of back. Tail white beneath, raised vertically in alarm. Young brown with white spots, white belly. Body shape and antlers like red brocket, but more slender and gracile.
Similar species. Red brockets *(M. americana)* are larger, red with red bellies; white-tailed deer *(Odocoileus virginianus)* are much larger, with facial markings, branched antlers, and a straight back.
Sounds. Snorts in alarm.
Natural history. Mostly diurnal; terrestrial; solitary. Feeds on browse and fallen fruits and flowers. Gray brockets seem to favor

Map 135

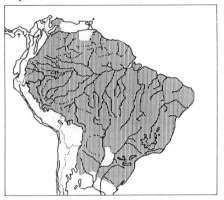

Map 136

Gray brocket deer, *Mazama gouazoubira*

White-tailed deer, *Odocoileus virginianus*

dense vegetation such as streamsides in rainforest, where they always seem rarer than red brockets. At night they lie down in sheltered spots such as thickets. Found in rainforest and more open, drier habitats such as chaco. They are the more common brocket in dry scrub of Paraguay and the only brocket in Uruguay.

Geographic range. Central and South America: in Central America known only from San José Island, Panama; east of the Andes from Colombia and Venezuela south to Uruguay and N Argentina.

Status. Uncommon but widespread in rainforest.

Local names. Venado plomo, venado cenizo (Span); corzuela parda (Ar); veado-birá (Br); matacán (Ve); guazú-birá (Gua); koeriakoe (Su).

References. Stallings, J. R. 1984. Notes on the feeding habits of *Mazama gouazoubira* in the Chaco Boreal of Paraguay. *Biotropica* 16:155–57.

White-tailed Deer
Odocoileus virginianus
Plate 18, map 136
Identification. Measurements: HB = 1,130–2,260; T = 130–140; HF = 350–380; E = 110–130; WT of males about 50 kg, females 30 kg.

Upperparts gray-brown to light brown with a reddish tinge grading to chestnut on rear of rump, or all chestnut. **Head gray-brown with distinct markings: tip of chin white, nose and muzzle dark brown bordered by a pale band from the corner of the mouth upward; dark spot below corner of mouth; eye surrounded by pale ring especially prominent above;** eyeshine bright yellow-white; ear lined with thick white hair on inner edges. **Antlers branched in mature males, simple spikes in yearling males,** none in females. **Tail white below, long white hairs protruding as a white bordering fringe,** raised and fanned in alarm to show white underside. Belly white. Legs slightly paler than back. Young red with white spots, belly white. **Large deer with straight, horizontal back when head is raised; head carried higher than back when walking.**

Variation. Color varies from gray-brown "winter" coat to red, fine-haired "summer" coat, depending on both habitat and season.

Similar species. For lowland rainforest only: red brockets *(Mazama americana)* are red, with no facial markings, spike antlers, and red belly; gray brockets *(M. gouazoubira)* are much smaller, with no facial markings and spike antlers.

Sounds. A whistling bleat in alarm.

Natural history. Diurnal and nocturnal; terrestrial; small groups or solitary. Feeds on browse and grass, with some fruit and fallen flowers. White-tailed deer are not rainforest animals. Their branched antlers and high head-and-shoulder carriage are ill suited to running through dense, viny forest. Whitetails are found in open and secondary habitats bordering the rainforest, old secondary forest, montane habitats in upper elevations and cleared midelevations of the Andes, and open woodlands and fields, from which they may sometimes enter rainforest habitats in Central America.

Geographic range. North, Central, and South America: S Canada to South America, Andes from Venezuela and Colombia south to N Bolivia; west of the Andes in appropriate lowland habitats from Panama to S Peru, and east of the Andes along the northern coast and in savannas from Colombia to French Guiana.

Status. CITES Appendix III (Gu). Widespread and often common. An important game animal for meat and sport; some local populations in South and Central America are at risk from overhunting.

Local names. Venado (Span); deer (Be); venado llanero, venado de cornamenta (Co); veado galeiro, cariacu (Br); cariacú (Su, FG, Gy); lluichu (Qui); venado cola blanca (Cent Am).

References. Halls, L. K., ed. 1984. *White-tailed deer.* Harrisburg, Pa.: Stackpole Press.

Manatees (Sirenia)

Trichechidae

Dental formula: 6/6 molarlike teeth usually present at one time; vestigial incisors and deciduous premolars are lost before adulthood. There are an indefinite number of molars that erupt in sequence throughout the life of the animal; as the anterior teeth are worn down, they fall out and are replaced by a new tooth at the back. Manatees are very large, hairless, cigar-shaped mammals, with forelimbs modified to flippers, no free hindlimbs, and the rear of the body in the form of a horizontal paddle. They are entirely aquatic and never leave the water. They are gentle herbivores that browse on aquatic vegetation. They give birth to a single young which accompanies its mother. There are two families, two genera, and four species worldwide, with one genus and two species in the New World.

Amazonian Manatee
Trichechus inunguis
Figure 7, map 137
Identification. Measurements: length to 2.8 m; WT = 350–500 kg.
Body large, cylindrical; hindend a single, flat, rounded, horizontal paddle; color gray. **Head small, upper lip modified into a large bristly surface; eye tiny; no external ear. Forelimbs short, rounded flippers with no nails on tips.** Chest and **abdomen with large, irregular white patch.**
Similar species. West Indian manatees *(T. manatus)* have nails on the flippers and no white patch on underparts; swimming otters *(Lutra, Pteronura* spp.), tapirs *(Tapirius* spp.), and capybaras *(Hydrochaeris hydrochaeris)* eventually surface and swim with the top of head and eyes out of water; dolphins *(Inia, Sotalia* spp.) roll forward and surface top of head and part of back to breathe.
Sounds. None usually heard; may snort or puff when coming up to breathe; makes underwater calls.
Natural history. Nocturnal and diurnal;

Figure 7. Manatees: *(a)* **Amazonian manatee,** *Trichechis inunguis; (b)* **West Indian manatee,** *Trichechis manatus.*

Map 137

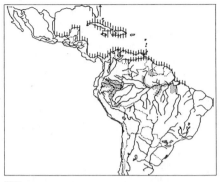

▨ Amazonian manatee, *Trichechis inunguis*
▥ West Indian manatee, *T. manatus*

aquatic; solitary or mothers with young. Feeds on aquatic vegetation such as grasses, water hyacinths, and water lettuce *(Pistia)*. Manatees live entirely under the water; only their nostrils break the surface when they rise to breathe. They favor areas of dense aquatic vegetation and are therefore extremely difficult to see. Their presence below is betrayed by their droppings, which are balls of macerated fibers, like horse droppings, that float to the surface. Manatees produce prodigious amounts of these, since they can eat up to 8% of their body weight in vegetation per day. Amazonian manatees do most of their feeding during the wet season, when they eat new vegetation in seasonally flooded backwaters. When the rivers shrink in the dry season, the manatees return to the main channels, where they may fast for weeks for lack of available food plants. Found in rivers and their associated lakes below any major rapids.

Geographic range. South America: the Amazon and lower reaches of its tributaries from Ecuador and N Peru to the estuaries at its mouth; and isolated populations on the Rupununi and Essiquibo rivers (Guyana).

Status. CITES Appendix I, US-ESA endangered. These huge animals once teemed in Amazonia, but they were ruthlessly hunted almost to extinction for their meat, oil, and hides. Few populations are known to remain; despite legal protection, they are still killed for meat in the remote areas where they survive.

Local names. Peixe-boi (Br); vaca marina, manatí (Span).

References. Best, R. C. 1984. The aquatic mammals and reptiles of the Amazon. In H. Sioli, ed., *The Amazon,* 371–412. Dordrecht: W. Junk.

Timm, R. M., and L. Albuja. 1986. Ecology, distribution, harvest, and conservation of the Amazonian manatee *Trichechus inunguis* in Ecuador. *Biotropica* 18:150–56.

West Indian Manatee
Trichechus manatus
Figure 7, map 137

Identification. Measurements: length = 2.5–4.5 m; WT = 200–600 kg.

Similar to Amazonian manatee except larger; flippers with large, flat nails on their tips; underparts gray or with pink blotches, without white patches.

Similar species. See Amazonian manatee.

Sounds. Underwater squeals, whines, and grunts; puffs when breathing.

Natural history. Diurnal and nocturnal; aquatic; solitary or females with young, but temporary congregations form in favored spots and during the mating season. West Indian manatees feed on both bottom and floating aquatic vegetation. Like Amazonian manatees, they spend their lives almost completely submerged. They surface only the nostrils to breathe two to four times in rapid succession about every four minutes. Found in both fresh and salt water.

Geographic range. North, Central, and South America and Caribbean islands: the coasts of Georgia and Florida, United States, Mexico and Central America; the north coast of South America from Colombia to the mouth of the Amazon in Brazil; and the drainages of the Rios Cauca and Magdalena in Colombia and the Orinoco in Venezuela.

Status. CITES Appendix I, US-ESA endangered. Rare or extinct over much of its former range, it has been overhunted for meat in South America and the Caribbean; in the United States many animals are killed or maimed by boat propellers.

Local names. Vaca marina (Span); peixeboi (Br); sea cow (Be); sekoe (Su).

References. Husar, S. L. 1978. *Trichechus manatus*. Mammalian Species, no. 93.

Rodents (Rodentia)

The rodents are most easily characterized by their teeth: they have a single large pair of chisellike, ever-growing incisors in the front of each jaw, no other incisors or canines, and three to five cheekteeth (or rarely, fewer) on each side in the rear of the mouth, separated from the incisors by a large gap (diastema). The incisors have hard enamel only on their front surfaces; the posterior part is softer dentine that wears down more quickly. This arrangement ensures that as the teeth are used, they continually keep a sharp, blade-like edge at the front surface. These incisors are remarkably versatile tools: they can be used to cut, pry, slice, gouge, dig, stab, or delicately hold like a pair of tweezers; they cut grass, open nuts, kill animal prey, dig tunnels, and fell large trees. To do these things, rodents have evolved several different complex systems of muscles of the jaws and structures of the skull that support these muscles. These differences are used to classify most of the rodents into three major groups (squirrellike, Sciuromorpha; mouselike, Myomorpha; cavylike, Caviomorpha). Most of the world's rodents are small (< 1 kg), ratlike animals. This body type is extremely versatile, and slight modifications allow rats to be terrestrial, arboreal, semiaquatic, or fossorial. A number of large species, chiefly those in the group of cavylike rodents, are not at all ratlike. The wide scope of activities and diets made possible by the rodent anatomy has made these by far the most diverse order of living mammals: there are about 1,750 species of rodents worldwide, or close to as many as all other species of mammals combined (roughly 2,300), and almost twice as many as the next largest order (bats).

The tails of ratlike rodents are covered with skin that is formed into patterns of tiny scales. The scales may be organized into even, prominent rings around the tail or staggered into diagonal rows so that no rings are evident. Hairs grow backward in a fixed pattern around each scale, either from beneath the rear edge of scales or from between them. A magnifying lens is needed to see the pattern. All rodent tails have hairs on them; when a tail is called "naked" it means that the hairs are short and thinly scattered and do not hide the scales beneath. "Hairy" tails may have few, but long, hairs that curl outward and make the tail look bristly or hairy, but with the scales still partly visible beneath, or many short flat hairs that cover the scales, or they may have a complete covering of long, dense hair. Hairy tails often have hairs that extend beyond the tip of the flesh in a tuft or "pencil" that may be either barely visible or large and brushlike. Because there is a continuous gradient of amount of hair on the tail, differences in the degree of hairiness are often hard to describe, yet they are one of the most easily seen external differences between some genera and species. Arboreal and semiaquatic species tend to have hairier tails than terrestrial species.

Squirrels (Sciuridae)

Dental formula: I1/1, C0/0, P1 or 2/1, M3/3 = 20–22. Forefeet with four long toes with claws, and thumb a short stump with a small nail; hindfeet with five long toes with claws. The squirrels of Neotropical rainforest are all tree squirrels; they have broad heads, large eyes, short ears, soft fur, long bushy tails, and quite long legs. They have flexible ankle-joints that can rotate to allow them to descend trees headfirst, with the hindfeet flat against the trunk. Most species have strong, chisellike incisors and large jaw muscles, which enable them to gnaw open the hardest nuts. They all are excellent climbers, but some species do much of their foraging on the ground. Squirrels are generally omnivorous and feed on nuts, soft fruits, insects, fungi, and sometimes leaves, flowers, and bark. However, many Neotropical rainforest species feed chiefly on the hard-shelled nuts of palms and a few other trees with large, hard nuts. Squirrels can often be located by the sounds of gnawing. Those for which nesting has been described either build a round ball of leaves and twigs in a vine tangle or on a branch or make a nest of leaves in a tree hole. Litter size is about two for those few Neotropical species for which there is information. There have been few studies of Neotropical squirrels; the natural history of most species is poorly known, and the systematics is unclear in several cases. Squirrels of the genus *Sciurus,* especially Central American species, tend to be geographically highly variable in color and color pattern, and many species have a tendency to produce melanistic or black-pigmented populations or individuals. There are about 65 genera and 377 species of squirrels worldwide, with 4 genera and 18 species in the rainforest region.

Northern Amazon Red Squirrel
Sciurus igniventris
Plate 19, map 138

Map 138

Identification. Measurements: HB = 240–295; T = 240–305; HF = 55–75; E = 30–40; WT = 500–900 g.
Back grizzled black and yellowish or orange; hindlegs and forelegs above pure chestnut, rust-red, or orange; or black. Crown usually blackish; cheeks orange or chestnut; ears large, protruding well above crown, thinly haired; sometimes yellow behind. **Tail usually longer than head and body, very bushy,** exceeds diameter of body, **base black or dark red-brown, orange to yellowish orange distally. Feet pure red or orange,** not mixed with black hairs except in melanistic individuals. **Underparts** thinly haired, sharply demarcated from sides, **contrasting pure pale orange, rust-red, or white,** or black in melanistic individuals. Mammae pairs = 4. A large, long-legged red or black squirrel, with very bushy tail which is often curled up against the back when the squirrel is at rest.
Variation. Animals from north of the Amazon in Venezuela and Brazil are often melanistic, with varying amounts of blackness; melanism is rare in Ecuador and Peru; animals from Ecuador often have a black lateral line separating sides from belly.
Similar species. The northern and southern (*S. spadiceus*) Amazon red squirrels are almost indistinguishable in the field. The northern species is coarsely grizzled yellowish black on the back; the southern appears red, more finely grizzled with black, and has black hairs mixed with red on the feet (this character visible only with the animal in the hand); Junín red squirrels are uniform chestnut red; all other squirrels in range are small and brown.
Sounds. Alarm call short, low-frequency chatters and chucks; rarely given. Gnaws loudly on palm nuts.
Natural history. Diurnal; arboreal and terrestrial; solitary, but several may feed together at a fruit tree. Feeds chiefly on large palm nuts (*Astrocaryum* spp., *Scheelea*

Northern Amazon red squirrel, *Sciurus igniventris*
Guayaquil squirrel, *S. stramineus*
Junín red squirrel, *S. pyrrhinus* (arrow)

spp.) and other nuts and fruits. These squirrels usually travel in the forest understory or on the ground and can be seen feeding in or near palm trees. They are usually wary and run quickly away, keeping on the ground or low in the undergrowth. They are easiest to locate and approach while they are gnawing on nuts, which they do on an elevated perch. Found in mature and disturbed rainforest.
Geographic range. South America: the Amazon Basin lowlands north of the Amazon from the Rio Negro west through Brazil, Venezuela, Colombia, and Ecuador, and south of the river to N Peru and to the Rio Juruá, Brazil; in S Peru found only in the Andes in upper tropical forest. To 3,300 m elevation.
Status. Common and widespread, sometimes hunted for meat.
Local names. Esquilo, quatipuru (Br); ardilla (Span); huayuashi (Ec).
References. Patton, J. L. 1984. Systematic status of the large squirrels (subgenus *Urosciurus*) of the western Amazon Basin. *Studies on Neotropical Fauna and Environment* 19:53–72.

Junín Red Squirrel

Sciurus pyrrhinus

Plate 19, map 138

Identification. Measurements: HB = 240–280; T = 208–210; HF = 59; E = 21.
Upperparts entirely uniform dark red, often with small white dots (from scars). **Tail dark chestnut-brown at base, orange distally. Underparts sharply contrasting pure rusty orange, white, or orange with white patches medially.** Feet like back.
Variation. The systematic status of this form is unclear; it seems closely related to, and may be a subspecies of, the northern Amazon red squirrel; specimens of the two appear often confused.
Similar species. Northern (*S. igniventris*) and southern (*S. spadiceus*) Amazon red squirrels are larger, have tails longer than head and body, and back and top of head grizzled with black.
Natural history. Found in montane forests at 600–2,500 M.
Geographic range. South America: Peru, on the eastern Andean slope in Junín, perhaps to San Martín, but range limits unclear because of confusion with Amazon red squirrels.
Status. Unknown.
Local names. Ardilla.

Southern Amazon Red Squirrel

Sciurus spadiceus

Plate 19, map 139

Identification. Measurements: HB = 240–290; T = 235–300; HF = 59–70; E = 30–37; WT = 600–650 g.
Upperparts dark chestnut-red or rusty orange mixed with black on head, neck, and shoulders, becoming pure red on hindquarters; or animal black or blackish in melanistic individuals. **Crown often black, cheeks orange;** ears thinly haired, long, protruding well above crown. **Tail very bushy, exceeds diameter of body, black or blackish at base, orange or rusty distally. Feet red with mixture of black hairs. Underparts thinly haired, sharply contrasting pure pale orange, white, or yellowish.** Mammae pairs = 4. **A large, long-legged squirrel,** often holds the tail up over the back.
Variation. Blackish or sometimes pure black individuals are common in some regions of Brazil; they are rare in Peru.

Map 139

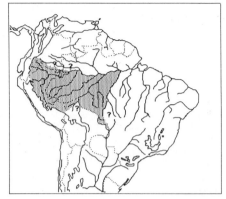

░░ Southern Amazon red squirrel, *Sciurus spadiceus*

Similar species. See northern Amazon red squirrel (*S. spadiceus*).
Sounds. Rarely calls; alarm a snort or sneeze followed by chatter of soft chucks; sneezing calls during mating bouts; gnaws loudly on nuts.
Natural history. Diurnal; terrestrial and arboreal; solitary, but several may feed at the same tree. Feeds chiefly on large hard nuts of palms and other trees, and other fruits. These squirrels are often on the ground and in low undergrowth, or in palm trees; they do not travel in the forest canopy. Loud rapid gnawing often betrays their location. When alarmed they run away in the undergrowth, sometimes with long leaps between vertical understory trunks. They bury nuts, and several individuals will quickly strip hundreds of nuts from a favored palm. They build a nest of leaves in a tree hollow. During a mating bout, many males will pursue the female. Mature and disturbed evergreen rainforest habitat.
Geographic range. South America: the Amazon Basin of Ecuador, Peru, Bolivia, and Brazil south of the Amazon and west of the Rio Tapajós.
Status. Common and widespread, sometimes hunted for meat.
Local names. Ardilla (Span); esquilo, quatipuru (Br); huayuashi (Ec).
References. See northern Amazon red squirrel.

Map 140

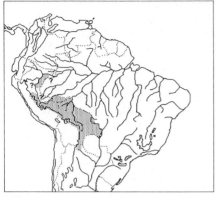

▨ Bolivian squirrel, *Sciurus ignitus*

Bolivian Squirrel
Sciurus ignitus
Plate 20, map 140
Identification. Measurements: HB = 180–195; T = 183–195; HF = 41–52; E = 20–26; WT = 225–240 g.
Upperparts uniform brown-olivaceous, finely grizzled. Ears protruding above crown of head, small buff patch behind ear near base; eye surrounded by indistinct pale ring. Tail quite slender, same color as back, hairs slightly tipped with yellowish. **Feet the same color as back.** Chin, throat, and chest buff to white washed with pale buff; pale color of chest usually increasingly mixed with gray posteriorly to **grayish inner thighs and inguinal region not sharply contrasting with sides.** Underparts well furred. Mammae pairs = 3.
Variation. Belly and inner thighs may be whitish. The small brown squirrels of the genus *Sciurus* of Peru and Bolivia are confusing and in need of taxonomic revision; they are often misidentified in museum collections.
Similar species. Dwarf squirrels (*Microsciurus* spp.) have short ears that do not protrude above crown; Sanborn's squirrels (*S. sanborni*) are smaller, have feet paler than back, sharply contrasting pale underparts, including inner thighs, and bright buff eye ring; Guianan squirrels (*S. aestuans*) are smaller and slightly darker, probably indistinguishable in the field.
Sounds. Soft chucks in alarm.
Natural history. Diurnal; arboreal and ter-

restrial; usually solitary. Feeds on nuts, fruits, mushrooms, and insects. These squirrels are most common in dense viny vegetation such as along riversides, where they use all levels from the ground to the midstory. They are not usually seen high in the canopy. They make a round nest of green leaves and twigs hidden about 6–10 m up in the top of a palm or dense vine tangle. Mature and disturbed lowland and montane rainforest.
Geographic range. South America: Amazon Basin of Peru, Bolivia, W Brazil, and N Argentina. To 2,700 m elevation.
Status. Locally common; too small to hunt.
Local names. Ardilla (Span); ardilla roja (Ar).

Guianan Squirrel
Sciurus aestuans
Plate 20, map 141
Identification. Measurements: HB = 160–202; T = 135–200; HF = 42–50; E = 16–25; WT = 159–218 g.
Upperparts uniform brown-olivaceous, finely grizzled. Ear thinly haired, protruding above crown, with or without pale buff patch behind; eye ring pale buff. **Tail frosted whitish, yellowish, or reddish,** and black. **Underparts orange, brightest on chest, mixed with gray posteriorly, or throat and inguinal region pale gray or white, or underparts gray.** Mammae pairs = 4.
Variation. There are several forms of this squirrel that are sometimes recognized as separate species: *S. a. aestuans,* mainly from north of the Amazon, has pale ear patches and a reddish tinge to body and feet; *S. a. gilvigularis,* mainly from south of the Amazon and Venezuela, has no ear patches; *S. a. alphonsei,* from coastal Pará south to Pernambuco, has a paler, more yellowish back and gray underparts; *S. a. ingrami,* from Bahía to Rio Grande do Sul, is olivaceous with whitish or whitish and buff underparts.
Similar species. These are the only squirrels in much of their geographic range; pygmy squirrels (*Sciurillus pusillus*) are much smaller, pale gray, with prominent white patches behind ears; see Bolivian squirrel (*Sciurus ignitus*).
Sounds. Alarm calls sharp chucks or chat-

Map 141

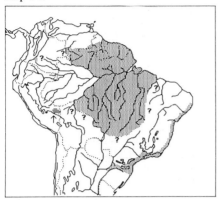

▓ Guianan squirrel, *Sciurus aestuans*
▆ Sanborn's squirrel, *S. sanborni*

ter, soft single chips, and high-pitched, twanging whines.

Natural history. Diurnal; arboreal and terrestrial; solitary, sometimes pairs. Feeds on small fruits and nuts and something it scrapes from tree bark. Guianan squirrels use all levels of the forest but are most often seen in the low to middle levels. When alarmed, they run up a tree from the ground with a chatter and continue to chatter or give sharp chucks from high in a tree as they hide or move off through the trees. They are most common around treefalls and viny vegetation but also use tall, open forest. They are active and agile and sometimes travel through the understory with long leaps from one vertical trunk to another. Mature and disturbed rainforest, secondary forest, gardens and plantations.

Geographic range. South America: north of the Amazon and east of the Rio Negro in Venezuela, the Guianas and Brazil, south of the Amazon west to at least the Rio Madeira, possibly to Bolivia, and Atlantic coastal rainforest to Rio Grande do Sul.

Status. Widespread and locally common, not hunted.

Local names. Caixé, serelepe, quatipuru, caxinguele (Br); eekhoorn (Su); bonboni (Sar).

Sanborn's Squirrel
Sciurus sanborni
Plate 20, map 141

Identification. Measurements HB = 152–175; T = 161–184; HF = 44–50; E = 20–21.

Upperparts brown-olivaceous. Eye ring and area around mouth bright pale buff; ears thinly haired, protruding above crown, bright white or buff patches behind. Feet yellowish, sometimes distinctly paler than back. Underparts including inner thighs entirely pure white or yellowish orange, sharply contrasting with sides.

Variation. An animal from the lowlands in Peru has buff underparts and ear patch; one from between the Rios Inambari and Tambopata has those parts white. These squirrels seem closely related to Guianan squirrels and may be a subspecies.

Similar species. Bolivian squirrels (*S. ignitus*) are larger, with gray inguinal region and three pairs of mammae.

Natural history. From lowland rainforest.

Geographic range. South America: Peru, Madre de Dios Department between the mouth of the Río Manu and The Río Tambopata, 300–580 m elevation. Probably also in NW Bolivia.

Status. Unknown; apparently rare and with a small geographic range; but some specimens of *S. "aestuans"* from Bolivia may belong to this species.

Red-tailed Squirrel
Sciurus granatensis
Plate 19, map 142

Identification. Measurements; HB = 200–285; T = 140–280; HF = 40–65; E = 16–36; WT = 212–520 g.

This squirrel has three basic color patterns; there are individuals and populations with intermediate colors between any of the following three. (1) Upperparts and tail entirely bright orange-red; underparts pure white; (2) upperparts brown-olivaceous; tail olive at base, heavily frosted rust-red or orange distally; underparts pure dark red to orange; (3) back and head blackish olivaceous, top of head and midback sometimes black; legs and feet red to orange; tail dark at base, orange for most of length, pitch black at tip; underparts dark red to bright rusty orange. Generally a

Map 142

▦ Red-tailed squirrel, *Sciurus granatensis*
▤ Neotropical pygmy squirrel, *Sciurillus pusillus*

medium-sized squirrel with large ears protruding above crown; a red or brown body; bushy, red and black, or red-tinged tail. Mammae pairs = 3.
Variation. Coloration (1) is found in N Colombia east of the Andes; (2) in Venezuela and Central America; and (3) on the west coast of Colombia and Ecuador. There are many intermediate colors; squirrels isolated on different sides of a river or mountain range may be different; at higher elevations of the eastern Andes of Colombia, these squirrels are completely olivaceous, with only the faintest reddish tinge to the tail; those on Trinidad are brown-olivaceous, with red-tinged tail; those on Tobago are grizzled brown-yellow-olivaceous, with pale rusty tail with a black tip. Squirrels from coastal Ecuador and Nariño, Colombia, may be speckled with many white spots due to scars.
Similar species. Dwarf squirrels (*Microsciurus* spp.) are smaller, with short ears, and no red on tail; variegated squirrels (*Sciurus variegatoides*) are larger, with black tails heavily frosted white; Deppe's squirrels (*S. deppei*) have blackish tails frosted with white and grayish bodies.
Sounds. Calls with short chucks; gnaws loudly on hard nuts.
Natural history. Diurnal; arboreal; solitary. Feeds chiefly on the large, hard nuts of palms and other trees, other fruits, and fungi. Red-tailed squirrels use all levels of the forest, including the ground, but most often the middle levels. They cache nuts by burying them and park fruits to store

them briefly in a branch fork or on a vine. They nest in tree holes or leaf nests in vine tangles but may not always sleep in a nest. During a mating bout several males follow the estrous female. Females defend small territories against other females, but males are nonterritorial and have overlapping home ranges. Found in mature and secondary evergreen and seasonally deciduous forests.
Geographic range. Central and South America: Costa Rica south through Panama; west of the Andes to N Ecuador; in the inter-Andean valleys of N Colombia and NW Venezuela; Trinidad and Tobago. To 2,500 m elevation.
Status. Often common. Widespread and adapts to disturbed vegetation.
Local names. Ardilla chisa, ardilla roja (CR); ardilla común (Co) ardilla colorada (Pn).
References. Nitikman, L. Z. 1985. *Sciurus granatensis.* Mammalian Species, no. 245.
Glanz, W. E., R. W. Thorington, Jr., J. Giacalone-Madden, and L. R. Heaney. 1982. Seasonal food use and demographic trends in *Sciurus granatensis.* In E. G. Leigh, Jr., A. S. Rand, and D. M. Windsor, eds., *The ecology of a tropical forest*, 239–52. Washington, D.C.: Smithsonian Institution Press.

Guayaquil Squirrel
Sciurus stramineus
Plate 19, map 138
Identification. Measurements: HB = 180–320; T = 250–330; HF = 50–65; E = 28–39.
Color variable: (1) Ecuador lowlands: **shoulders coarsely grizzled dirty white and blackish; rump and tail base faint or strong dull orange grizzled black; head and ears black; feet black or white; tail black frosted with white;** underparts dull brown or reddish. (2) Peru, S Ecuador highlands: **upperparts and tail** black heavily frosted with white, **appears pale gray; rump washed with faint or bright orange or buff; neck behind ears pure white or pale yellow; ears and feet pitch black;** underparts gray. Mammae pairs = 4.
A large squirrel with a long, slender, gray tail, and long black ears.
Variation. There is much individual variation in color within populations; the body

is often spotted with white scars.
Similar species. These are the only squirrels in their range.
Natural history. Diurnal; arboreal. Guayaquil squirrels are seen in the trees in both mature and secondary forest and coffee plantations. Found in humid, dry, and montane forests; in the southern part of its range occupies humid montane forest along the western Andean slope at 1400–2,000 m; in the northern part of its range it occurs both in highlands and humid and dry forests at sea level.
Geographic range. South America: SW Ecuador and NW Peru, around the Gulf of Guayaquil and south along the Andean slope to Cajamarca. To 2,000 m elevation.
Status. Locally common.
Local names. Ardilla parda, ardilla nucha blanca, ardilla mora, ardilla negra (Pe).

Variegated Squirrel
Sciurus variegatoides
Plate 21, map 143
Identification. Measurements: HB = 220–337; T = 226–325; HF = 45–70; E = 20–35; WT = 428–909 g.
Color highly variable: **upperparts either unpatterned, coarsely grizzled black and gray, yellowish, tawny, or rufous; or patterned, with sides and feet grizzled pale gray or yellowish, center of back from neck to tail a sharply demarcated band of pure dark brown or jet black.** There are pure brown and brownish black populations and individuals. **Ear with prominent pale patch behind,** white in pale animals, tawny to orange in dark animals. **Tail long and bushy,** above **black heavily frosted white,** below with median strip of yellowish to reddish. Feet and limbs in patterned individuals like back or contrasting red; feet white, buff, blackish, or orange. **Underparts pure white, pure rust red, or red with white patches** on throat and chest and lower belly.
Variation. The 14 named subspecies attest to the great color variation in this species. There is also much variation between individuals within some populations: squirrels from the wetter eastern coasts are generally blackish; those from drier areas are paler gray; those with bright red legs are found in W Costa Rica.
Similar species. Red-tailed squirrels

Map 143

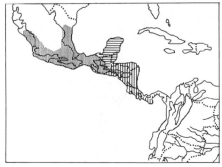

▓ Red-bellied squirrel, *Sciurus aureogaster*
▥ Variegated squirrel, *S. variegatoides*
▤ Yucatan squirrel, *S. yucatanensis*

(*S. granatensis*) are smaller, with orange tails; Deppe's squirrels (*S. deppei*) are smaller, with slender tail and gray belly and feet; Yucatan squirrels (*S. yucatanensis*) are gray, with a gray belly; Mexican red-bellied squirrels (*S. aureogaster*) are more finely grizzled gray, without prominent pale ear patches; the last two species may not occur together with variegated squirrels.
Sounds. Chucks in alarm.
Natural history. Diurnal; arboreal; solitary. Feeds mainly on soft, juicy fruits, other types of fruits, and flowers. Variegated squirrels are usually seen high in the trees, rarely on the ground. They are most active in the early morning. Found in deciduous forest, open woodland, scrub, and plantations of fruit trees; uncommon in evergreen rainforest.
Geographic range. Central America: Chiapas, Mexico, south to Panama on the east side of the canal. To 600 m elevation.
Status. Locally common; hunted for meat, but adapts to disturbed vegetation.
Local names. Chiza (CR); ardilla negra (Pn).
References. Harris, W. P., Jr. 1937. Revision of *Sciurus variegatoides*, a species of Central American squirrel. *Misc. Pub. Mus. Zool. Univ. Mich.* 38:7–39.
  Glanz, W. E. 1984. Food and habitat use by two sympatric *Sciurus* species in central Panama. *J. Mammal.* 65: 342–47.

Yucatan Squirrel
Sciurus yucatanensis
Plate 21, map 143
Identification. Measurements: HB = 208–260; T = 220–258; HF = 53–65; E = 17–28.
Upperparts coarsely grizzled or variegated black and gray, olivaceous or tawny. Ear with a pale patch behind. **Tail black frosted with white. Feet often blackish, darker than body, or gray. Underparts like back, or pale gray or whitish,** or black in melanistic individuals.
Variation. General appearance from blackish brown to gray.
Similar species. See variegated (*S. variegatoides*) and Deppe's squirrels (*S. deppei*).
Natural history. Found in rainforest, lowland broadleaf forest, pine-oak forest, and plantations of coffee and cacao.
Geographic range. Central America: the Yucatán Peninsula, Chiapas, Mexico, Guatemala, and Belize.
Status. Probably hunted for meat.

Deppe's Squirrel
Sciurus deppei
Plate 21, map 144
Identification. Measurements: HB = 163–220; T = 152–190; HF = 46–58; E = 22–26; WT = 190–220 g.
Upperparts finely grizzled **brown-olivaceous or reddish brown; forelegs from shoulder clear gray.** Ears sometimes with small pale patch behind. **Tail slender,** black frosted with white above; brown-olive medially beneath. Underparts pale gray to whitish, sometimes buff. Mammae pairs = 4. **A small brown or gray squirrel.**
Variation. The forelegs in some populations are colored like back.
Similar species. All other squirrels in its range are much larger and have bushy tails.
Sounds. Calls throughout the day with a series of high-pitched, closely spaced notes or birdlike trills.
Natural history. Diurnal; arboreal and terrestrial; solitary; sometimes several call near each other. Feeds on berries, acorns, and fungi. Deppe's squirrels are usually seen on the ground and in lower vegetation levels of dense forest, especially in vine-covered trees. They move quietly and are

Map 144

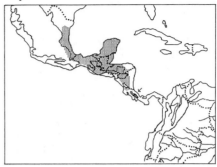

 Deppe's squirrel, *Sciurus deppei*
Montane squirrel, *Syntheosciurus brochus*

difficult to observe. They make a leaf nest on a large branch 6 m or more above the ground. Found in dense broadleaf forests and subtropical cloud forests.
Geographic range. Central America: Veracruz, Mexico, south to N Costa Rica. To 2,800 m elevation.
Status. CITES Appendix II (CR). Locally threatened by deforestation.
Local names. Ardilla montañera, ardilla chica (Me).
References. Hall, E. R., and W. W. Dalquest. 1963. The mammals of Veracruz. *Univ. Kansas Pub. Mus. Nat. Hist.* 14:165–362.

Richmond's Squirrel
Sciurus richmondi
Map 145
Identification. Measurements: HB = 160–210; T = 130–184; HF = 45–55; E = 15–25; WT = 235–268 g.
Upperparts entirely uniform brown, tinged olivaceous; fur short and slightly glossy. Ears large, no pale patches behind. **Tail shorter than head and body, fairly slender, brown** or blackish frosted buff. **Underparts orange** or orange washed. Mammae pairs = 3. **A small squirrel.**
Variation. Similar to and may be a subspecies of the red-tailed squirrel, *S. granatensis*.
Similar species. Deppe's squirrels (*S. deppei*) have gray or white underparts, usually gray forelegs.
Natural history. Diurnal; terrestrial and arboreal. This squirrel forages on the ground or on the main trunks and lower branches

Map 145

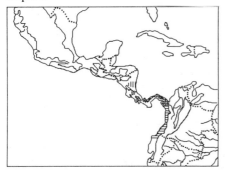

Central American dwarf squirrel, *Microsciurus alfari*

Richmond's squirrel, *Sciurus richmondi*

Western dwarf squirrel, *Microsciurus mimulus*

of trees, rarely high in the canopy. Found in mature and secondary rainforest, gallery forest, and plantations with trees.
Geographic range. Central America: Nicaragua only, the Caribbean coast and drainage.
Status. Unknown, but geographic range small and much deforested. Likely to be the most endangered Neotropical squirrel.
References. Jones, J. K., Jr., and H. H. Genoways. 1975. *Sciurus richmondi*. Mammalian Species, no. 53.

Red-bellied Squirrel
Sciurus aureogaster
Plate 21, map 143
Identification. Measurements: HB = 220–310; T = 206–315; HF = 65–74; E = 27–35; WT = 432–680 g.
Highly variable, **upperparts finely grizzled gray, brownish, black,** or intermediate colors; **nape of neck, and/or shoulders, and/or rump sometimes with patch of orange or chestnut,** or darker black in melanistic individuals. Tail black frosted with white above, gray or reddish medially below. **Underparts pure white, deep chestnut-red, or any intermediate color.**
Variation. Animals from the Isthmus of Tehuantepec northward up the Pacific coast and to Veracruz have bright chestnut rump or nape patches. Animals from Veracruz and Puebla northward are often melanistic.
Similar species. See Deppe's squirrel (*S. deppei*) and variegated squirrel (*S. variegatoides*).

Sounds. A resonant, harsh, trilling chatter.
Natural history. Diurnal; arboreal; solitary. Feeds on fruits, seeds including conifer seeds, conifer leaves, and sometimes corn. Red-bellied squirrels are usually seen in the trees, and they rarely come to the ground. They make leaf nests in trees. Found in a wide range of habitats from lowland tropical evergreen broadleaf forest to pine-oak woodland and thorn scrub on the Mexican plateau, but may be absent from intermediate cloud forests, where Deppe's squirrel is found.
Geographic range. Central America: Mexico from Nayarit and Nuevo León south to Chiapas and the highlands of Guatemala. To about 3,000 m elevation.
Local names. Ardilla, ardilla negra (melanistic individuals), ardilla pinta, ardilla gris (Me).
References. Musser, G. G. 1968. A systematic study of the Mexican and Guatemalan gray squirrel, *Sciurus aureogaster* F. Cuvier (Rodentia: Sciuridae). Misc. Pub. Mus. Zool. Univ. Mich., no. 137.

Amazon Dwarf Squirrel
Microsciurus flaviventer
Plate 20, map 146
Identification. Measurements: HB = 120–160; T = 96–150; HF = 35–43; E = 10–17; WT = 60–128 g.
Upperparts brown finely grizzled reddish to olivaceous. **Ears short, not protruding above crown of head,** pale yellow behind with short pale hairs not visible above rim of ear. **Tail slightly shorter than head and body,** relatively slender, tapered toward tip, slightly frosted with dirty white. Underparts thickly haired, deep to pale orange, brightest on chest, not sharply demarcated from sides. Mammae pairs = 3.
Variation. Some animals from Ecuador, Colombia, and Brazil have grayish underparts washed with orange.
Similar species. Pygmy squirrels (*Sciurillus pusillus*) are much smaller and gray, with prominent white ear patches; Guianan (*Sciurus aestuans*), Sanborn's (*S. sanborni*), and Bolivian (*S. ignitus*) squirrels have long ears that protrude well above crown.
Sounds. Soft chucks in alarm; soft, low-frequency, descending, birdlike trill.
Natural history. Diurnal; arboreal; solitary and in pairs. Amazon dwarf squirrels use

Map 146

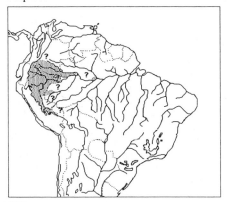

▨ Amazon dwarf squirrel, *Microsciurus flaviventer*

all levels of the forest from the ground to the canopy. They can often be seen foraging, apparently for arthropods, by searching actively over large trunks, vines, and treefalls. One nest was a ball of leaves lined with fibers at 3.5 m in the top of a small palm. Found only in evergreen rainforest.

Geographic range. South America: Amazon Basin of Colombia, Ecuador, Peru, and Brazil west of the Rios Negro and Juruá. To 2,000 m elevation.

Status. Locally common to rare; not hunted.

References. Allen, J. A. 1914. Review of the genus *Microsciurus. Bull. Amer. Mus. Nat. Hist.* 33:145–65.

Western Dwarf Squirrel
Microsciurus mimulus
Plate 20, map 145
Identification. Measurements: HB = 128–150; T = 92–150; HF = 33–42; E = 12–17; WT = 120 g.
Upperparts dark brown finely grizzled reddish to olivaceous, with or without distinct broad black stripe down midback; top of head the same as back or blackish or black; ears short, not protruding above crown, with or without pale patch on back. **Tail distinctly shorter than head and body,** in black-striped forms black toward tip, frosted with gray. Underparts thickly haired, dull to bright, dark rust-orange brightest on chest, not sharply demarcated from sides.

Variation. Animals from Nariño, Colombia, south through Ecuador have a black stripe down midback; those from El Valle, Colombia, north to Darién, Panama, have uniform brown backs; those from eastern Panama are brown-olivaceous and have white behind the ear tips. Specimens from lowlands often speckled with small white spots (probably from scars of botflies). *Microsciurus similis* (not described in this book) inhabits the higher elevations of Cauca and Nariño in the Andes of Colombia.

Similar species. These are the only lowland dwarf squirrels in the southern half of their range. In Panama, where they occur with Central American dwarf squirrels (*M. alfari*), they apparently live at higher elevations, but some overlap occurs. The species are almost indistinguishable in the field: the back of Central American squirrels is more finely ticked, and the underparts are grayish or yellowish.

Natural history. Diurnal; arboreal and terrestrial; solitary and in pairs. Forages on tree trunks and around bush piles, probably searching for arthropods. Found in evergreen forest. In Panama only at elevations above 800 m; in Colombia and Ecuador it is found in the lowlands.

Geographic range. Central and South America: Chiriquí, Panama south to NW Ecuador, and N Colombia. To 1,600 m elevation.

Status. Not hunted, but parts of range extensively deforested.

Central American Dwarf Squirrel
Microsciurus alfari
Plate 20, map 145
Identification. Measurements: HB = 123–159; T = 100–125; HF = 32–42; E = 12–18; WT = 75–100 g.
Upperparts dark brown, very finely grizzled. Ear short, not protruding above crown, with or without pale white or buff patch behind. **Tail shorter than head and body, tapered toward tip,** like back or more reddish. **Underparts well furred, dull gray-tinged, or gray frosted yellowish, not sharply demarcated from sides.**
Feet like back or slightly more reddish.
Variation. Animals from Darién, Panama, have white ear patches and yellow-gray underparts; those from Chiriquí, Panama,

have no ear patches and gray underparts.
Similar species. See western dwarf squirrel (*M. mimulus*).
Natural history. Diurnal; arboreal and terrestrial. Central American dwarf squirrels use all levels of the forest. Found in cloud forest and dense rainforest.
Geographic range. Central America: S Nicaragua and Costa Rica south to W Colombia. To 1,700 m elevation.
Status. Uncommon.
Local names. Chiza (CR); ardilla voladora (Pn).

Neotropical Pygmy Squirrel
Sciurillus pusillus
Plate 20, map 142
Identification. Measurements: HB = 89–115; T = 89–113; HF = 24–29; E = 10–15; WT = 33–45 g.
Upperparts of body and tail **pale gray** with faint yellowish tinge, but may look brown in poor light in field. **Head slightly reddish; ears short, do not protrude above crown, with prominent white patches behind, white hairs protruding over rim.** Tail slightly longer than head and body. Underparts gray or buff-gray, not sharply demarcated from sides, only slightly paler than back. A **tiny squirrel with slender legs.**
Variation. Animals from French Guiana have more conspicuously reddish heads than other populations.
Similar species. These are the smallest New World squirrels and the only pale gray squirrels in the Amazon Basin. Amazon dwarf squirrels (*M. flaviventer*) are larger and brown; all other brown squirrels in the region have ears that protrude well above crown and sharply contrasting paler underparts.
Sounds. A soft to amazingly loud cricket-like chirp, "tsick"; more rarely a soft chatter or trill.
Natural history. Diurnal; arboreal; solitary and females with young; up to at least four may feed on the same tree, or a group of males may chase the female during a mating bout. Feeds chiefly on a substance scraped from the inner surface of tree bark. These tiny squirrels forage from near the ground to so high in the canopy that they disappear from sight. They pull chips of outer bark from certain large trees (often

Inga spp.), leaving the trunk densely pockmarked, and are most easily seen by waiting near such a tree (but some birds likewise chip bark). They move rapidly, running up and down trunks and hopping and flitting from branch to branch in the canopy like small birds. They always travel from tree to tree high in the canopy. On flat trunks they have a highly characteristic habit of making almost instantaneous 180-degree switches of direction. As they forage on a trunk, the tail is carried stiffly out behind and given little jerks from time to time, but is not brought up over the back. Young foraging near their mothers frequently play by dashing back and forth around the trunk. Guianan squirrels (*S. aestuans*) sometimes forage on the bark of the same trees used by pygmy squirrels. Their tiny size makes these squirrels difficult to spot, but they are not wary of humans and are easy to observe. Found with a patchy distribution in mature, lowland, evergreen rainforest only.
Geographic range. South America: known only from three widely separated areas of the Amazon Basin of Peru (near Iquitos); Surinam and French Guiana and adjacent Brazil, and Brazil on the lower Rio Tapajós.
Status. Locally common, but the distribution is extremely patchy even within its known geographic areas and in individual forests, suggesting specialization to some as-yet-unknown feature of the habitat, such as particular species of trees.
Local names. Quatipuru-zinho (Br); n'gue, ngua (Sar).
References. Olalla, A. M. 1935. El genero *Sciurillus* representado en la Amazonia y algunas observaciones sobre el mismo. *Rev. Museu Paulista* 19:425–30.

Anthony, H. E., and G. H. H. Tate. 1935. Notes on South American Mammalia. No. I. *Sciurillus. Amer. Mus. Novitates,* no. 780.

Montane Squirrel
Syntheosciurus brochus
Map 144
Identification. HB = 153–185; T = 120–160; HF = 41–48; E = 8–21.
Upperparts uniform finely grizzled brown-olivaceous. Ears tiny, well furred, and inconspicuous, not protruding above

crown. **Tail thickly furred, shorter than head and body,** frosted with olivaceous-buff above, blackish below. Underparts grayish orange. **Fur thick and soft, giving squirrel a roly-poly look.**
Similar species. Central American dwarf squirrels (*Microsciurus alfari*) are smaller, dark brown, slender-tailed, and with a grayish white belly in area of overlap.
Natural history. Crepuscular and diurnal; terrestrial; in pairs or family groups. This rarely seen squirrel spends most of its time on the ground and in the underbrush, where it hides and is difficult to see. One nest was in a tree hollow about 7 m up.

Found in wet montane palm and oak forests.
Geographic range. Central America: Costa Rica and Panama in the Cordillera de Talamanca; scattered localities in a strip less than 250 km long. At 1,920–2,300 m elevation.
Status. Apparently rare or extremely difficult to see; geographic range small and narrow.
Local names. Ardilla (Span).
References. Enders, R. K. 1980. Observations on *Syntheosciurus:* Taxonomy and behavior. *J. Mammal.* 61:725–27.

Pocket Mice (Heteromyidae)

Dental formula: I1/1, C0/0, P1/1, M3/3 = 20. This family is characterized by external cheek pouches; small, simple cheekteeth, and a variety of skull features. Many species have large, strong hindlegs and small, weak forelegs and travel by hopping on their hindlegs. All species are small (< 200 g). The family includes kangaroo rats, kangaroo mice, and pocket mice. These rodents are generally seed eaters that use their pockets to transport seeds, which they often store in caches. Most live in arid habitats, a few in humid forest. There are 5 genera and 65 species, all in the New World, with 1 genus and about 3 species in rainforest.

Spiny Pocket Mice
Heteromys spp.
Plate 22, map 147
Identification. Measurements: HB = 108–148; T = 119–175; HF = 27–39; E = 13–22; WT = 49–103 g.
Upperparts and limbs dark brown-black; fur a mixture of **spiny** guard hairs and softer underhairs; spines sometimes partly pale, giving grizzled look; spines not apparent visually from a distance, detectable by rubbing fur backward. **Tail bicolored dark above, pale below, usually longer than head and body. Underparts pure white, sharply demarcated from sides. Cheek pouches large, externally opening pockets** extend back to shoulders. All feet white or hindfeet black. Scrotum of adult males greatly elongated behind.
Species. Three of about 11 species in the genus occupy different regions of Neotropical rainforest—*H. anomalus, H. australis, H. desmarestianus.* All are similar in appearance. The genus needs revision, and species definitions are likely to change.
Similar species. No other mice in region

Map 147

Spiny pocket mice, *Heteromys*

have cheek pouches; spiny mice (*Neacomys* spp.) are tiny, with relatively short tails; spiny and armored rats (*Proechimys* spp., *Hoplomys*) are much larger, with tails shorter than the head and body; all other mice or rats in range are soft-furred. Another genus of spiny pocket mouse (*Liomys*) occupies arid habitats in Central America; they are paler and have white forelegs.

Natural history. Nocturnal; terrestrial; solitary. Feeds on seeds and berries, including small palm nuts and coffee beans, and on fungi. The cheek pouches are used to transport seeds and other food. Spiny pocket mice are often found near streams and around logs and the bases of trees. They den in burrows and have a home range of 1–2 ha. They are associated with very wet habitats, which occur at sea level in some regions, but mostly in montane forests in Central America. Found mostly in mature forest, but also in coffee plantations and other secondary vegetation.
Geographic range. Central and South America: rainforest species from Oaxaca, Mexico, through Central America to N Ecuador west of the Andes and to northern Colombia and Venezuela east of the Andes; Trinidad and Tobago. *H. anomalus* east of the Andes in Colombia and Venezuela; *H. australis* west of the Andes in Colombia and Ecuador; and *H. desmarestianus* in Central America. To about 2,200 m elevation.
Status. Often common.
Local names. Ratón semiespinosa (CR); ratón de bolsa (Co).
References. Rood, J. P., and F. H. Test. 1968. Ecology of the spiny rat, *Heteromys anomalus* at Rancho Grande, Venezuela. *Amer. Midl. Nat.* 79:89–102.

Fleming, T. H. 1974. The population ecology of two species of Costa Rican heteromyid rodents. *Ecology* 55:493–510.

Murid Rodents (Muridae)

Dental formula: I1/1, C0/0, P0/0, M3/3 = 16, or sometimes fewer molars. Forefeet with four long toes and a short thumb; hindfeet with five toes. This family includes most of the familiar rats and mice. They are characterized mainly by skeletal and internal features; they range in size from about 4 to 2,000 g; most are small rats and mice in the 15–300 g range. (There is no taxonomic basis, real difference, or dividing line between rats and mice, which are simply traditional words for larger and smaller species.) Murids show a wide array of lifestyles and reproductive traits. A knowledge of the general morphological features associated with different ecologies is one of the most useful aids for identifying small rodents. Generally, arboreal species tend to have broad, pink-soled feet with sharp curved claws; short legs; long, often robust, hairy tails; smallish ears; large eyes; dense and long whiskers; and somewhat blunt muzzles. Terrestrial grazers tend to have short, fine whiskers; short tails; hispid fur; and short rotund bodies. Mice that tunnel and dig have short tails and ears and long claws. Terrestrial omnivores have long, narrow, black-soled hindfeet; long legs; large ears; well-developed whiskers; long, naked tails; and pointed snouts. Semiaquatic species have large, paddle-shaped hindfeet with webs between the toes and a fringe of hairs on the side of the foot, slightly to very hairy tails, small ears and eyes, stiff whiskers, and dense, glossy fur. Species that feed on terrestrial invertebrates tend to have long snouts, short round ears, long claws on the forefeet, and short, slightly hairy tails. These are not rigid rules but general tendencies; some species have mixed ecologies and features; other factors such as habitat and climate are also associated with certain morphologies. Most murids make nests of fibers to shelter themselves and their young. Many species have large litters of altricial young that grow quickly and mature early. The life expectancy of individuals of most species seems less than a year. This is the largest family of mammals, with about 1,140 species worldwide. These are divided into 15 subfamilies: all of the native genera described in this book belong to the subfamily Sigmodontinae; the introduced house mouse and black and Norway rats belong to the Murinae.

Rice Rats
Oryzomys spp.
Plate 22, map 148
Identification. Measurements: about HB = 100–190; T = 79–210; HF = 25–40; E = 15–25; WT = 40–120 g.

Medium to small rats: brown, tawny, reddish, or gray; fur fine and soft, dark gray at base. Ears large, thinly haired; whiskers long or short; eyes large; eyeshine bright yellow. Tail usually slightly shorter than head and body,

Map 148

Rice rats, *Oryzomys*

but may be slightly longer, naked, slender; solid or bicolor pale below, dark above, not white tipped. Hindfeet narrow, usually white above, with black soles. Underparts grayish white. Mammae pairs: 1 pectoral, 1 postaxial, 1 abdominal, 1 inguinal = 4.
Species. A large genus of about 20 species. Up to 4 species may occur in one locality. A few of the most common rainforest species are described below, but the species cannot be identified with confidence without examining the skull.
O. bolivaris. Dark brown to tawny, long velvety fur, short ears; distinguished from all others by an exceptionally long whisker above the eye that reaches well back behind ear. Costa Rica, Panama, the west coast of Ecuador and Colombia.
O. capito (plate 22). Uniform brown, with yellowish sides, darker midback, and gray belly. Young are gray, young adults brown, old adults reddish. The most widely distributed and common species in the Amazon Basin and the Guianas.
O. talamancae is similar, from Central America, coastal Colombia, Ecuador, and Venezuela. Several other species are also externally similar.
O. macconnelli (plate 22). Back a mixture of tawny and blackish, sides bright cinnamon, fur long and soft; distinctive shape with long hindlegs and feet; long pointed muzzle, long, slender bicolored tail. Found in very dense vegetation such as viny bamboo thickets and large treefalls, apparently

always on the ground. Throughout Amazonia and the Guianas.
O. nitidus, O. intermedius. Color like *O. macconnelli,* body shape normal like *O. capito. O. nitidus* in the Amazon Basin; *O. intermedius* in coastal forests of SE Brazil.
O. caliginosus (subgenus *Melanomys*). Distinctive and sometimes placed in a separate genus. It is blackish, with reddish tints; underparts about the same as upperparts; it has tail much shorter than head and body and short ears, unlike other species. Honduras south to W Ecuador.
Similar species. Arboreal rice rats (*Oecomys* spp.) have slightly hairy tails, short ears, and pink-soled feet; climbing rats (*Rhipidomys* spp.) have tufted tails and pink-soled feet; pygmy rice rats (*Oligoryzomys* spp.) are much smaller, with relatively longer tails; Atlantic forest rats (*Delomys* spp.) sometimes have a black stripe on midback, extreme tip of tail white, and pink soles on toes; may not be distinguishable by external features.
Sounds. No sounds are usually heard in field.
Natural history. Nocturnal (except *O. caliginosus* diurnal); terrestrial; solitary. Feed on fruits, seeds, and insects. The diet of *O. capito* includes about 30% insects and 67% fruits and seeds. Rice rats are generally most common in areas of dense undergrowth and around fallen logs. They are usually a dominant part of the rainforest small mammal fauna and are easy to see. They run about actively on the ground or on fallen logs; if frightened they may bounce away with a series of high jumps. After the initial activity of the night, they may climb onto a branch at 1–2 m and rest motionless, head propped on the incisors against the branch. *O. capito* makes a small nest of leaves and twigs in a hollow log or on the ground surface under the litter. Rice rats occupy a wide range of climates and humid habitats. Some rainforest species (e.g., *O. capito*) adapt well to secondary vegetation and plantations and invade camps or buildings in the forest; others do not.
Geographic range. North to South America: S United States to N Argentina.
Status. Common to rare.
References. Guillotin, M. 1982. Rythmes d'activité et régimes alimentaires de *Proe-*

chimys cuvieri et d'*Oryzomys capito velutinus* (Rodentia) en forêt Guyanaise. *Rev. Ecol.* 36:337–71.

Musser, G. G., and M. M. Williams. 1985. Systematic studies of oryzomyine rodents (Muridae): Definitions of *Oryzomys villosus* and *Oryzomys talamancae*. Amer. Mus. Novitates, no. 2810.

Pygmy Rice Rats
Oligoryzomys spp.
Plate 23, map 149

Map 149

Identification. Measurements: HB = 70–110; T = 85–155; HF = 20–30; E = 11–17; WT = 9–40 g.
Upperparts tawny yellow-brown, mixed with black hairs on back producing hispid look; muzzle, sides of head and neck, and sometimes forequarters tawny gray; fur short in warm lowlands, long in cold climates. Whiskers usually short, sometimes reach ear tip but not shoulder; eyeshine bright. **Tail very slender, much longer than head and body, naked, with a few hairs protruding beyond tip, gray. Underparts gray, with hairs gray to base.** Hindfeet narrow. **Tiny mice** with pointed muzzles, flat fur, and oval, laid-back ears. Mammae pairs: 1 pectoral, 1 postaxial, 1 abdominal, 1 inguinal = 4.
Species. There are about 10–15 species, perhaps fewer; all are similar in external appearance and difficult to distinguish in the field. Only one species usually seems to be found at any one lowland rainforest site; some of these include: *O. fulvescens,* in Central America; *O. microtis,* in Brazil, Bolivia, and Peru; *O. chacoensis,* in Bolivia, Paraguay, S Matto Grosso, Brazil; *O. nigripes,* in SE Brazil. The systematics of this group has not yet been completely worked out. This is often considered a subgenus of *Oryzomys*.
Similar species. Bicolored arboreal rice rats (*Oecomys bicolor*) are reddish brown, with a pure white belly and slightly tufted tail; house mice (*Mus musculus*) have a shorter tail and large, rounded ears; spiny mice (*Neacomys* spp., *Scolomys*) have spiny fur; pocket mice (*Heteromys* spp.) have spiny fur, cheek pouches, and white bellies. Above 1,000 m in the Andes there are other diminutive rice rats (*Microryzomys* spp).
Natural history. Nocturnal; terrestrial; but

Pygmy rice rats, *Oligoryzomys*

climb well; solitary. Probably feed on seeds, fruits, and insects. These little mice prefer dense and brushy habitats; they are rare in mature, terra firme rainforest but do occur there. They are more numerous in open areas such as gardens, secondary brush, rice fields, and plantations. Pygmy rice rats commonly occupy houses and camps in rainforest clearings. They can be agricultural pests in rice fields and storage barns. *O. fulvescens* builds a nest in the rice plants, 50–80 cm above the ground. Found in a wide range of habitats and climates from rainforest to grassland and from sea level to the high Andes.
Geographic range. Central and South America: Mexico to Tierra del Fuego. To over 4,000 m elevation.
Status. Locally common.
References. Olds, N., and S. Anderson. 1987. Notes on Bolivian mammals 2: Taxonomy and distribution of rice rats of the subgenus *Oligoryzomys. Fieldiana Zool.,* n.s., 39:261–81.

Veiga-Borgeaud, T. 1982. Données écologiques sur *Oryzomys nigripes* (Desmarest, 1819) (Rongeurs, Cricétidés) dans le foyer naturel de peste de Barracão dos Mendes (Etat de Rio de Janeiro, Brésil). *Mammalia* 46:335–59.

Map 150

Arboreal rice rats, *Oecomys*

Arboreal Rice Rats
Oecomys spp.
Plates 23, 25, map 150
Identification. Measurements: HB = 95–175; T = 90–190; HF = 20–32; E = 11–20; WT = 21–88 g.
Upperparts uniform warm brown to reddish sometimes with gray tones; fur soft, dense, dull. Ears medium length, brown, set low on head; muzzle short, quite blunt; whiskers dense but not long or coarse, standing up in a fan above nose; **eyes large and bulging;** eyeshine bright. **Tail long, brown, lightly haired, with a few hairs extending beyond tip in an inconspicuous tuft. Feet short, broad, with strongly curved claws, dusky or white above, often with a dark patch across hindfeet above toes, soles pink.** Underparts pure white, pale gray, or gray with orange wash; area around mouth sometimes white. Mammae pairs: 1 pectoral, 1 postaxial, 1 abdominal, 1 inguinal = 4.
Species. There are about 10 species, up to 4 of which can be found together in lowland Amazon Basin localities. Bicolored arboreal rice rats (*O. bicolor*) are distinctive and widespread: small (20–35 g), reddish, with white underparts, with hair white to the roots. Most other species are larger (40–80 g) and difficult to distinguish in the field.
Similar species. Small species of climbing rats (*Rhipidomys*) are similar and difficult to distinguish externally; they are usually grayer, with a longer, more robust tail with

more pronounced tuft, and long, coarser whiskers that reach to shoulder. Females have six mammae.
Natural history. Nocturnal; arboreal; solitary. Feed on fruits and green seeds. These mice use all levels of the forest, including occasionally the ground. They are most numerous in dense, viny vegetation. At night they can be seen running rapidly along pencil-thin vines and slanting branches or sitting motionless in leafy thickets above the ground. They nest in tree holes, in dense masses of vines or epiphytes, and among palm leaves. They often invade houses in the forest and seem particularly fond of thatched roofs. Mature and secondary rainforest, dry forest, savanna, scrub, and gardens and plantations.
Geographic range. Central and South America: Costa Rica south to Argentina. To at least 2,200 m elevation.
Status. Often common.

Spiny Mice
Neacomys spp.
Plate 23, map 151
Identification. Measurements: HB = 71–100; T = 70–111; HF = 20–24; E = 13–19; WT = 15–32 g.
Upperparts brown or red-brown finely streaked with black, or head and midback blackish streaked with orange; sides of head, neck, shoulders, and lower sides orange, often bright; **fur spiny,** with narrow spines mixed with hairs, **spines not visually apparent, detectable by rubbing fur backward.** Whiskers fine and long, reaching beyond ear. **Tail slender, naked,** brownish, as long as head and body. Forefeet white, hindfeet dusky, or legs and feet gray. **Underparts pure white, pale orange, or white along midline, bordered by orange laterally. Tiny, sleek-looking mice.**
Species. There are at least three similar species: *N. guianae* is generally brown above, with underparts whitish medially, orange laterally; *N. spinosus* is more blackish above, with gray tones on the head, and white or pale orange underparts; *N. tenuipes* is rich, reddish brown above.
Similar species. Pocket mice (*Heteromys* spp.) have cheek pouches; gray spiny mice (*Scolomys melanops*) are gray above and below, with tail much shorter than head

Map 151

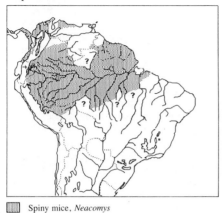

Spiny mice, *Neacomys*

Gray spiny mouse, *Scolomys melanops*

and body; no other small mice have spiny fur.

Natural history. Nocturnal; terrestrial; solitary. Diet of *N. guianae* is about 60% insects and 40% fruits, seeds, and other plant material. These mice favor areas of dense ground cover and travel on the ground or on logs or vines near the ground. They are restricted to rainforest regions and live in mature rainforest, but can sometimes be found in secondary forest, grassy clearings, or plantations.

Geographic range. Central and South America: E Panama to Ecuador west of the Andes; east of the Andes to the southern Amazon Basin of Brazil. *N. guianae* north of the Amazon and east of the Rio Negro; *N. spinosus* from the rest of the Amazon Basin; and *N. tenuipes* from Panama, Colombia, Ecuador, and Venezuela west of the Andes. To 1,875 m elevation.

Status. Rare in some regions (i.e., Panama), common in others.

Gray Spiny Mouse
Scolomys melanops
Plate 23, map 151
Identification. Measurements: HB = 146–160; T = 55–67; HF = 20–22; E = 15; WT = 20–33 g.
Upperparts dark gray finely ticked with brown or fulvous; **fur spiny, spines not apparent visually, detectable by rubbing fur backward.** Whiskers fine, reaching ear. **Tail much shorter than head and body,**

naked. **Legs and feet entirely gray. Underparts clear medium gray, paler than back.**
Variation. There is individual variation in the amount of brown on the upperparts.
Similar species. See spiny mice (*Neacomys* spp.).
Natural history. In lowland rainforest this mouse was found in a disturbed patch of forest with dense undergrowth.
Geographic range. South America: Ecuador, known from two localities 200 km apart on the eastern side of the Andes at 1,150 m and 250 m elevation.
Status. Apparently patchy or rare in a tiny geographic range. Known only from a total of nine individuals, taken in two series 50 years apart.

Water Rats
Nectomys spp.
Plate 24, map 152
Identification. Measurements: HB = 135–232; T = 152–234; HF = 37–53; E = 17–26; WT = 133–288 g.
Upperparts glossy dark brown, a finely grizzled mixture of black and yellowish; center of back darkest, sometimes blackish, **paling to yellowish sides; fur soft and dense, medium to long. Ears well furred around base, naked toward tip;** whiskers short, few or none extending beyond tip of ear when flattened. **Tail robust,** usually slightly longer than head and body, **lightly haired with brown hairs densest midventrally, sometimes forming a slight "keel," and extending slightly beyond tip in a small tuft (but looks naked from a distance). Hindfeet wedge-shaped, with narrow heel and broad palm** at base of toes, **toes partially webbed; sides of feet with fringe of downcurling silvery white hairs; sole of foot to heel covered with roundish scales.** Underparts gray washed dirty yellow or buff, with silvery sheen, grading gradually from sides with no line of demarcation. Young gray-brown. Mammae pairs: 1 pectoral, 1 postaxial, 1 abdominal, 1 inguinal = 4. Large, plain brown rats.
Species. The two species are similar: *N. squamipes* is larger, *N. parvipes* is smaller (the lower measurements given above are from the type and only specimen). *N. squamipes* varies from blackish to

Map 152

Water rats, *Nectomys*

Map 153

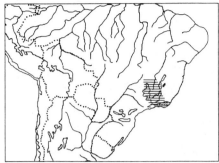

Rio de Janeiro arboreal rat, *Phaenomys ferrugineus*
Brazilian arboreal mouse, *Rhagomys rufescens*

yellowish or grayish brown; it is likely that these rats represent more than one species.
Similar species. Marsh rats (*Holochilus* spp.) are externally similar; they usually have tails slightly shorter than head and body, short round ears hairy inside and outside at tips, no scales on sole of heel, very short whiskers, and often orange on sides and underparts. Large climbing rats (*Rhipidomys* spp.) have tail much longer than head and body, short, broad feet with no webs or fringe, and long, coarse whiskers. Spiny rats (*Proechimys* spp.) and fishing rats (*Ichthyomys* spp.) have white bellies sharply demarcated from sides; Norway rats and black rats (*Rattus* spp.) may be similar in color but have coarse, sparse fur, naked ears and tails, and no webs between toes or fringe on foot.
Natural history. Nocturnal; terrestrial and seimaquatic, solitary. Feeds on arthropods and other invertebrates, fruits, and fungi. Water rats are adapted for swimming and are almost always found near water. In captivity they have been shown to catch small aquatic prey by probing with the forefeet. They make nests under logs or roots or in dense vegetation; an *N. squamipes* nest of grass in a tree stump in a swamp had several underwater entrances. These rats seem most common in grassy areas, but are also found in closed-canopy forest. In rainforest, gallery forest, and disturbed forest, gardens, plantations, and clearings.
Geographic range. South America: east of the Andes from Colombia and Venezuela south to N Argentina (*N. squamipes*). *N. parvipes* is known only from French Guiana. To 1,100 m elevation, but rare above 500 m.
Status. Widespread and sometimes common.
Local names. Rata de agua (Span); rato d'agua, rato paca (Br): ratón nativo (Ec).
References. Ernest, K. A. 1986. *Nectomys squamipes*. Mammalian Species, no. 265.

Rio de Janeiro Arboreal Rat
Phaenomys ferrugineus
Map 153
Identification. Measurements: HB = 148; T = 187; HF = 33; E = 17.
Upperparts brilliant rusty orange, slightly darker on crown and midback; fur long, soft, dense, and straight, slaty at base. **Ear short,** both sides covered with hair, **rustorange;** whiskers fine and long, longest to shoulder. **Tail robust, much longer than head and body,** lightly covered with fine dusky hairs, dark brown. Feet dark reddish brown, soles pink; **hindfeet with a broad white patch across upper foot and ankle. Underparts** from chin **cream,** sharply demarcated from sides, hairs cream to base. Mammae pairs: 1 pectoral, 1 postaxial, 1 abdominal, 1 inguinal = 4.
Similar species. The only other bright red mice in its range are Brazilian arboreal mice (*Rhagomys rufescens*), which are smaller, with tail equal in length to the head and body, an orange belly, and a

round nail, not a claw, on first hindtoe.
Natural history. Unknown. The morphology of this rat suggests that it is arboreal.
Geographic range. South America: Brazil, known only from the state of Rio de Janeiro.
Status. Extremely rare; known from five individuals collected in the last century.

Brazilian Arboreal Mouse
Rhagomys rufescens
Map 153
Identification. Measurements: HB = 94; T = 93; HF = 20; E = 15.
Upperparts and underparts entirely rich rust-orange; hairs slate blue at base, rusty at tips over whole body; fur long and dense. **Ears short, barely projecting above fur, thickly haired, red-brown;** whiskers fine, longest to shoulder. **Tail equal in length to head and body,** thinly covered with dusky hairs, slight tuft at tip. Feet pinkish-yellowish, soles pink; first toe of hindfoot with a rounded nail, not claw. Underparts only slightly paler than back. Mammae pairs: 1 pectoral, 1 abdominal, 1 inguinal = 3.
Similar species. See Rio de Janeiro arboreal rat (*Phaenomys ferrugineus*).
Natural history. Unknown. The morphology of this species suggests that it is arboreal.
Geographic range. South America: SE Brazil; known only from Minas Gerais and Rio de Janeiro.
Status. Rare; known from only a few individuals.

Climbing Rats
Rhipidomys spp.
Plate 25, map 154
Identification. Measurements: HB = 110–190; T = 130–230; HF = 24–36; E = 18–23; WT = 35–170 g.
Upperparts dull yellow-brown, fulvous brown, or gray-brown; usually slightly grizzled or streaked with black; fur soft, dense, sometimes long. **Eyes large, often narrowly ringed with dark fur; whiskers** beside nose **long, dense, and coarse,** when pressed back reach beyond ear to shoulder. **Tail 20–40% longer than head and body, robust, thinly covered with coarse, dark hair that forms a small to prominent tuft at tip. Feet short and**

Map 154

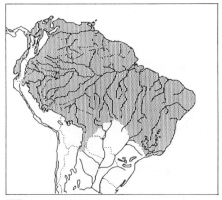

Climbing rats, *Rhipidomys*

broad, with sharp, curved claws, conspicuously black or dusky across top, usually with paler toes, soles pink or dark with pink toes. Underparts gray, white with gray base, pure white, or yellowish white. Mammae pairs: 1 postaxial, 1 abdominal, 1 inguinal = 3.
Species. There are about ten species, ranging from medium to large. Widespread forms include: *R. couesi*, a large, yellowish brown rat with tail tuft poorly developed, from Venezuela to Argentina along the eastern base of the Andes; *R. leucodactylus*, a large variegated tawny rat with long fur and prominent tail tuft, from Venezuela, Colombia, and Ecuador; *R. mastacalis*, a medium-sized fulvous rat with little tail tuft, from the eastern Amazon Basin to SE Brazil.
Similar species. The smaller climbing rats, such as *R. mastacalis*, are very similar externally to the larger arboreal rice rats, such as *Oryzomys trinitatus;* climbing rats have longer, thicker tails with coarser, longer hairs and terminal tufts, and black, thick, coarser whiskers, several of which reach well behind ears if pressed back. Members of the closely related genus of Thomas's paramo rats (*Thomasomys*) are found at 1,000–3,500 m in cloud forest or paramo.
Sounds. None usually heard in field.
Natural history. Nocturnal; arboreal; solitary. Climbing rats use all levels of the forest, sometimes including the ground, and seem to favor areas of dense, viny vegetation, or rocks and caves. Found in mature

and secondary rainforest, montane and cloud forest, deciduous forest and scrub, and gardens and plantations; also invades rural camps and houses.
Geographic range. Central and South America: barely enters Central America in extreme E Panama; west of the Andes south to NW Peru; east of the Andes widespread in all countries south to N Argentina, but not Paraguay. To 2,400 m elevation.
Status. Widespread and often common.

Atlantic Forest Rats
Delomys spp.
Plate 22, map 155
Identification. Measurements: HB = 114– 138; T = 90–135; HF = 26–30; E = 20–23. Upperparts dark to tawny brown, sides paler; with or without a dark midback stripe (see species below); fur soft and medium length. Whiskers fine, reaching just behind ear tip, no long whisker from behind eye; ear finely haired, appears naked, moderately large. **Tail about equal to or shorter than head and body,** naked, tip sometimes white. **Hindfeet long and narrow, with long toes, white above** with white hairs extending beyond claws, soles black with pink toetips. Mammae pairs: 1 pectoral, 1 postaxial, 1 abdominal, 1 inguinal = 4; or lacking pectoral pair = 3.
Species. There are two or three species.
Delomys dorsalis. **Upperparts dark brown, midback with a black stripe from shoulder to rump;** sides slightly paler, tawny brown; ear dark gray; eye sometimes with dark ring or darker area between eye and nose; **tail equal to or slightly longer than head and body, gray-brown above, whiter below.** *D. d. collinus* is similar; with black stripe indistinct, it may be a subspecies or a species.
Delomys sublineatus. **Upperparts tawny brown, finely streaked with blackish; sides washed with orange, brightest on cheeks and elbows;** ear brown; **tail slender, shorter than head and body;** underparts gray washed with white or orange.
Variation. This genus seems related to paramo mice (*Thomasomys*) and climbing rats (*Rhipidomys*) and has generally been included in the former.
Similar species. Rice rats (*Oryzomys* spp.)

Map 155

Atlantic forest rats, *Delomys*

are extremely similar and may not be distinguishable in the field from Atlantic forest rats that lack a backstripe; most rice rats are smaller, with longer tails and longer whiskers; climbing rats (*Rhipidomys* spp.) have hairy tails much longer than head and body and broad, dark-banded hindfeet; grass mice (*Akodon* spp.) have short whiskers that do not reach ear and short tails.
Natural history. Terrestrial. These rats can be found in holes around fallen logs and roots. The striped back suggests diurnal activity. *D. sublineatus* is found in secondary forest, *D. dorsalis* in primary forest. Found in Atlantic coastal forests.
Geographic range. South America: SE Brazil to NE Argentina; Rio de Janeiro to Paraná (*D. dorsalis*), Espírito Santo, and Rio de Janeiro (*D. sublineatus*). To at least 1,600 m elevation.
Status. May be locally common.
References. Davis, D. E. 1947. Notes on the life histories of some Brazilian mammals. *Bol. Mus. Nac.* (Rio de Janeiro), n.s., zool. 76:1–8.

Crab-eating Rats
Ichthyomys spp.
Plate 24, map 156
Identification. Measurements: HB = 110– 171; T = 110–155; HF = 28–40; E = 8–11; WT = 127 g.
Upperparts glossy dark brown, protruding black overhairs give rump a slightly bristly look; fur soft, with dense woolly underfur. **Head with blunt, rounded muzzle; eyes small, ears small,** hairy, incon-

Map 156

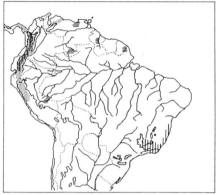

▒ Crab-eating rats, *Ichthyomys*
▬ South American water mice, *Neusticomys*
▥ Brazilian shrew mouse, *Blarinomys breviceps*

spicuous, **sometimes white hairs on inside show as a prominent white spot on side of head. Whiskers coarse, stiff, and straight,** longest reaching just beyond ear. Tail equal to or shorter than head and body, **completely covered with flat hairs, dark brown or sharply bicolored white below, brown above. Hindfeet wedge-shaped,** with narrow heel and broad palm; **outer edges of foot and toes fringed with downcurling silvery hairs;** toes with partial webs near base; forefeet with five pads on palm. **Underparts silvery gray washed with brown, to pure white,** sharply demarcated from sides. Mammae pairs: 1 postaxial, 1 abdominal, 1 inguinal = 3.
Species. Three of the four species occur in rainforest below 1,000 m: *I. tweedii, I. stolzmanni, I. pittieri;* the first two are also found at higher elevations.
Similar species. Central American water mice (*Rheomys* spp.) are similar; they have forefeet with four pads on palm; South American water mice (*Neusticomys* spp.) are smaller, with dark underparts not sharply contrasting with sides and hindfeet not wedge-shaped. Marsh and water rats (*Holochilus, Nectomys* spp.) are usually larger, with naked, scaly tails.
Natural history. Probably nocturnal; terrestrial and semiaquatic. Feeds on crabs, other aquatic invertebrates such as insect larvae, and occasionally small vertebrates. Wild-caught captives kill and eat crabs

with a stereotyped behavior: first they bite off the claws, then the other legs, then they bite open the body at the back of the shell and eat the contents. The empty dorsal shell and legs or leg tips remain. The presence of crab-eating rats in an area may be shown by the discarded empty shells of small crabs left on rocks in streams. Crab-eating rats are always found near fresh (not salt or brackish) water; they appear to require clear, fast-flowing streams. Such streams occur in hilly terrain and generally have sandy or rocky beds, and they may be small and shallow and shaded within tall rainforest, cloud forest, or secondary forest.
Geographic range. Central and South America: *I. stolzmanni* on the eastern slope of the Andes in Peru and Ecuador; *I. tweedii* on the western slope of the Andes from E Panama to Ecuador; *I. pittieri* in the coastal cordillera of Venezuela.
Status. Unknown; rarely encountered.
References. Voss, R. S., J. L. Silva L., and J. A. Valdes L. 1982. Feeding behavior and diets of Neotropical water rats, genus *Ichthyomys* Thomas, 1893. *Z. Saugetierk.* 47:364–69.
    Voss, R. S. 1988. Systematics and ecology of Ichthyomyine rodents (Muroidea): Patterns of morphological evolution in a small adaptive radiation. *Bull. Amer. Mus. Nat. Hist.* 188:259–493.

Central American Water Mice
Rheomys spp.
Plate 24, map 157
Identification. Measurements: HB = 102–140; T = 109–160; HF = 30–41; E = 8–12.
Upperparts glossy dark brown, overhairs protrude to give rump a bristly look; **rump and sides often with a sprinkling of silvery overhairs;** fur soft and dense, underfur woolly. **Muzzle blunt; eyes tiny; ears small, inconspicuous,** covered with hair; whiskers thin and stiff, longest reaching well behind ear. Tail equal to or longer than head and body, **completely covered with flat hair, dark brown or bicolored,** white below, dark above. **Hindfeet wedge-shaped, sides and toes fringed with silvery hair; forefeet with four pads on palm. Underparts contrasting silvery**

Map 157

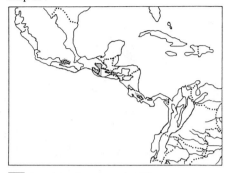

Central American water mice, *Rheomys*

gray or white. Young slate gray.
Species. Two of the five species can occur
in the lowlands. *R. thomasi* is smaller,
with tail < 137 mm; *R. mexicanus* is larger, with tail > 140 mm; all species are primarily montane.
Similar species. See crab-eating rats
(*Ichthyomys* spp.)
Natural history. Semiaquatic. Feeds on
aquatic insects and other invertebrates and
occasional small vertebrates. These mice
are found near clear, tumbling mountain
streams that are shaded by vegetation.
Mainly a genus of montane and cloud forest species, but also found where suitable
streams occur below 1,000 m.
Geographic range. Central America: Oaxaca, Mexico, south to Darién, Panama. *R.
thomasi* in Chiapas, Mexico, Guatemala,
and El Salvador; *R. mexicanus* in Oaxaca,
Mexico. To about 2,000 m elevation.
Status. Unknown; rare or difficult to
capture.
References. Hooper, E. T. 1968. Habitats
and food of amphibious mice of the genus
Rheomys. J. Mammal. 49:550–53.

South American Water Mice
Neusticomys spp.
Plate 24, map 156
Identification. Measurements: HB = 100–
128; T = 82–111; HF = 24–30; E = 9–
12; Wt = 75 g.
**Upperparts glossy blackish brown or buff
brown;** fur soft and smooth, pale gray at
base. **Muzzle blunt; eyes small; ears
small, but conspicuous,** hairy inside and
out. **Whiskers fine and long, several extending well behind ear. Tail** equal to or

shorter than head and body, **completely
covered with flat dark brown hair. Feet
not strongly wedge-shaped, sides with
slight fringe of brown or silvery hair,**
blackish or cream above. **Underparts dark
gray or brown with silvery sheen, not
sharply contrasting with sides.**
Species. Three of the four species
(*N. oyapocki, N. peruviensis, N. venezuelae*) occur in the lowlands; the other species is montane.
Similar species. Crab-eating rats (*Ichthyomys* spp.) are similar but have contrasting
pale underparts and strongly wedge-shaped
feet; no other small soft-furred mice have
tails fully covered with smooth hairs that
completely hide underlying scales.
Natural history. Semiaquatic. The montane species feeds on aquatic insects; lowland species may also eat small crabs.
Lowland water mice have been captured
near clear streams in rainforest.
Geographic range. South America: lowlands east of the Andes and Andes of Colombia and Ecuador; *N. venezuelae* in Venezuela and Guyana; *N. peruviensis* in SE
Peru; and *N. oyapocki* in French Guiana.
Range extensions are likely.
Status. Rare or difficult to capture in lowlands; two lowland species are known from
single individuals.
References. Musser, G. G., and A. L.
Gardner. 1974. A new species of the Ichthyomyine *Daptomys* from Perú. *Amer. Mus.
Novitates,* no. 2537.
    Voss, R. S. 1988. Systematics and ecol-
ogy of Ichthyomyine rodents (Muroidea):
Patterns of morphological evolution in a
small adaptive radiation. *Bull. Amer. Mus.
Nat. Hist.* 188:259–493.

Brazilian Shrew Mouse
Blarinomys breviceps
Plate 22, map 156
Identification. Measurements: HB ~ 100
mm; T = 30–49; HF = 16–21; E = 8–
10.
**Upperparts uniform glossy chocolate-
brown. Muzzle blunt, point of nose,
muzzle, and chin white; eyes and ears
tiny, lost in fur.** Forefoot with fifth digit
greatly reduced; fore and hindfeet tiny,
claws long. **Tail very short, less than half
the length of head and body.** Underparts
gray-brown, hardly differentiated from

back. A small cylindrical blunt-nosed mouse with short legs.

Similar species. All similar mice and short-tailed opossums have conspicuous eyes and ears.

Natural history. Terrestrial. Brazilian shrew mice are probably semifossorial; they are said to live in burrows under the leaf litter. Their diet is unknown, but their tooth morphology suggests that they feed on invertebrates. From montane rainforest 650 m and above.

Geographic range. South America: Brazil; coastal mountains from Bahía to Rio de Janeiro and mountains of Minas Gerais.

Status. Apparently rare in a small geographic range. Known from about 40 specimens.

References. Matson, J. O., and J. P. Abravaya. 1977. *Blarinomys breviceps.* Mammalian Species, no. 74.

Naked-tailed Climbing Rats
Tylomys spp.
Plate 25, map 158

Identification. Measurements: HB = 184– 260; T = 191–290; HF = 32–43; E = 20–28; WT = 156–326 g.

Upperparts dull slate gray to warm gray-brown, head often slightly darker than back; fur soft, dense and slightly woolly. **Ears large, naked, set low, held semihorizontally, with funnel pointed downward;** muzzle pointed; whiskers black, long, reaching well past ear. **Tail robust, naked, with prominent scales in rings, longer than head and body, dark and sometimes shiny for basal half to two-thirds, long tip sharply contrasting white or yellow. Feet** short and broad, **dark above, usually with sharply contrasting white toes.** Underparts entirely white, or throat, chest, and usually inguinal region white or yellowish and midriff gray. Young slate gray. Large, robust rats.

Species. There are about three species, all quite similar in size and external appearance. There is individual color variation within a population, from gray to brown: adults seem to bleach browner and paler as they age.

Similar species. No other large rats in geographic range have naked, white-tipped tails. The ear posture and white tail tip give these rats a superficial resemblance to some opossums.

Map 158

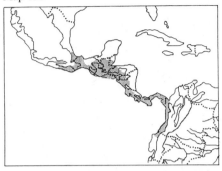

Naked-tailed climbing rats, *Tylomys*

Natural history. Nocturnal; arboreal and terrestrial; solitary. Diet described as greenish plant material, possibly lichen or bark. These rats frequent areas of rocks and caves and fallen logs in dense forest, and they are often on the ground as well as in the trees. Found in wet evergreen broadleaf forests.

Geographic range. Central and South America: Mexico to Panama and west of the Andes in Colombia and N Ecuador. To about 1,100 m elevation.

Status. Apparently uncommon.

References. Hall, E. R., and W. W. Dalquest. 1963. The mammals of Veracruz. *Univ. Kansas Pub. Mus. Zool.* 14:165–362.

Big-eared Climbing Rat
Ototylomys phyllotis
Plate 25, map 159

Identification. Measurements: HB = 152– 170; T = 127–178; HF = 22–29; E = 22–27; WT = 70–130 g.

Upperparts dark to pale gray-brown; fur long and lax, parting easily to give a disheveled look. **Ears very large, naked; eyes large; head and muzzle long, narrow;** whiskers medium to long, to ear or beyond. **Tail** slightly shorter than head and body, robust, **naked, shiny** as if varnished, **with prominent scales in rings, dark brown or black above, sometimes white below,** especially near base. Forefeet dark above, toes white; hindfeet short and broad, dark above with white toes and edge of foot, or all whitish. Underparts pure white, or white on chin and chest, pale gray posteriorly. A small, slender rat.

Map 159

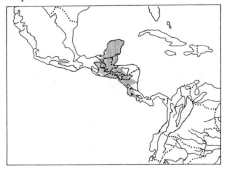

Big-eared climbing rat, *Ototylomys phyllotis*

Map 160

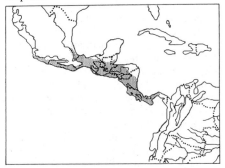

Vesper rat, *Nyctomys sumichrasti*

Variation. Animals from higher elevations tend to be large, with gray bellies; those from the lowlands are smaller, with white bellies, but there are exceptions.

Similar species. The combination of large ears and robust dark, shiny tail with large scales in rings distinguishes this species from other small rats; naked-tailed climbing rats (*Tylomys* spp.) are larger, with white tail tips.

Natural history. Nocturnal; arboreal and terrestrial; solitary. Feeds on seeds, fruits, and browse. Big-eared climbing rats usually frequent areas of rocks, caves, or sinkholes, and around fallen trees. They use the lower levels of the forest, generally below 10 m, where they run along vines and branches or the ground. Their nests are at ground level among rocks or under logs, or in low tree holes. This species probably only marginally inhabits rainforest; it is found in a wide variety of habitats from coastal lowland rainforest to montane forest and arid scrub; and in mature forest and secondary vegetation such as roadsides and hedgerows.

Geographic range. Central America: Tabasco and Chiapas, Mexico, south to central Costa Rica. To about 2,000 m elevation.

Status. Locally common.

References. Lawlor, T. E. 1969. A systematic study of the rodent genus *Ototylomys*. *J. Mammal.* 50:28–42.

Lawlor, T. 1982. *Ototylomys phyllotis*. Mammalian Species, no. 181.

Vesper Rat
Nyctomys sumichrasti
Plate 25, map 160

Identification. Measurements: HB = 99–140; T = 102–157; HF = 20–27; E = 15–21; WT = 38–67 g.

Upperparts uniform bright rust-orange or red-brown, sides bright orange; fur short or long, straight and slightly stiff. Ears short, rounded, broad at base; eyes very large, rimmed with narrow black rings, sometimes faint, that extend forward as a dusky smudge at base of nose and whiskers; whiskers very long, sometimes reaching to midbody. Tail usually slightly longer than head and body, robust, thickly covered with dull, red-brown hairs that curl outward in a sparse brush and extend beyond tip in a small tuft. Feet short and broad; forelegs and hindfeet dusky to top of foot; toes white. Underparts pure white, including area around mouth and lower cheeks. Young red or gray-brown with fur soft and straight, or distinctly woolly.

Variation. Animals from Panama and Costa Rica are sometimes red-brown, and their young are gray-brown, with woolly fur, until they reach almost adult size; those from Mexico are bright rust, with young the same color as adults.

Similar species. Bicolored arboreal rice rats (*Oecomys bicolor*) and harvest mice (*Reithrodontomys* spp.) are smaller, with slender, smooth tails; rice rats (*Oryzomys* spp.) have naked tails and slender feet. Yucatan vesper rats (*Otonyctomys hatti*; not described in this book) are almost identical externally but brighter red. They are extremely rare, known from only a half-dozen specimens from the eastern half of the Yucatán Peninsula; their natural habitat is unknown.

Natural history. Nocturnal; arboreal; solitary. Feeds on fruits, seeds, and flowers. Vesper rats use the middle and upper levels of the forest, above 3 m, and rarely descend to the ground. They nest in tree hollows and favor those with small entrances 2.5–8 cm in diameter. The nest is a mass of shredded bark within the hollow. When disturbed in her nest by day, a mother vesper rat will flee along horizontal branches, with her young clinging tightly by their mouths to her teats. Found in evergreen lowland and lower montane forests, old secondary and riparian forests, and semideciduous forests.

Geographic range. Central America: Veracruz, Mexico, south to central Panama. To 1,600 m elevation.

Status. Usually uncommon or rare; common in a few localities.

References. Genoways, H. H., and J. K. Jones, Jr. 1972. Variation and ecology in a population of the vesper mouse (*Nyctomys sumichrasti*). Occ. Pap. Mus. Texas Tech. Univ., no. 3.

Grass Mice
Akodon spp.
Plate 22, map 161

Identification. Measurements (rainforest species): HB = 90–136; T = 32–98; HF = 17–28; E = 11–21; WT = 19–89 g.
Upperparts uniform dark brown to brown-olivaceous; fur soft, dark gray at base. Ears rounded, often small; muzzle pointed; **whiskers fine and short, not reaching tip of ear. Tail short,** 50–66% of head and body length, robust at base tapering to thin tip; black. Feet blackish, narrow, soles black, claws sometimes long, with little curve. **Underparts dark,** about the same color as back, slightly washed with buff, not strongly differentiated from sides. **Body short and rotund.** Mammae pairs: 1 pectoral, 1 postaxial, 1 abdominal, 1 inguinal = 4.
Species. The genus is one of the largest in the Neotropics, with over 35 species. Most live in open grasslands; only a few species are regularly found in rainforest, including *A. urichi, A. cursor, A. dayi, A. nigrita,*

Map 161

Grass mice, *Akodon* (excluding southern part of range)

and possibly others. The description above applies to species in the rainforest region; some species from south temperate latitudes are chestnut, or washed with pale gray, and may have gray underparts and hairy, bicolored tails.

Similar species. Long-nosed mice (*Oxymycterus* spp.) are similar but have very long muzzles and strongly undershot jaws and are usually larger. In old-field habitats may be confused with cane mice (*Zygodontomys brevicauda*); these have pale-gray frosted underparts. See short-tailed opossums.

Natural history. Terrestrial; nocturnal. Probably feed on insects, seeds, and browse. Some grass mice nest in burrows and travel in tunnels under the leaf litter. Found in grassy areas, especially with tall grass, young second growth, gardens, and clearings, but also sometimes in evergreen forest, in coastal regions of Venezuela and SE Brazil.

Geographic range. South America: all countries from N Venezuela to Tierra del Fuego. To 5,000 m elevation.

Status. Locally common.

References. Reig, O. A. 1987. An assessment of the systematics and evolution of the Akodontini, with description of a new fossil species of *Akodon* (Cricetidae: Sigmodontinae). *Fieldiana Zool.*, n.s., 39:347–99.

Map 162

▨ Long-nosed mice, *Oxymycterus*

Long-nosed Mice
Oxymycterus spp.
Plate 22, map 162
Identification. Measurements: HB = 130–164; T = 70–120; HF = 26–32; E = 17–23; WT = 70–125 g.
Upperparts blackish on midback, grading to olivaceous, or faint to bright fulvous mixed with black hairs on sides of body and head; fur long and soft, dark gray or blackish at base. **Muzzle extremely long and narrow; chin greatly undershot, usually pure white; whiskers fine and short,** usually not reaching to base of ear. **Tail short,** about two-thirds of head and body, thinly covered with short blackish hair. **Feet small and narrow, soles black, claws very long, white,** with little curve. Underparts gray washed with yellowish or dull to bright fulvous, not sharply demarcated from sides. **Body short, rotund, and vole-like.**
Species. There are at least 10 species; most occur in temperate climates in lowland grasslands or above 1,000 m, from cloud forest to Puna, but *O. inca* (and perhaps other species) occur in lowland Amazonian rainforest. In SE Peru the lowland rainforest form is bright fulvous, almost rust-red on underparts and sides; the form south of the Amazon in Brazil is faint fulvous on sides; the forms from E coastal Brazil are olivaceous on sides and larger than Amazonian forms. It is not currently clear how many species there are or to what species the lowland rainforest long-nosed mice belong.

Similar species. Grass mice (*Akodon* spp.) are similar; they have a shorter muzzle and claws, are usually smaller, and might occur only in the same localities as long-nosed mice in SE Brazil; Brazilian shrew mice (*Blarinomys breviceps*) have inconspicuous eyes and ears; other rainforest mice have long tails. See short-tailed opossums.
Natural history. Nocturnal and/or diurnal; terrestrial. Feed mainly on insects and other invertebrates and may use their long claws to dig these from the litter. Long-nosed mice are rare in rainforest, where they seem to be found in swampy areas, at least some of which are covered with sedges and have no forest canopy above. Also found in secondary growth and plantations in Amazon Basin rainforest areas.
Geographic range. South America: south of the Amazon from Pará to Peru; south to Argentina and Uruguay. To over 4,000 m elevation.
Status. Some species are locally common.
References. Barlow, J. C. 1969. Observations on the biology of rodents in Uruguay. Roy. Ontario Mus. Life Sci. Contrib., no. 75.

Marsh Rats
Holochilus spp.
Plate 24, map 163
Identification. Measurements: HB = 150–230; T = 140–190; HF = 35–49; E = 15–22; WT = 90–320; males reach larger size than females.
Back dull brown sometimes streaked black and olivaceous, tawny, or orange or sprinkled with gray hairs; **sides of head, neck, forearms, and body lightly to brightly tinged orange;** fur short and woolly or long and silky. **Ears short and round, hairy inside and out to tips; whiskers short and fine, none usually reaching to ear tip. Tail robust, shorter than head and body, slightly hairy but without tuft** or usually, ventral keel of hairs. **Hindfeet wedge-shaped, with narrow heel and broad palm, partial webbing between first, second, third, and fourth toes; sides of feet with downcurling fringe of whitish hairs; sole of heel without prominent scales.** Underparts usually gray washed with orange, not sharply demarcated from sides, may also be white. Young like adults, with orange on sides. Mammae

Map 163

Map 164

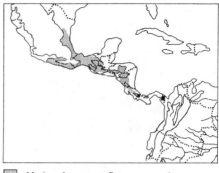

▒ Mexican deer mouse, *Peromyscus mexicanus*
■ Isthmus rats, *Isthmomys*

▒ Marsh rats, *Holochilus*

pairs: 1 pectoral, 1 postaxial, 2 abdominal, 1 inguinal = 5.
Species. There are about four species, two in rainforest regions: *H. sciureus* and *H. brasiliensis*. More species are likely to be described from this group. Some animals from Bolivia have pure white bellies.
Similar species. See water rats (*Nectomys* spp.)
Natural history. Nocturnal and possibly diurnal; terrestrial; solitary. Feed on grass and other green plants. Marsh rats are not found in forest but frequent wet grasslands, marshes, and cultivated areas within the rainforest region. They make oval, two-chambered nests of woven strips about 30 by 20 cm, either on the ground in crevices or under brush piles, or among grass and cane stems up to 2 m above ground. Marsh rats are prodigious breeders, and in some agricultural areas they undergo periodic plaguelike outbreaks and become serious agricultural pests in rice, sugarcane, and oil-palm plantations.
Geographic range. South America: east of the Andes from Colombia and Venezuela south to N Argentina and Uruguay: *H. sciureus* from the Amazon Basin, and *H. brasiliensis* from SE Brazil south to Argentina. To about 2,000 m elevation.
Status. Widespread and often common.
References. Twigg, G. I. 1965. Studies of *Holochilus sciureus berbicensis*, a cricetine rodent from the coastal region of British Guiana. *Proc. Zool. Soc. London* 145:263–83.

Mexican Deer Mouse
Peromyscus mexicanus
Plate 22, map 164
Identification. Measurements: HB = 116–131; T = 115–135; HF = 25–28; E = 18–22; WT = 29–57.
Upperparts brown, tawny in older individuals, slate in younger; **midback dark, paling to cinnamon-brown on sides** in tawny individuals; fur long and soft, slate gray at base on all of body. **Eye surrounded by dark ring;** nose at base of whiskers with darker spot; **ears very large,** rounded, gray-brown; **whiskers long, reaching shoulder. Tail slightly longer than head and body,** naked, **bicolored gray-brown above, with paler blotches below.** Hindfeet narrow and delicate, dark to ankles, then whitish to toes. **Underparts pale gray washed with dirty white or yellowish, sharpy demarcated from sides.** Mammae pairs: 1 postaxial, 1 abdominal, 1 inguinal = 3. Young gray-brown with whitish belly. This is the largest deer mouse in the region.
Variation. Animals from higher elevations are generally darker and more blackish on the back, with longer fur. Those from the lowlands are more tawny. There are about 30 species of deer mice in Mesoamerica; most occupy either high elevations above 1,000 m or habitats drier than rainforest.
Similar species. Rice rats (*Oryzomys* spp.) usually have the tail slightly shorter than the head and body, smaller ears, no black eye ring, eight mammae, and often dusky

feet or short, fine whiskers that do not reach ear tip.

Natural history. Nocturnal; terrestrial. Feeds on fruits and seeds and makes caches of seeds in crevices. This is chiefly a montane and cloud forest mouse. Found in evergreen forests and thickets, especially near watercourses.

Geographic range. Central America: Mexico, Oaxaca and Veracruz south to the highlands of Costa Rica and Panama. To about 2,400 m elevation.

Status. Uncommon in lowland rainforest; can be very common in humid forests above 1,000 m.

References. Hall, E. R., and W. W. Dalquest. 1963. Mammals of Veracruz. *Univ. Kansas Pub. Mus. Nat. Hist.* 14:165–362.

Isthmus Rats
Isthmomys spp.
Plate 22, map 164

Identification. Measurements: HB = 130–175; T = 174–214; HF = 31–37; E = 24–29.

Upperparts cinnamon-brown, sometimes nearly orange, or top of neck and/or shoulders gray-brown; sides paler, bright cinnamon; fur long and soft, dark slate gray at base over whole body. Ears large, naked; whiskers long, reaching shoulder. Tail much longer than head and body, robust, naked, bicolored pale below, dusky above. Feet dusky brown to below wrist or ankle, sometimes to base of toes, then pure white to toe tips, with a sharp straight line across foot where color changes; hindfeet quite broad and robust. Underparts gray heavily washed with white or yellowish white; chest often orange. Young dark brown with velvety fur, can be recognized by their bicolored feet. Large, mouselike rats.

Species. The two species, *I. flavidus* and *I. pirrensis,* are quite similar externally but are separated geographically. The position on the foot of the sharp change from dusky to white varies individually; in a few individuals the line is indistinct.

Similar species. Mexican deer mice (*Peromyscus mexicanus*) have dark eye rings, narrow feet, and shorter tails; vesper rats (*Nyctomys sumichrasti*) have hairy tails, dark eye rings, and snow-white underparts;

arboreal rice rats (*Oecomys* spp.) have smaller ears and no sharp line separating colors on feet; rice rats (*Oryzomys* spp.) are smaller and do not have bicolored feet. There are no other large, soft-furred, cinnamon-colored rats where they occur.

Natural history. Terrestrial. Feed on plants and insects. These rats are found under and around fallen logs and tree roots. Occur in evergreen montane forests and cloud forest from about 500–1,500 m, most often above 1,000 m.

Geographic range. Central and South America: Panama and Colombia; *I. flavidus* from Chiriquí, Panama only; *I. pirrensis* around Cerro Pirre, including the Colombian side of the Serrania del Darién.

Status. Each species is uncommon in a small geographic range.

Harvest Mice
Reithrodontomys spp.
Plate 23, map 165

Identification. Measurements: HB = 43–100; T = 66–142; HF = 16–25; E = 12–19; WT = 8–29 g.

Upperparts tawny brown to warm fulvous brown, darkest on midback; sides of head, neck, and body paler, often orange or cinnamon; fur usually long and soft, dark gray at base. Ears relatively large, lightly haired to tips behind; whiskers long, several reaching to shoulder; eyes large; muzzle pointed; upper incisors grooved. Tail longer than head and body, lightly haired, solid or bicolor dark above, paler below. Underparts pure white, white with gray base, gray, or gray washed with orange. Hindfeet usually dusky above, with white toes. Mostly very tiny, delicate, mice. Mammae pairs: 1 postaxial, 1 abdominal, 1 inguinal = 3.

Species. There are about 19 species; only a few partially enter the lowland rainforest region. *R. mexicanus,* from Mexico to N Ecuador, is warm fulvous; *R. sumichrasti,* from Mexico to Panama, is tawny brown; *R. paradoxus,* known only from Nicaragua, is ochraceous brown; *R. creper,* from elevations generally above 1,000 m, is blackish brown with red tints and sometimes has a white or spotted tail tip.

Similar species. The following species only partially overlap the geographic range; none has grooved incisors: bicolored ar-

Map 165

Harvest mice, *Reithrodontomys*

boreal rice rats (*Oecomys bicolor*) are reddish with white bellies, usually weigh more than 20 g, and have short, broad, pink-soled feet; pygmy rice rats (*Oligoryzomys* spp.) have grayish heads and necks; house mice (*Mus musculus*) have tails the same length as head and body.

Natural history. Nocturnal; terrestrial and arboreal; solitary. Feed on small seeds and browse. Harvest mice climb well and may be seen running around in the undergrowth above the ground. On the ground they use networks of runways. Their nests are round balls of grass or fibers, 13–20 cm in diameter, with a small chamber inside. Nests are hidden at ground level under rocks, logs, old boards, and such, or may be aboveground hidden in a small tree or shrub. These mice are characteristic of temperate lowland or montane tropical habitats. In the rainforest region, most are found above 1,000 m, but a few species occasionally live lower. They are most common in overgrown secondary habitats such as old fields, dense roadside weeds, coffee plantations, and orchards, but they may sometimes be found in evergreen forests.

Geographic range. North, Central, and South America: Canada south through Central America to the Andes of Colombia and Ecuador. To 3,100 m elevation.

Status. Common to rare.

References. Jones, J. K., Jr., and H. H. Genoways. 1970. Harvest mice (genus *Reithrodontomys*) of Nicaragua. Occas. Pap. W. Found. Vert. Zool., no. 2.

Hooper, E. T. 1952. A systematic re-

view of the harvest mice (genus *Reithrodontomys*) of Latin America. Misc. Pub. Mus. Zool. Univ. Mich., no. 77.

Black or Roof Rat
Rattus rattus
Plate 24

Identification. Measurements: HB = 140–200; T = 165–235; HF = 35–40; E = 20–30; WT = 100–240.

There are two colors: (1) **upperparts tawny yellow-brown streaked with black and usually white hairs;** underparts pure white, yellow-white, yellow-brown, or gray; **or** (2) **upperparts black or dark gray streaked with white hairs,** especially on rear half of body; underparts gray. Both forms have **fur long and coarse, but not spiny or bristly, quite sparse, not dense or velvety,** a few long guard hairs often protrude to give bristly look; hairs pale gray at base. **Ears medium, naked; whiskers long and strong, reaching shoulders. Tail longer than head and body, robust, naked and scaly,** with short, stiff hairs. Feet robust and broad. Young brown or dark gray. Mammae pairs: 1 pectoral, 1 postaxial, 2 abdominal, 1 inguinal = 5 (sometimes more).

Variation. The number of black and white hairs mixed in the fur is variable, making individuals more blackish, brownish, or yellowish.

Similar species. Norway rats (*Rattus norvegicus*) are larger (usually more than 200 g), have tails shorter than head and body, short whiskers, and short ears; water rats (*Nectomys* spp.) and marsh rats (*Holochilus* spp.) have soft, dense fur, usually with a gloss or sheen both dorsally and ventrally, ears hairy at base or entirely hairy, short whiskers, paddle-shaped hindfeet with a fringe of hairs along sides; spiny rats (*Proechimys* spp.) have bristly or spiny fur, oblong ears, bicolored tails pale below, and long narrow hindfeet with black soles; they are often glossy chestnut.

Sounds. Gallops and gnaws noisily in the roofs of houses at nights.

Natural history. Nocturnal; mostly terrestrial. Feeds on grains, garbage, fruits, carrion, and almost anything remotely edible, such as soap, candles, or leather. This introduced species was probably brought to the New World by the European exploring ships. It is a serious pest worldwide. In the rainforest region it is found only in and around permanent human settlements. It does not appear to invade Neotropical rainforests, although it occupies such forests in other regions. In some parts of the world, black- and brown-colored forms occupy different habitats. Often found in buildings and graneries. These rats carry disease and often destroy native vertebrate species. They should be eliminated whenever possible.

Geographic range. Worldwide.

Local names. Rata negra, rata doméstica, rata casera (Span); rato-doméstico, caseiro (Br); zwarte rat, ton-alata (Su); rat noir (FG).

Norway Rat
Rattus norvegicus
Plate 24

Identification. Measurements: HB = 196–270; T = 170–235; HF = 41–49; E = 18–24; WT = 200–505 g.

Upperparts tawny yellow-brown streaked with black and/or white hairs; blackest streaks on head and midback; fur texture like black rat. **Ears short, naked; whiskers medium length, reaching ear but not shoulder. Tail shorter than head and body, robust and scaly, with short stiff hairs, sometimes indistinctly bicolored,** dark above, pale below. Feet large and broad, white above, usually with pink soles. Underparts gray washed with yellow or whitish. Young brown. Mammae pairs: 1 pair pectoral, 1 postaxial, 2 abdominal, 2 inguinal = 6. Very large, robust, ugly rats.

Similar species. See black rat.

Sounds. Squeals and often fights with other rats.

Natural history. Nocturnal and diurnal; terrestrial. Diet like black rat; also will sometimes bite sleeping humans or livestock. An introduced pest that arrived in the Americas much more recently than the black rat. This rat lives in large colonies and has a complex social organization. It does not invade the Neotropical rainforest but is mainly found in cities and towns, where it reaches high numbers in sewers, canals, waterfronts, fish markets, and other wet and trashy places. It often lives in complex burrow systems in gardens and parks and likes drainage pipes. It can often be seen by day, dashing along a gutter and diving down a drain or scuttling from one hiding place to another. Extremely aggressive, a carrier of disease, and a danger to native fauna, it should be eliminated when possible. Much less common than the black rat in farms, country towns, or plantations.

Geographic range. Worldwide.

Local names. Rata de alcantarilla, rata doméstica (Span); rato-doméstico, caseiro (Br); bruine rat, rioolrat, ton-alata (Su).

House Mouse
Mus musculus
Plate 23

Identification. Measurements: HB = 66–108; T = 60–92; HF = 14–20; E = 11–17; WT = 10–21 g.

Upperparts uniform tawny to slate brown, paling to tawny or grayish sides; fur soft. **Ears large, rounded, naked;** whiskers fine, reaching to base of ear. Tail slender, naked, about the same length as head and body. **Underparts not sharply demarcated from sides, same color as sides, gray, or gray washed with whitish orange.** Hindfeet narrow, dusky or whitish. Mammae pairs: 1 pectoral, 1 postaxial, 2 abdominal, 1 inguinal = 5. Tiny mice.

Similar species. Pygmy rice rats (*Oligoryzomys* spp.) have longer feet, and these and harvest mice (*Reithrodontomys*) have tails much longer than head and body. See those species for other comparisons.

Natural history. Nocturnal; terrestrial but climbs well. Feeds mainly on grains and cereals stored by humans, such as rice and corn, supplemented by other stored foods and insects. These mice are an introduced pest. In the rainforest region, house mice are rare and restricted to the vicinity of buildings. Most small mice in houses in isolated camps, settlements, and Indian villages in the forest are native species, such as pygmy rice rats and bicolored arboreal rice rats. House mice build a round nest of fibers. Areas contaminated by house mice and trapped individuals have a strong, musky, "mousy" odor stronger than the odor of native species. In agricultural areas they may be found in grain fields and hedgerows as well as buildings. They cause much damage to stored grain.

Geographic range. Worldwide.

Local names. Ratón casero (Span); cachita (Br); huismuis, moismoisi (Su).

Cavylike Rodents (Suborder Caviomorpha)

Dental formula: I1/1, C0/0, P1/1, M3/3 = 20. The caviomorphs include most of the world's largest rodents and are all of relatively large size (120 g to 60 kg). Their distinctive features include an outward flaring structure of the jaw, to which the masseter muscles are attached. Externally, they usually have relatively large heads and large eyes, long, stout whiskers, and relatively small ears; many have bristly or spiny fur (many also do not). Most have a sheet of muscle tightly attached under the skin of the back, which raises the spines, bristles, or fur. Most have some lateral mammae, on the sides of the body rather than the belly. They have long gestation periods and few young in a litter (usually one to three), which are born with eyes and ears open, fully furred, and able to run around. Many species are grazers or browsers. This suborder of rodents underwent a large evolutionary radiation in South America while it was isolated from other continents. Most living caviomorphs are found in Central and South America; only the porcupines have spread naturally to North America. The species that occur in the rainforest region can be divided into three main groups: the porcupines, the large cavylike rodents, and the spiny rats. There are 13 families, about 40–45 genera, and 200 or more species in the New World. It is not yet clear how many species there are in many genera.

Porcupines (Erethizontidae)

All the rainforest porcupines are similar in body shape and many physical features. Their upperparts are covered with thick, sharp, stiff, dangerous spines that are circular in cross section and have barbed tips. The spines detach easily when touched (porcupines cannot "throw" their spines) and embed themselves in the skin of an enemy; the barbs then cause them to work inward into the flesh. The feet all have four large toes with long, strong, curved claws; the hindfoot also has a reduced hallux. The feet are highly modified with a broad naked pad that flares sideward and covers the area where the first digit would be; the toes and claws flex in a tight, pincerlike grip over this pad. The tail is muscular and prehensile, thick at the base and tapering to a thin, flexible tip that is naked on its upper surface and has stiff bristles on its underside. The tail is wrapped "backward" in a spiral twist around branches. The tail is not lashed in defense like that of North American porcupines. The nose and muzzle are greatly swollen and bulbous, naked and pinkish. The head is relatively small and round, with small eyes and inconspicuous ears buried in the spines on the head. The body is thick and heavy, with short legs. Some small dwarf species have long, soft, woolly hair between or over the spines; others do not. Porcupines give birth to a single young. The young of all species are apparently hairy as juveniles and subadults; young of large, hairless species are therefore easily confused in the field with hairy dwarf porcupines. In this book species that have no soft fur are described first, followed by hairy species. The porcupines need taxonomic revision; in particular, no two authorities treat the hairy dwarf porcupines (often placed in their own genus, *Sphiggurus*) in the same way. The accounts in this book are descriptions of specimens from known areas, including many type specimens. The animals described occur in the regions noted, but the scientific names and species delimitations given here may change when the systematics are better understood. In particular, all hairy porcupines from Paraguay and from Bahía to Rio Grande do Sul, Brazil may intergrade and be a single, extremely variable species. There has been no in-depth field study of any species, and for most there is no published natural history information at all. Many of the dwarf species are known from only a few specimens collected long ago. All Neotropical porcupines except two are described in this book, including three high-elevation species. The family is restricted to the New World and includes 4 genera and about 15 species.

Bristle-spined porcupines (*Chaetomys subspinosus*) are morphologically distinctive animals with no apparent close living relatives. They share some features with both the true porcupines (Erethizontidae) and the spiny rats (Echimyidae) but also have unique features. Their evolutionary relationships are unclear. There is one species.

Brazilian Porcupine
Coendou prehensilis
Plate 28, map 166
Identification. Measurements: HB = 444–560; T = 330–578; HF = 80–95; E = 20–29; WT = 3.2–5.3 kg.
Upperparts covered with strong, barbed spines, many spines tricolor with roughly equal width, white or pale yellow base and tip, black or dark brown center; general color black or dark brown heavily overlaid with white or yellowish, from a distance animals look gray or yellow; no soft fur between spines; spines on head, legs, and tail short; spines on back long and thick. Head round, face usually white; nose and lips large, soft, bulbous, pink; ears small, inconspicuous, lost in spines of head; whiskers stout and long, reaching shoulder; eyes small, black, eyeshine very faint dull red. Tail shorter or longer than head and body, robust, prehensile, whitish at base, with short spines except distal one-third of top surface naked; tail often curled around branches in an upcurling coil or a twisted, longitudinal spiral. Feet with broad, expanded pads opposing four long, strong, inward-flexing claws; pale gray-brown above. Underparts covered with short, soft spines, pale gray-brown or whitish. Young hairy, with red-brown or brown, long, soft fur partially covering spines. Strong, peculiar odor. A large, whitish porcupine.
Variation. Light parts of spines vary from white to bright yellow, dark parts from brown to black.
Similar species. Bicolor-spined porcupines (*C. bicolor*) are blackish overall; white tips on spines, if present, are short; tail always longer than head and body. Black dwarf porcupines are smaller and blackish overall; other dwarf porcupines have soft fur covering most of spines as adults. Bristle-spined porcupines (*Chaetomys subspinosus*) have short, soft, brown spines without white tips.
Sounds. No calls are usually heard, but porcupines rattle the leaves when they travel through the trees and drop a patter of fruit fragments when they eat.
Natural history. Nocturnal; arboreal; solitary. Feeds on seeds from immature fruits, green or ripe fruits, palm fruits, and occasionally bark and probably leaves. Brazilian porcupines appear to favor old second-

Map 166

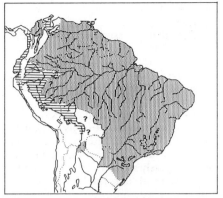

Brazilian porcupine, *Coendou prehensilis*
Bicolor-spined porcupine, *C. bicolor*

growth or exceptionally viny areas of forest, where they use the middle and upper levels of the vegetation, but they can also be seen in the high canopy of tall, open, mature forest. They do not jump and must descend to the ground to cross a gap between trees. They are usually slow-moving, often immobile and difficult to spot, but they nonetheless are surprisingly agile and can climb swiftly. Their tightly gripping feet and tail allow them to negotiate amazingly thin vines and branches. By day they den in hollow trees (which smell strongly of their odor) or crouch, head between the forelegs, in a shaded place in the branches of a tree. Found in mature and secondary rainforest, deciduous forest, gallery forest, and gardens and plantations.
Geographic range. South America: east of the Andes in Colombia, Venezuela, the Guianas, Amazonian and SE Brazil, Bolivia, and Trinidad. To 1,500 m elevation.
Status. Rare in many areas but common, dominant mammals in others. Widespread, sometimes hunted for meat.
Local names. Puerco-espín (Span); ouriço, porco-espinho (Br); erizo (Co); boomstek-elvarken, djiendjamaka (Su); cuandu (FG).
References. Montgomery, G. G., and Y. D. Lubin. 1978. Movements of *Coendou prehensilis* in the Venezuelan llanos. *J. Mammal.* 59:887–88.

Charles-Dominique, P., et al. 1981. Les mamifères frugivores arboricoles nocturnes d'une forêt guyanaise: Inter-relations plantes-animaux. *Rev. Ecol.* 35:341–435.

Bicolor-spined Porcupine
Coendou bicolor
Plate 28, map 166
Identification. Measurements: HB = 378–
493; T = 460–540; HF = 92–100; E =
27–30; WT = 3.4–4.7 kg.
Generally like Brazilian porcupine except:
**upperparts blackish; shoulders and back
black or dark brown speckled with some
yellow or white tips of spines; face, sides,
tail, and legs usually strongly speckled
pale yellow or white, but may be entirely
blackish or dark brown; large spines on
back relatively narrow, often bicolored
white for basal half, black or brown for
distal half, pale tip, if present, much
shorter than black section of spine. Tail
longer than head and body.** Feet and underparts pale gray-brown, mixed with
white on feet.
Variation. The length of pale tip and number of pale-tipped spines vary from none to
most of spines pale-tipped. Animals from
Iquitos appear intermediate between this
and Brazilian porcupines, and the two may
intergrade (the geographic ranges do not
seem to make biogeographic sense). If
C. rothschildi is a valid species, the *C.
"bicolor"* west of the Andes are probably
that species. There is another, distinctive
porcupine, *C. quichua*, in the Andes of
Ecuador. It is the same color as bicolor-
spined porcupines but is smaller, with a
short tail (about 180–200 mm, measured
on skins), conspicuously slender spines,
and short whiskers about reaching ear; at
close range, sparse, soft black hair can be
seen to be sprinkled throughout the dorsal
spines (map 168).
Similar species. See Brazilian porcupine
(*C. prehensilis*); white-fronted hairy dwarf
porcupines are brown with a white blaze
on face and a very short tail.
Natural history. Nocturnal; arboreal. Adult
pairs have been found denning together in
tree holes. Mature and disturbed lowland
and montane rainforest.
Geographic range. South America: east of
the Andes in the upper Amazon Basin of
Colombia, Venezuela, Ecuador, Peru, and
Bolivia, and west of the Andes in Colombia, Ecuador, and Peru to at least Cajamarca. To 2,500 m elevation.
Status. Can be locally common,
widespread.

Map 167

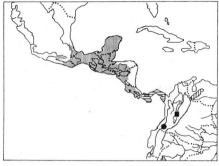

▨ Mexican hairy porcupine, *Coendou mexicanus*
▤ Rothschild's porcupine, *C. rothschildi*
▥ Frosted hairy dwarf porcupine, *C. pruinosus*
◉ White-fronted hairy dwarf porcupine, *C. sneiderni*
■ Brown hairy dwarf porcupine, *C. vestitus*

Local names. Puerco-espín (Span); casha
cuchillo, erizo, pululo (Pe); puchan (Ec).

Rothschild's Porcupine
Coendou rothschildi
Plate 28, map 167
Identification. Measurements: HB =
332–438; T = 260–413; HF = 61–78; E
= 23–26; WT = 2 kg.
**Similar to bicolor-spined porcupine but
somewhat smaller: upperparts black uniformly speckled with small white tips on
some spines; entirely spiny, with no fur
covering spines;** spines mostly bicolored
white at base and black for distal half,
some spines with a white tip much shorter
than black part. Tail 70–90% of head and
body length, black, or speckled with white
on sides, black below. Underparts gray-
brown. Feet dark brown. Young with
spines covered by soft hair.
Variation. Degree of speckling with white
varies from almost none to considerable.
Sometimes considered a subspecies of
C. bicolor, but also resembles *C. quichua*.
See comments for *C. bicolor*.
Similar species. These are the only heavily
spined animals in Central America with no
fur covering spines in adults; Mexican
hairy porcupines (*C. mexicanus*) have soft,
blackish hair, pale spines showing on head.
Natural history. Nocturnal; arboreal. Feeds
on fruits and perhaps leaves. Found in lowland evergreen forests.

Geographic range. Central America: Panama (but see comments for *C. bicolor*).
Local names. Puercoespín, gato de espinas (Pn).

Black Dwarf Porcupine
Coendou sp. nov.
Plate 28, map 168
Identification. Measurements: HB = 319–364; T = 280–326; HF = 50–65; WT = 950 g.
Upperparts entirely spiny; no fur covering spines; black overall with yellow or white patches where base of spines visible, especially on head and forequarters; spines may have small white or pale red-brown tips, giving finely speckled appearance; if spine tips are red-brown, these are concentrated around the shoulder and midback; spines generally white or yellow at base, black for two-thirds of distal part, pale tip if present, short. Spines on rump above base of tail short and upstanding, black, with sharp tips worn off (area is black). Ear with naked area behind. Tail 80–90% of head and body length, prehensile, black. Feet black. Underparts dark brown or gray-brown. Young with rust-red hair between spines. **A very small, blackish porcupine.**
Variation. Amount of pale tipping on spines varies. A specimen of dwarf, hairless porcupine from Ecuador is quite similar but has long, fine white tips on spines, so that the animal is heavily speckled with white or yellow overall; it may be an undescribed species or closely related to the black dwarf porcupine, or quichua or black-tailed hairy dwarf porcupines.
Similar species. Brazilian porcupines (*C. prehensilis*) are much larger and have thicker, heavily white-tipped spines and look gray overall; giant tree rats (*Echimys grandis*) have no spines, nonprehensile tails, nonbulbous noses, and gold-black backs.
Natural history. Unrecorded. The short, worn spines at the tail base suggest that this animal rests with its rump wedged against a surface (this feature is not present in all individuals and is also found in several other species). Known only from lowland evergreen rainforest areas.
Status. Unknown; apparently much rarer than Brazilian porcupine, with which it oc-

Map 168

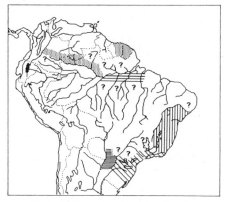

▦	Black-tailed hairy dwarf porcupine, *Coendou melanurus*
▤	Black dwarf porcupine, *Coendou* sp. nov.
▤	Paraguay hairy dwarf porcupine, *C. paragayensis*
▥	Bahía hairy dwarf porcupine, *C. insidiosus*
▧	Orange-spined hairy dwarf porcupine, *C. villosus*
■	Quichua hairy dwarf porcupine, *C. quichua*

curs. Known from fewer than 50 specimens.
Geographic range. South America: Brazil south of the Amazon from the Rio Madeira east to Belém.
Local names. Cuandu, ouriço.

Black-tailed Hairy Dwarf Porcupine
Coendou melanurus
Plate 28, map 168
Identification. Measurements: HB = 280–380; T = 220–363; HF = 52–75 (su); E = 7–24; WT = 0.5–1.3 kg.
Upperparts clothed with soft black hairs overlaid with scattered long, thin, tricolored bristles with long yellow-white tips; fur pale at base, thickest over shoulders, rump with thick yellow spines emerging from fur, or fur may completely cover all dorsal spines. Head finely grizzled, whiskers fine. **Tail long, about 80% to longer than head and body, pitch black except above base, where colored like back.**
Legs and feet grizzled dark gray-brown or blackish. **Underparts pale gray-brown frosted with whitish.**
Variation. This species is sometimes incorrectly called *C. insidiosus*.
Similar species. These seem to be the only hairy dwarf porcupines in their range.

Natural history. Found in lowland rainforest habitat.
Geographic range. South America: north of the Amazon in the Amazon basins of Colombia, probably Venezuela, the Guianas, and Brazil.
Status. Very rare, known from only a few specimens, but widespread.
Local names. Ouriço cacheiro (Br); listig stekelvarken; djiendjamaka (Su).

Bahía Hairy Dwarf Porcupine
Coendou insidiosus
Plate 28, map 168
Identification. Measurements: HB = 290–350; T = 180–222; HF = 50; E = 16.
Entirely covered with thick, soft, smoky brown fur which completely hides spines except on head, fur unicolor to base, or base faintly darker; thick, sharp, hidden spines of back bicolor, basal half whitish, distal half dark brown, or faintly tricolor with tips faintly orange. Face dark brown with some white showing at bases of spines; whiskers black. Tail short, 50–75% as long as head and body, haired above like body for basal third, dark brown-black for about one-third, naked for terminal third. Feet dark brown. Underparts dark to medium brown.
Variation. Pale, almost white or smoke-gray animals from E Brazil probably belong to this species. A gray animal resembling *C. insidiosus* from Espírito Santo has dark hair base, white-tipped whiskers, a long tail, and faintly tricolor spines (characters of *C. villosus*); this suggests that the forms may intergrade.
Similar species. Orange-spined hairy dwarf porcupines (*C. villosus*) have black fur frosted with dirty yellow, and strongly orange-tipped spines.
Natural history. Has been found at over 900 m elevation in Bahía.
Geographic range. South America: Brazil, Bahía and probably Espírito Santo.
Status. Unknown.
Local names. Ouriço cacheiro.

Orange-spined Hairy Dwarf Porcupine
Coendou villosus
Plate 28, map 168
Identification. Measurements: HB = 300–538; T = 200–378; HF = 46–70; E = 11–25.
Upperparts except head entirely covered with soft black fur with long pale yellow or dull pale orange tips, looks washed overall with those colors, or with little or no hair; spines on head and shoulders tricolor with yellow base, black center, orange tip, sometimes with a fourth black band at tip; spines on rump bicolor, pale at base with short black tip, spines long, to 3.5–6 cm; orange spines visible on head. Whiskers long, reaching behind ear, bicolored dark at base, pale at tip. Tail long or short, 41–100% as long as head and body, rusty orange below and on sides for basal half, naked part of top tip long (10 cm). Feet colored like back. Underparts thickly or thinly furred, with no spines, dirty yellow-orange with sharply demarcated contrasting black or dark brown hair base.
Variation. Extremely variable (or possibly more than one species is involved): animals from Paraná, Brazil (*C. v. roberti*), can be almost hairless or densely haired; they have short black whiskers. An animal from Santa Catarina is dark brown, with only faint yellow tipping of hairs, and long spines (6 cm) on midback.
Similar species. See Bahía hairy dwarf porcupine (*C. insidiosus*); Paraguay dwarf porcupines (*C. paragayensis*) have a cape of long, yellow exposed spines over head and shoulders.
Natural history. From Atlantic coastal forests.
Geographic range. South America: SE Brazil, Rio de Janeiro and probably Minas Gerais, to Rio Grande do Sul. To 1,150 m elevation.
Status. Apparently uncommon.
Local names. Ouriço cacheiro.

Paraguay Hairy Dwarf Porcupine
Coendou paragayensis
Plate 28, map 168
Identification. Measurements: HB = 288–340; T = 228–260; HF = 53–65.
Upperparts blond; strong, pale yellow spines showing on whole back, top of head to shoulders with a "cape" of long (5–6 cm on shoulder) spines; thinly overlaid with blond-tipped soft hairs; rump dark brown black, with short black, upstanding spines hidden in dark, yellow-tipped fur. Tail long, 73–90% of head and body length, yellow at sides and be-

low base, dark brown at tip. Feet brown grizzled with pale yellow. Underparts without spines, hair gray-yellow with contrasting dark brown base. **A small to tiny dwarf porcupine.**
Similar species. See orange-spined hairy dwarf porcupine (*C. villosus*).
Geographic range. South America: eastern Paraguay.

Brown Hairy Dwarf Porcupine
Coendou vestitus
Plate 28, map 167
Identification. Measurements: HB = 290; T = 130.
Upperparts thickly covered with soft, warm smoky brown hair covering spines except on head and face, where white-tipped spines emerge through hair. Long, thin bristles emerging through fur are bi-colored pale for basal half, brown for distal half, which blends almost invisibly with dorsal fur; spines on back long, to 3.7 cm, pale yellow or white for most of length, with short, dark brown tips; spines on face thick, sharp and robust. Whiskers fine. **Tail short, about 50% of head and body.** Feet colored like back. Underparts thickly haired, with no spines, warm brown, slightly paler than back.
Similar species. See frosted hairy dwarf porcupine (*C. pruinosus*); stump-tailed porcupines (*Echinoprocta rufescens*), only at high elevations, are hairless, with red-brown spines and a stumpy, nonprehensile tail.
Natural history. Has been collected at about 2,600 m elevation, but is said to be found in the warm lowlands.
Geographic range. South America: Colombia, Cudinamarca, the upper Magdalena Valley west of Bogotá (San Juan de Rio Seco), and possibly the lower eastern Andean slopes. Poorly known.
Status. Apparently extremely rare, known only from a few individuals; the region where the specimens described above were collected has been almost entirely deforested.

Frosted Hairy Dwarf Porcupine
Coendou pruinosus
Plate 28, map 167
Identification. Measurements: HB = 330–380; T = 190; HF = 50–60.
Upperparts thickly haired with soft, gray-brown, **almost blackish fur that completely hides spines of body, fur sprinkled throughout with prominent, long, thin, white-tipped bristles, fur frosted with whitish, especially on sides.** Head gray-brown finely ticked with white of thin white tips of hairs and bristles; whiskers stiff, black. **Spines on back very short, less than 2.5 cm,** most spines on back bicolor, pale yellow for most of length, with black tips; spines on head and a few on shoulder tricolor pale-dark-pale. **Tail short, 50% as long as head and body,** naked area on top of tip short (~5 cm); base pale yellow or whitish, grizzled with fine white spines, tip and undersurface black. Feet brown, grizzled whitish. **Underparts thickly furred, with no spines,** fur dark brown at base, **heavily frosted with silver,** looks whitish with dark brown showing through parts of fur.
Similar species. Brown hairy dwarf porcupines (*C. vestitus*) are warm brown, without white frosting on hairs above or below and longer spines; white-fronted hairy dwarf porcupines (*C. sneiderni*) have a white blaze on face, and soft fur does not cover spines; black-tailed hairy dwarf porcupines (*C. melanurus*) have a long tail and long spines.
Natural history. Four frosted porcupines were found together in a nest. They nest in hollows in rotten tree trunks and perhaps in the branches of trees. Found at 2,500–2,600 m elevation.
Geographic range. South America: Venezuela, known only from the Andes near Merida.
Status. Apparently rare, known only from a few specimens.

White-fronted Hairy Dwarf Porcupine
Coendou sneiderni
Map 167
Identification. Measurements: HB = 335; T = 125; HF = 47 (su).
Upperparts blackish brown sprinkled with reddish tips of spines; fine, soft, brownish gray **fur inconspicuous,** mostly hidden below spines, **animal appears spiny except on rump near tail base, where fur covers spines;** spines long, about 4 cm, tricolor pale yellow at base, black, and rust at tip, which is often prolonged into a long, thin bristle. **Head with a sharply**

defined, pure white blaze down center from crown to upper lip, bordered on each side by a blackish brown band. Tail short, less than 40% of head and body length, prehensile, naked on upper surface of tip. Underparts sharply demarcated from sides, clothed with stiff, flat, bristly hairs white at base with brown tips.
Variation. The description above is taken from the original description of the type, which may be the only specimen. It suggests close alliance to the stump-tailed porcupines (*Echinoprocta rufescens*). It is perhaps an intermediate form between prehensile-tailed and stump-tailed species.
Similar species. Stump-tailed porcupines (*Echinoprocta rufescens*), from the Colombian Andes near Bogotá, have similar color, often including white spots on the face, but they have no soft fur, and the short, stumpy tail is nonprehensile and well haired to its tufted tip; brown hairy dwarf porcupines (*C. vestitus*) have thick fur covering bicolored spines of back.
Natural history. The type specimen was collected at 2,000 m. This is the only hairy dwarf porcupine known to occur on the western slope of the Andes.
Geographic range. South America: Colombia, Cauca Department, on the western slope of the Andes, known from only one locality.
Status. Apparently known from a single individual.

Mexican Hairy Porcupine
Coendou mexicanus
Plate 28, map 167
Identification. Measurements: HB = 350–460; T = 200–360; HF = 63–85; E = 19–25; WT = 1.4–2.6 kg.
Upperparts from neck to tail base thickly covered with long, soft, pitch black or dark brown fur that covers spines; appears black; scattered parts in fur show yellow spines; spines beneath fur of back long (3–3.5 cm), bicolor, yellow with short black tip. Head and neck with bright yellow spines showing through thin hair, or thickly haired, hair covering spines; ear buried in fur but marked by tuft of pale whitish brown fur behind ear; whiskers robust and long, reaching to behind ear or to shoulder; eyeshine dull red. Tail 50–80% of length of head and body, cov-

ered with stiff black bristles below, above like back for basal third, black bristles for middle third, naked for distal third. Underparts with fur and no spines, brown or brown mixed with whitish. Strong, peculiar odor. A small porcupine that looks black with a pale head.
Variation. Animals from Chiriquí, Panama, are very small, dwarf porcupines and have been considered a separate species (*C. laenatus*); some animals, especially from the Yucatán, have sparse fur and many yellow spines on body exposed.
Similar species. Rothschild's porcupines (*C. rothschildi*) have no soft fur (adults) and are black sprinkled with white.
Natural history. Nocturnal; arboreal; adult pairs have been found together in dens. Feeds on ripe and green fruits and seeds and probably browse when fruit is scarce. By day dens in hollow trees or sits on a tree branch among dense vines or foliage. Most common in montane forest; uncommon in the lowlands. Rainforest and deciduous forest.
Geographic range. Central America: Veracruz, Mexico, south to W Panama. To 3,000 m elevation.
Status. Unknown, unlikely to be threatened. Hunted for meat.
Local names. Puerco espín (Span); kixpach och, citam (May).

Bristle-spined Porcupine
Chaetomys subspinosus
Plate 28, map 169
Identification. Measurements: HB = 380–450; T = 260–275; HF = 67–70; E = 9; WT = 1.3 kg.
Upperparts pale brown; head and shoulders densely and evenly covered with short (1.5 cm), sharp, upstanding, kinky spines; back behind shoulder to rump, legs, and tail base thickly covered with long (to 5 cm), stiff, slightly wavy, dry bristles like thin broom straws, these tricolored pale yellow at base, dark brown, then pale brownish yellow at tip; no sharp spines on lower back. Head round, ears buried in spines of head, marked by a small tuft of soft brown fur, the only soft fur on entire body; whiskers medium, reaching ear; muzzle almost naked, brown, nose not greatly swollen and bulbous. Tail prehensile, curling dorsally,

Map 169

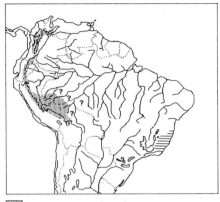

Pacarana, *Dinomys branickii*
Bristle-spined porcupine, *Chaetomys subspinosus*

60–70% of head and body length, thick at base tapering to slender tip, above like back for basal fifth, then thinly haired with brown bristles, becoming almost naked near tip, below rusty at base then dark brown to tip. Feet brown, similar to those of true porcupines. Underparts thickly covered with flat bristles, yellow-brown slightly paler than back. **A thick-bodied, cylindrical animal with a long back and short legs.**

Variation. Color overall varies from dark brown to pale whitish brown, sometimes with white patches.
Similar species. Other porcupines all have sharp spines on lower back; Bahía hairy dwarf porcupines (*C. insidiosus*) have soft, smoky fur covering spines of back.
Sounds. A raucous puffing call.
Natural history. Nocturnal; arboreal. Feeds on fruits, including cacao. Bristle-spined porcupines move slowly and use their tails like true porcupines. They are excellent climbers that can scale stone walls. Found in mature and secondary Atlantic coastal forest and plantations, especially cacao plantations.
Geographic range. South America: Brazil, SE Bahía and N Espírito Santo.
Status. US-ESA endangered. Very few specimens are known, but the animals were apparently relatively common in some areas at least until 1950. Only one individual has been reported in recent years. This unique animal deserves a special effort for protection because it is the only member of its genus and perhaps subfamily.
Local names. Ouriço preto.
References. Moojen, J. 1952. *Os roedores do Brasil.* Biblioteca Ciêntifica Brasileira. Ser. A. II.

Large Cavylike Rodents (Agoutidae, Dasyproctidae, Hydrochaeridae, Dinomyidae)

This mixed group includes the capybara, agoutis, acouchys, pacas, and pacarana. All but the last are long-legged, ungulatelike rodents. All have four toes on the front foot; the weight-bearing toes of the hindfoot are reduced to three in capybaras, acouchys, and agoutis, four in pacaranas, and three, with two much smaller side toes, in pacas. The hair is straight, stiff, and bristlelike. They all have large heads with bulging jaw muscles, short ears, short to miniscule tails, and cylindrical and sometimes piglike bodies. All have the typical rodent posture of sitting on the haunches; all but the capybara can hold their food in the forepaws. They have precocial young in litters of one or two, except the capybara, which has litters of up to eight. Most of the genera include only one or two species and do not present taxonomic problems. Agoutis (*Dasyprocta* spp.), however, are highly variable geographically and have not received any modern taxonomic revision. They are here divided into the most generally accepted set of species, but there appear to be intermediate forms where species meet geographically, and there are many skins that would be difficult to identify without knowing where they came from. There are other named species. It is likely that the species limits given here will change in the future, perhaps radically, reducing the number of species to two or three.

Map 170

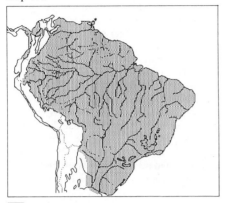

Capybara, *Hydrochaeris hydrochaeris*

Capybara
Hydrochaeris hydrochaeris
Plate 18, map 170
Identification. Measurements: HB =
1,070–1,340; T = 10–20; HF = 220–
250; E = 60–70; SH = 500–620; WT =
35–65 kg.
**Entirely uniform tan or yellowish brown,
or rarely dark red-brown;** midback sometimes darker than sides; hair coarse and
sparse, long on midback. **Head large,
rectangular, muzzle square; ears small;
eyes small, set high on the head,** giving a
supercilious look; eyeshine moderate, eyes
look small and yellow or reddish. **Tail rudimentary, not visible. Feet partly
webbed,** four toes on front, three on hindfoot; legs robust. Young like adults but
more thickly furred. **Large, stocky, piglike
rodents.** Often sit on haunches like a dog.
The world's largest rodents.
Variation. Dark red-brown animals (in
mixed groups with tan ones) seem to be
found more often in rainforest than in open
grasslands.
Similar species. Peccaries (*Tayassu* spp.)
are black, with a narrow snout and two-
toed hooves; pacas (*Agouti paca*) are red
with lines of white spots.
Sounds. Sharp yelps in alarm; also call to
each other with a variety of small talk
(growls, whinnies, twitters) not usually
heard from a distance.
Natural history. Naturally diurnal with
some feeding at night, but has become
nocturnal where intensively hunted; terrestrial; family groups of two to six, or in

open grasslands in herds of up to dozens.
Feeds on grass and browse, especially
aquatic vegetation. Capybaras are always
found near water, in rainforest along rivers
and lakes large enough to have open sky
above and a margin with aquatic plants or
grassy vegetation. The one to six young
accompany the mother. When alarmed,
capybaras usually plunge into the water
and swim under the surface to escape.
They leave distinctive piles of smooth oval
droppings (like giant rabbits) and star-
shaped tracks along the water's edge.
Found in lowland rainforest, dry forests,
scrub, or grassland near water; most numerous in open habitats of seasonally
flooded grasslands with permanent water
sources (llanos, pantanal), where they are a
major prey of jaguar.
Geographic range. Central and South
America: Panama to NE Argentina.
Status. Locally common and widespread,
but uncommon or rare in Panama and in
populated areas of Amazonia; extensively
hunted for meat and easily found with
dogs. Populations in rainforest are small
and narrowly restricted to open watersides.
Because rivers are the main travel routes of
hunters, capybaras are no longer seen on
river trips in populated areas. Large populations still exist in remote forests and
some wet grassland regions (llanos, pantanal). Capybaras are "farmed" for meat
and leather in some areas, with manage-
ment of wild herds for sustainable yield;
their herbivorous diet, large size, high reproductive rate, and herding behavior make
them ideal for this purpose.
Local names. Capibara (Span); carpincho
(Ar, Pa); capivara, cupido (Br); capihuara
(Ec); chigüire (Co, Ve); ponche, lancho
(Co, Pa); ronsoco (Pe); cabiaï, kapoewa
(Su, FG).
References. Ojasti, J. 1973. *Estudio biolo-
gico del chigüire o capibara.* Caracas: Re-
publica de Venezuela fondo Nacional de
Investigaciones Agropecuarias.

Paca
Agouti paca
Plate 29, map 171
Identification. Measurements: HB = 616–
775; T = 12–32; HF = 107–130; E =
43–56; SH = 27–32; WT = 5–13 kg.
**Upperparts chestnut-red or dark brown,
sides from neck to rump with three or**

Map 171

Paca, *Agouti paca*

four lines of large white spots sometimes
coalescing into stripes. Hair coarse,
sparse, and flat. **Head large, cheeks swollen, ears short,** brown; whiskers long;
eyes large; eyeshine bright yellow to orange, eyes widely spaced. **Tail a tiny
stump** hidden beneath rump hair. Lower
cheeks, throat, chest, and **underparts
white.** Forefeet with four toes, hindfeet
with three large toes and two tiny toes that
usually do not touch the ground. **Body
heavy, piglike, with large rump.** Young
like adults.
Similar species. Pacaranas (*Dinomys branickii*) are black, with a conspicuous tail and
spots only on the rear of the body; baby
tapirs (*Tapirus* spp.) have spotted legs and
head; baby deer (*Mazama, Odocoileus*
spp.) have thin necks and legs, and distinct
tails. Another species (*Agouti taczanowskii*)
occurs at higher elevations in the Andes
from Venezuela to Peru; it has spots on the
top of the back and long, dense hair.
Sounds. Pacas walk heavily and noisily in
dry leaves. In alarm they usually do not
call; rarely they give a hoarse bark or make
a deep rumbling sound by grinding their
teeth. The head is modified with expanded
cheek bones and pockets in the cheeks that
probably act to resonate this sound, which
seems to be a social signal.
Natural history. Nocturnal; terrestrial; solitary or rarely in pairs. Feeds on fallen
fruits, browse, and some tubers. Pacas are
most common near water, from large rivers
to small seeps, and in swampy areas and
dense thickets, but they are also sometimes
seen in open forest far from water. They
live in monogamous pairs that share small

territories, but they forage alone. By day
the male and female usually den separately,
in large burrows with a main entrance and
several hidden entrances that are plugged
with leaves. The burrows are usually in
permanently dry banks near water and may
be marked by mounds of dirt. At night between bouts of foraging, pacas sometimes
lie down to rest in the open. Found in mature, disturbed, and secondary rainforest,
montane forest and gallery forest, gardens
and plantations.
Geographic range. Central and South
America: SE Mexico to S Brazil and
N Paraguay. To 2,000–3,000 m elevation.
Status. CITES Appendix III (Ho). Pacas
are the most prized Neotropical game animals for their tender, veallike meat; they
are easily hunted by day with dogs, or at
night with headlights, and are now scarce
or locally extinct in overhunted areas.
Where not hunted they are common and
easily seen. Because of their huge geographic range and liking for inaccessible
swamps, thickets, and second growth, lowland pacas are not threatened except locally, especially in Central America. Efforts
should be made to manage this species for
sustainable hunting yield from intact
forests.
Local names. Paca (Ar, Br, Su); gibnut
(Be); guagua, lapa (Co); guanta (Ec); pak
(FG); tepezcuintle, haleb (Cent Am); ma-
jaz, picuru (Pe); conejo pintada (Pn); water
haas (Su); hei (Sar); acutipá (Gua).
References. Marcus, M. 1983. Population
density, home range, and foraging ecology
of paca (*Agouti paca*) on Barro Colorado
Island. Unpublished Report. Smithsonian
Tropical Research Institute.
    Collett, S. F. 1981. Population charac-
teristics of *Agouti paca* (Rodentia) in Colombia. Pub. Mich. State Univ. Biol. Ser.
5:489–601.

Pacarana
Dinomys branickii
Plate 29, map 169
Identification. Measurements: HB = 475–
513; T = 145–174; HF = 114–123; WT
= 13 kg.
**Upperparts and legs black to brown-
black sprinkled with white hairs; sides
with two wide white stripes grading to
rows of spots from behind shoulder to
rump, with two disorganized rows of**

white spots below these. Head large, grizzled gray-black, muzzle blunt; ears short; eyes small; whiskers long and stout. **Tail about 25% as long as head and body, thick and stumpy, thickly furred,** black or dark brown. Underparts grizzled brown-black variably mixed with gray. Feet with four toes; claws long. **Thick-bodied, short-legged, thick-necked rodents that move slowly and often adopt a sitting posture.** Young like adults.

Similar species. These are the only black animals with white stripes and rows of spots and a conspicuous tail; pacas (*Agouti paca*) are red or brown with no tail.

Sounds. Sounds noted in captivity include foot stamping, tooth chatter, hiss, grunt, and growl.

Natural history. Unknown. Pacaranas are herbivorous, and they are thought to be nocturnal and to live in burrows. In captivity they climb well and like to rest on elevated platforms. Found in lowland and montane rainforest.

Geographic range. South America: the eastern foothills of the Andes from Colombia and Venezuela to Bolivia, and the Amazon lowlands of Peru and W Brazil. To 2,000 m elevation.

Status. Appears to be rare throughout its range; hunted for meat.

Local names. Pacarana, paca-com-rabo (Br); guagua loba (Co); machetero, pacarana, picuru maman (Pe).

References. Tate, G. H. H. 1931. Random observations on habits of South American mammals. *J. Mammal.* 12:248–56.

Red-rumped Agouti
Dasyprocta agouti
Plate 29, map 172

Identification. Measurements: HB = 490–640; T = 13–30; HF = 118–148; E = 40–47; SH = 270–360; WT = 3–5.9 kg.

Head and forequarters finely grizzled olivaceous; rump dark red to brilliant orange, covered by long, straight hairs, which overhang rump in a fringe and are usually paler yellow or orange at base, this color visible when the hairs are erect. **Top of head, neck, and midback between shoulders sometimes blackish, or with crest of longer, pure blackish hairs.** Whiskers stiff, black, reaching to base of

Map 172

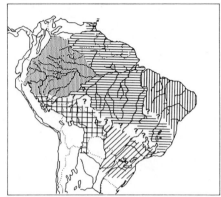

▨ Black agouti, *Dasyprocta fuliginosa*
☰ Red-rumped agouti, *D. agouti*
⫟ Black-rumped agouti, *D. prymnolopha*
▧ Azara's agouti, *D. azarae*
⊞ Brown agouti, *D. variegata*

ear; chin naked; ear short and roundish, mostly naked; eyes large. **Tail naked, a tiny, inconspicuous stub.** Feet dark brown, forefeet with four strong toes, **hindfeet with three long, strong toes and hooflike nails.** Underparts grizzled brownish, orange, or orange with white midline. **A large, ungulatelike rodent with long slender legs, large rump, and strongly humped back.** Runs with a bounding, bouncing gait; walks or trots slightly crouching, head held low. In alarm or excitement may raise long rump hairs in a fan.

Variation. Forequarters vary from dull brown-olivaceous to bright orange-olivaceous; some agoutis from Guyana have black crests on neck and shoulders. Those on the Lesser Antillies, Trinidad, and Tobago are generally smaller than mainland animals and are dark brown; on the Lesser Antilles they have long, chocolate-brown hairs forming a cape over the neck and shoulders. There are about 10 names (species and subspecies) that have been used for these agoutis. Some recent works use the species name *D. leporina*.

Similar species. Red acouchys (*Myoprocta acouchy*) are smaller, with black rump and a prominent little tail with a white tip. See range maps and descriptions of neighboring agouti species.

Sounds. Alarm often heard: a series of low grunts or loud hoarse squeals or barks, often two individuals call; hindfeet may also be stamped on the ground. Gnaws loudly on hard nuts.

Natural history. Diurnal; terrestrial; often in pairs, sometimes solitary. Red-rumped agoutis feed on fruit and nuts or seeds. They can be found throughout the forest, but often in areas of good undergrowth cover, around large treefalls, and near streams, rivers, and swampy areas. These agoutis live in monogamous pairs with their offspring on a territory. Pairs travel together; they are wary and are often heard fleeing in alarm but more rarely seen. The best way to observe them is from a blind near a fruit tree. They are most active in the early morning and late afternoon and are easiest to approach at dusk, when they either do not see well or are less wary. They may sometimes be active on bright, moonlit nights. They bury nuts singly in the ground for retrieval when food is scarce, and this behavior makes them important seed dispersers for many tree species. Found in mature, disturbed, and secondary evergreen, deciduous, and gallery forest, and in gardens and plantations.

Geographic range. South America: the Guianas and Brazil north of the Amazon and east of the Rio Negro, and south of the Amazon east of the Madeira.

Status. Much hunted for meat but usually very common; widespread.

Local names. Cutia (Br); goudhaas, konkoni (Su); picure, acure (Ve).

References. Dubost, G. 1988. The ecology and social life of the red acouchy, *Myoprocta exilis;* Comparison with the orange-rumped agouti, *Dasyprocta leporina. J. Zool., Lond.* 214:107–23.

Black Agouti
Dasyprocta fuliginosa
Plate 29, map 172

Identification. Measurements: HB = 542–760; T = 20–39; HF = 120–143; E = 36–49; WT = 3.5 kg.

Upperparts entirely black, finely grizzled with white (hairs black with tiny white tips); **rump hairs black with inconspicuous white tips on hairs,** not usually overhanging rump in fringe; nape with a slight crest of longer, black hair. Tail

black. Feet black. Throat white or strongly grizzled white; underparts dark brown and more or less grizzled with white, midline of belly sometimes white. General shape and posture like red-rumped agouti. Looks black or dark brown from a distance.

Variation. Some animals from Venezuela have rump hairs heavily frosted with white; some from Venezuela, Brazil, and Peru have hair tips tawny yellow or orange instead of white on foreparts, appearing slightly washed with orange; on close view some animals are dark brown rather than true black.

Similar species. Peccaries (*Tayassu* spp.) are much larger, with thick necks, large heads, and sharply tapering muzzles; acouchys (*Myoprocta* spp.) are smaller and either olivaceous or red with black rump, and have thin, white-tipped tails.

Sounds. In alarm stamps feet and gives a short series of grunts or, more rarely, whistlelike calls. Gnaws loudly on nuts.

Natural history. Diurnal; terrestrial; solitary and in pairs. Feeds on fruits and nuts. Found in mature and disturbed rainforest, deciduous forest, and montane forest.

Geographic range. South America: the western Amazon Basin west of the Rios Negro and Madeira in Brazil, Venezuela, Colombia, Ecuador, and central and N Peru. To above 1,000 m elevation.

Status. Intensively hunted for meat but geographic range large and sparsely inhabited.

Local names. Cutia (Br); guatín, agutí, picur, conejo negro, ñeque (Co); guatusa (Ec); añuje (Pe); picure, acure (Ve).

Brown Agouti
Dasyprocta variegata
Plate 29, map 172

Identification. Measurements: HB = 445–540; T = 11–38; HF = 94–120; E = 41–45.

Upperparts finely grizzled black and tawny yellow, brown and yellowish, or black and orange; from a distance appears blackish washed with tawny, plain brown, or orange; head often blackish, midline of back usually darker than sides. Feet dark brown except in orange animals, when colored like back. **Chin and throat and often midline of venter white, rest of underparts brownish yellow in brown in-**

dividuals, to bright pale orange in orange animals. Shape and posture like red-rumped agouti.

Variation. These agoutis seem to become progressively paler and more orange from north to south and east to west: those from Junín (Peru) are blackish; from the Rio Urubamba plain brown; from Madre de Dios yellowish to reddish brown; from Santa Cruz, Bolivia, bright orange, with orange bellies. This group of agoutis is considered by some to belong to the same species as the Central American agouti, by others to a group with the black agouti. It may intergrade with the black agouti to the north and Azara's agouti to the east of its range. The brown and orange forms have also been considered distinct species.

Similar species. See black agouti (*D. fuliginosa*)

Sounds. In alarm rushes away stamping feet and emitting a short series of low grunts or repeated nasal barks, sound between a bleat and woof. Gnaws loudly on hard nuts, with a slower frequency and deeper resonance than a squirrel.

Natural history. Diurnal; terrestrial; solitary and in pairs. Feeds on fruits and nuts. Brown agoutis favor areas of the forest with dense understory thickets, especially around fallen trees, *platanillos* in swampy areas and on watersides. They are wary and often heard calling in alarm, but they are difficult to see since they will run in alarm many yards from a danger. They are most easily approached when they are gnawing on a nut and are least wary at dawn and dusk, when the light is poor and they seem to venture more out into the open forest. At night they den under a dense pile of brush, usually a fallen tree with many vines. If the den is disturbed at night (such as by someone walking by), they will rush away calling. In mature, disturbed, and secondary forest and gardens and plantations.

Geographic range. South America: east of the Andes from central Peru (Junín) south to Bolivia and N Argentina. To at least 1,500 m elevation.

Status. Common and widespread; hunted for meat.

Local names. Agutí rojizo, akútí (Ar); cutia (Br); añuje (Pe).

Azara's Agouti
Dasyprocta azarae
Plate 29, map 172

Identification. Measurements: HB = 452–575; T = 14–33; HF = 100–120; E = 35–46; WT = 2.4–3.2 kg.

Upperparts gray washed with dull tawny or olivaceous to bright orange, hairs finely banded black, white, tawny yellowish, and orange; **top of head to behind shoulder blackest; midbody broadly washed with orange, flanks and thighs often bright orange, rump hairs black frosted with white or pale orange (looks gray); area around tail orange (viewed from rear).** Sides of head from nose to ear all or partly washed with orange. Legs orange; color ending abruptly at wrist and hind toes, extremities of feet with blackish "socks." **Underparts bright to pale orange,** inguinal region often white. Body shape and tail like red-rumped agouti. Young like adult but rump hair short. One of the smallest agoutis.

Variation. This is one of the most distinctive agoutis; the finely banded hair varies from blackish to grayish looking, with orange parts from dull tawny yellow or olivaceous to bright orange. Animals from drier regions (Paraguay, Argentina) tend to be gray or olivaceous.

Similar species. Rabbits (*Sylvilagus brasiliensis*) are smaller, with long ears; green acouchys (*Myoprocta acouchy*) have conspicuous tails.

Natural history. Diurnal; terrestrial. Probably similar to other agoutis. Found in rainforest, cerrado, and chaco; the southernmost agouti.

Geographic range. South America: east of the Andes in a belt from Santa Cruz, Bolivia, east to São Paulo, Brazil, including rainforests of SE Brazil and Matto Grosso, south to N Argentina and Paraguay.

Status. Often common in a large geographic range.

Local names. Agutí amarillento (Ar); cutia (Br); acutí (Pa).

Black-rumped Agouti
Dasyprocta prymnolopha
Plate 29, map 172

Identification. Measurements: HB = 450–525; T = 18–30; HF = 95–106; E = 36–43.

Foreparts yellow-orange grizzled with black becoming dark red-orange from midback aft; rump top covered by contrasting wedge of long, pitch black hairs, these hairs often pale yellow at base, visible when hairs erect. Crown blackish; neck with crest of longer, black hairs. Feet dark brown. Underparts pale orange sprinkled with white hairs down midline. Young like adult but darker and duller overall. Shape like red-rumped agouti.

Similar species. Red acouchys (*Myoprocta acouchy*) are similar but smaller, with a thin, white-tipped tail; they may not have any geographic overlap. Red-rumped agoutis (*D. agouti*) have an orange rump.

Natural history. Found in deciduous forest and scrub, including cerrado and caatinga, and probably coastal rainforest habitats.

Geographic range. South America: E Brazil from E Pará south of the Amazon to Bahía. The original locality given for this species was "Guiana"; this was probably erroneous, since there appear to be no good records from that region.

Local names. Cutia.

Central American Agouti
Dasyprocta punctata
Plate 29, map 173

Identification. Measurements: HB = 480–600; T = 20–55; HF = 120–156; E = 36–47; WT = 3.2–4.2 kg.

There are two basic color patterns: (1) **Upperparts uniformly warm red-brown, yellow-brown, or gray-yellow; fur banded with fine black striations throughout when viewed closely;** nape and rump hairs not different from rest of upperparts; chin and inguinal region clear orange, or white in gray-yellow animals; chest grizzled like back. (2) **Foreparts brown to blackish, finely grizzled with tawny or olivaceous; crown and nape often blackish; midbody forward of rump with a band of brighter, orange-banded hairs; rump hairs long, black, with long yellow to white tips on hair** overhanging rump in fringe; chin and inguinal region whitish; belly brown. Feet of both are black or brown.

Variation. This species is highly variable, with intermediate colors and patterns linking the extremes. Pattern (1) is found on the Pacific slope from Mexico to Panama and Ecuador; gray-yellow animals are from

Map 173

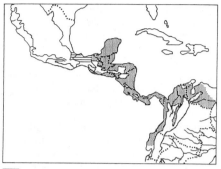

▨ Central American agouti, *Dasyprocta punctata*
▤ Mexican black agouti, *D. mexicana*

the Yucatán. In pattern (2) the foreparts range from blackish with a paler, more tawny area around midbody just forward of rump, to distinctly orange-olivaceous; some animals from Ecuador and W Colombia have pale yellow frosting on rump, tawny orange foreparts, and necks with grayish sides; this color pattern is found along the Atlantic slope from Costa Rica to Panama and Colombia, and NW Venezuela. Animals from Coiba Island, Panama, have been considered a separate species (*D. coibae*); they are small, pale brown agoutis with color pattern (2)

Similar species. Rabbits (*Sylvilagus* spp.) are smaller, with long ears; capybaras (*Hydrochaeris hydrochaeris*) are much larger and uniform yellow-brown. There are no other medium-sized tailless brown mammals in range.

Sounds. In alarm stamps feet and emits a short series of calls each consisting of a few low grunts followed by a sharp bark. Gnaws loudly on nuts.

Natural history. Diurnal; terrestrial; solitary or more rarely in pairs. Feeds chiefly on seeds, fruits, and cotyledons of seedlings, supplemented by fungi, flowers, leaves, and insects. These agoutis favor areas of the forest with dense undergrowth. Although wary where hunted, they are tame and easy to see in some protected reserves with many human visitors. They are most active in the late afternoon just before nightfall. Adults den at night in hollow logs, under fallen brush, or sometimes in burrows. A monogamous pair lives together on a territory, but the couple does not usually travel together. Newborn young live by themselves in a burrow that the mother does not enter: she calls them out for nursing and care. Agoutis bury nuts singly in the ground for retrieval when fruit is scarce, and this activity is important for the dispersal of seeds. Found in mature and secondary lowland and montane rainforest and deciduous forest and in gardens and plantations.

Geographic range. Central and South America: Chiapas and Campeche, Mexico, southeast through all countries of Central America; NW Venezuela, N and W Colombia, and Ecuador west of the Andes. To at least 1,500 m elevation.

Status. CITES Appendix III (Ho). Intensively hunted for meat and threatened by conversion of forest to pasture, but large populations remain in many areas.

Local names. Indian rabbit (Be); guatusa (CR, Ec); gautín (Co); guaqueque alazán, aguti (Me); ñeque, cuinco (Pn); picure (Ve).

References. Smythe, N. 1978. The natural history of the Central American agouti (*Dasyprocta punctata*). Smithsonian Contrib. Zool., no. 257.

Hallwachs, W. 1986. Agoutis (*Dasyprocta punctata*): The inheritors of guapinol (*Hymenaea courbaril*: Leguminosae). In A. Estrada and T. H. Fleming, eds., *Frugivores and seed dispersal*, 285–304. Dordrecht: W. Junk.

Mexican Black Agouti
Dasyprocta mexicana
Plate 29, map 173

Identification: Measurements: HB = 515–560; T = 20–30; HF = 116–127.

Upperparts entirely black or dark brown very finely grizzled with white (hairs have tiny white tips) except rump hairs long and entirely black, nape hairs usually long and entirely black, forming a crest. Chin and throat white, rest of underparts dark brown grizzled white or with white stripe down midbelly. Feet entirely black or with pure white toes on hindfeet. Young like adult.

Variation. An animal from Tabasco is grizzled rich brown; possibly a hybrid with the Central American agouti.

Similar species. Peccaries (*Tayassu* spp.) are much larger, with large heads, narrow snouts, and thick necks.

Sounds. Calls in alarm, probably like other agoutis.

Natural history. Diurnal; terrestrial. Found in wet coastal lowland rainforests.

Geographic range. Central America: Mexico, the lowland Atlantic coastal forests of Veracruz, Oaxaca, and Tabasco. To 500 m elevation.

Local names. Cerreti, cuacechi, guaqueque negro.

Red Acouchy
Myoprocta acouchy
Plate 29, map 174

Identification. Measurements: HB = 335–390; T = 51–78; HF = 90–104; E = 25–40; SH = 170–200; WT = 1.05–1.45 kg.

Upperparts dark, chestnut-red or orange on sides and legs; grizzled with black and some yellow on crown and neck; **midback and rump glossy black or very dark red; rump hairs long, straight, overhanging rear of body in a straight fringe, entirely dark, not banded.** Eyes large, ears quite large, naked, tips high above crown; areas behind ears, around mouth and eyes, and under chin almost naked. Whiskers well developed, black, reaching to behind ear. **Tail short, more slender than a pencil, white below and with a small white tuft at tip, often wagged or flipped,** or held up, showing the white. **Legs long and thin;** forefeet with four toes and a vestigial thumb with a claw, hindfeet with three large, elongated toes with hooflike claws, soles black. Underparts thinly haired, pale to dark rust-red or orange. Mammae pairs = 4. A small animal with large hindquarters, humped back; often runs in a slight crouch.

Variation. Some animals are quite grizzled with yellow and appear olivaceous, but they retain the long, dark rump hairs of this species; some hybridization with the green acouchy may have occurred in Colombia in the headwaters of the Río Uaupés (this form has been given the name *M. milleri*). Both red and green acouchys are said to occur in the Serranía de Macarena, Colombia. The red acouchy is sometimes given the name *M. exilis* and the green acouchy the name *M. acouchy*. Because some individuals are quite olivaceous, there seems no good basis for as-

Map 174

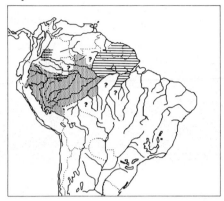

▨ Green acouchy, *Myoprocta pratti*
▤ Red acouchy, *M. acouchy*

signing the name *M. acouchy* to the green species.
Similar species. Green acouchys
(*M. pratti*) are green-olivaceous, with shorter rump hairs that are distinctly banded with different colors and are usually the same color as the shoulders, or only slightly darker, not black; red-rumped agoutis (*Dasyprocta agouti*) are much larger, with bright orange rump, and tail a short, thick, naked stump usually not visible; rabbits (*Sylvilagus* spp.) have long ears, tawny black-streaked backs, and buttonlike tails and are nocturnal.
Sounds. In alarm stamps hindfeet and emits a squeal or whistle followed by a short series of quickly repeated high-pitched chirps; also a trill-like squawk. Gnaws loudly on nuts.
Natural history. Diurnal; terrestrial; solitary. Feeds on seeds and fruits, and the cotyledons of seedlings. Acouchys bury nuts singly in the forest floor for use when food is scarce. Red acouchys favor zones with dense undergrowth, such as around tree falls. They are nervous and wary and are more often heard calling in alarm than seen, but they generally flee only a few yards and hide in a thicket or behind a log, where they can often be spotted standing motionless. After fleeing in alarm, they will often circle silently and approach a motionless observer from behind. They sometimes run along the tops of fallen logs or sit on them, although they are highly

terrestrial and not climbers. They are most active in the early morning and are least wary and most easily seen at dawn or dusk in half daylight. At night red acouchys rest in a nest of leaves within a hollow log, or rarely in a burrow made by another species; entrances not in use are plugged with leaves. If their den is disturbed at night (e.g., by stepping on the log) they run out, calling in alarm. Found in mature lowland rainforest, not usually in extensive areas of secondary forest.
Geographic range. South America: the Guianas and Brazil north of the Amazon and east of the Rio Negro, and south of the Amazon east of the Tapajós; perhaps an isolated population in Colombia in the upper Rió Uaupés.
Status. Common and widespread.
Local names. Cutiara (Br); acoechi, maboela (Su).
References. Dubost, G. 1988. The ecology and social life of the red acouchy, *Myoprocta exilis:* Comparison with the orange-rumped agouti, *Dasyprocta leporina. J. Zool., Lond.* 214:107–23.

Green Acouchy
Myoprocta pratti
Plate 29, map 174
Identification. Measurements: HB = 298–383; T = 40–58; HF = 74–98; E = 31–37; WT = 0.8–1.2 kg.
Upperparts and legs finely grizzled olivaceous, each hair with many narrow, alternating bands of black and yellow; **back and sides usually uniform, or rump sometimes darker, almost blackish, but hairs always banded;** rump hairs long, but usually not overhanging rump in a fringe; back sometimes washed with reddish tones. Sides of muzzle, cheeks, and patch behind ear rusty orange. Chin and area around eye almost naked. Underparts pale orange, thinly haired over yellow skin; **throat, chest, and midline of belly often white.** Small juveniles like adults but with more orange tints. Shape like red acouchy.
Variation. Acouchys from the Venezuelan Amazon may have gray-brown sides and thighs grizzled with white hairs. Sometimes the name *acouchy* is used for this species.
Similar species. See red acouchy; the agoutis that occur with this species are

blackish with black or yellow rumps, red, brown, or yellow-orange.

Sounds. In alarm stamps hindfeet and gives a short series of quickly repeated nasal squawks like a high-pitched party horn. Gnaws loudly on nuts.

Natural history. Diurnal; terrestrial; solitary or more rarely in pairs. Feeds on seeds and fruits. Green acouchys are most common in mature terra firme forest with dense undergrowth. They nest in hollow logs and sometimes in burrows. Their behavior when encountered is similar to that of red acouchys. Mature, lowland rainforest habitats.

Geographic range. South America: east of the Andes and west of the Rio Negro, north of the Amazon in southern Venezuela, Brazil, Colombia, Ecuador, Peru, and south of the Amazon in Peru and Brazil to at least the Rio Madeira.

Status. Often common in a huge geographic range; hunted for meat.

Local names. Cutiara, cutia de rabo (Br); guatusa pequeña, papali (Ec); punchana, añuje menor (Pe).

Spiny Rats and Tree Rats (Echimyidae)

Members of this family are all large (about 130–900 g) rodents with ratlike shapes. A few have soft fur, but in most it is spiny or bristly. The spines are flat, without barbs, always flexible, and have a longitudinal channel on the upper side; they usually lie flat and look like normal fur from a distance. Most species have tails constructed so that they break off easily if pulled, and as a result many individuals have shortened or entirely missing tails. The ears are short and naked, often with a group of hairs growing from the inside rim, and they have a characteristic irregular rear margin as if they had been damaged. All species have large eyes and long whiskers, including a large group of genal whiskers (from the rear corner of the eye). They have pairs of lateral mammae buried in the fur of the back, along the sides, never on the abdomen, except in the inguinal region. There is a long, narrow skin gland running down the midline of the sternum, especially well developed in males. The spiny rats (*Proechimys* spp.) are often the most numerous terrestrial mammals in a rainforest; they give birth to about two or three young as often as every two to three months. The tree rats (*Echimys* spp.) are all arboreal; in these the nail of the second toe of the hindfoot is flared inward and cup-shaped. They have small litters of one or two young, probably only once or twice a year. Tree rats are difficult to see or capture, and little is known about them, including their precise geographic ranges. Several very beautiful species with no apparent close living relatives seem to occur in relict populations, where they may be rare and perhaps threatened, including the painted tree rat, the red crested tree rat, and perhaps the dark tree rat. All but one species in SE Brazil belong to a distinct group, the Atlantic tree rats, *Nelomys* (sometimes placed in the genus *Echimys* or *Phyllomys*); the taxonomy of this group is still unclear; there may be other species in addition to those described below. This is the largest family of caviomorph rodents, with about 15 genera and 60–100 species; all but 4 genera are found in rainforest. There are certain to be future changes to the taxonomy, with additional species recognized beyond those described below.

Spiny Rats

Proechimys spp.

Plate 24, map 175

Identification. Measurements: HB = 180–275; T = 124–180; HF = 38–56; E = 20–26; WT = 150–550 g.

Upperparts glossy **chestnut-brown** to gray-brown; **fur stiff, bristly,** or with narrow flexible spines, spines not visually obvious except in SE Brazilian species; sides usually grayish or yellowish brown. **Un**derparts usually **snowy white**, occasionally partly gray or buff, **sharply demarcated from sides. Head long and narrow,** with pointed snout; **ears naked, oblong,** dark gray, and medium-sized, **protruding above crown;** eyes large, dark brown; eyeshine bright yellow; whiskers long, to shoulder. **Tail naked, bicolored dark gray above, whitish below;** breaks off easily and is often shortened or entirely missing. **Large, handsome rats.** Hindfeet usually long and

Map 175

 Spiny rats, *Proechimys (Proechimys)*
Spiny rats, *Proechimys (Trinomys)*

narrow, soles black; feet usually all or partly white above. Young of most species are gray-brown with white underparts, but young *P. brevicauda* are similar to adults.

Species. Over 70 forms have been described, but there are probably no more than 20–30 species; the taxonomy is still under revision. Color of back varies from dark blackish brown to cinnamon, or grayish or yellowish in species from drier areas. Museum specimens lose their gloss and are often faded to a cinnamon color unlike that of living animals.

P. semispinosus is the dominant species in Central America and N South America west of the Andes. Other common species include: *P. steerei, P. brevicauda, P. simonsi* (western Amazonia), *P. cuvieri* and *P. guyannensis* (Guianas and eastern Amazon Basin), *P. guairae* (N Venezuela). All species in E Brazil belong to the subgenus *Trinomys;* these have prominent spines on the lower back, are often grayish brown, and have the first and fifth toes of the hindfoot greatly reduced; some have hairy or tufted tails. In Amazonia, up to four species can occur in the same locality. Only experts can distinguish between them.

Similar species. Spiny rats are the only large, chestnut rats with bristly fur, ears that protrude above crown, and sharply demarcated snowy underparts; armored rats *(Hoplomys gymnurus)* are similar, with strong, upstanding spines in a geometric pattern on lower back; other large terrestrial rats (water rats [*Nectomys* spp.], marsh

rats [*Holochilus* spp.], crab-eating rats [*Ichthyomys* spp.], black and Norway rats [*Rattus* spp.]) have gray-brown backs, solid-color tails, or grayish bellies, and soft, rather than stiff, fur; tree rats (*Echimys* spp.) have short ears that do not protrude above crown.

Sounds. When alarmed sometimes calls with a musical twittering (especially females with young); almost always flees by suddenly bounding away with a series of rapid taps of the hindfeet on the ground (the only common small nocturnal mammal to do this).

Natural history. Nocturnal; terrestrial; solitary. Feeds on seeds, fruits, and fungi (especially mycorrhizal fungi), and a few leaves and insects. Spiny rats are found throughout dryland forest but are most common in dense, viny undergrowth and around fallen trees and complex tree roots. They do not climb except onto low fallen logs. They forage slowly, walking around in a small area, and are often seen sitting motionless near logs or roots or under shrubs. Piles of palm-nut shells show where they take nuts to eat under roots, logs, or brush. During the day they shelter under fallen brush, in hollow logs, or in holes in the ground. From an early age (1 week), the young follow the mother as she forages. Spiny rats are usually the most numerous terrestrial mammals in any rainforest habitat where they occur, and they are easy to see at night. They are the major prey of ocelots, bushmasters, and probably many other predators, and they are prolific breeders. They do not nest in houses, nor do they usually enter them, but they may live near them. Found in many forest types throughout the entire Neotropical rainforest area, including some dry forests and gallery forests extending into drier habitats.

Geographic range. Central and South America: Honduras to Paraguay. To maximum elevation of 1,800 m, but upper limit is usually much lower.

Status. Almost always common. Spiny rats have excellent meat and, although small, are sometimes trapped for food and may be sold in markets.

Local names. Sauiá, rato-de-espinho (Br); rata, sacha cuí (Pe); stekelrat, maka-alata (Su); mocangué (Pn).

References. Patton, J. L. 1987. Species groups of spiny rats, genus *Proechimys* (Rodentia: Echimyidae). *Fieldiana Zool.*, n.s., 39:305–45.

Emmons, L. H. 1982. Ecology of *Proechimys* (Rodentia, Echimyidae) in southeastern Peru. *Trop. Ecol.* 23:280–90.

Armored Rat
Hoplomys gymnurus
Plate 24, map 176
Identification. Measurements: HB = 212–300; T = 114–240; HF = 47–61; E = 20–29; WT = 218–790 g; males reach larger size than females.

Upperparts cinnamon-brown or almost black with prominent, geometrically spaced, strong, sharp, back-slanting black spines sticking up through fur on back and rump; rump almost black with dense spines; spines on sides may be pale-tipped, or older animals may have a sprinkling of white spines; sides cinnamon paler than back. **Head long and narrow; ears naked, narrow, standing up above crown;** whiskers long, black, reaching shoulder or beyond; eye quite large; eyeshine bright yellow. **Tail shorter than head and body, naked, robust, bicolored dark above, pale below,** especially marked toward tip; **often broken short or entirely missing. Underparts pure white, sharply demarcated from sides,** often with a dark collar across throat, **or** in black individuals may be **dark brown with variable amounts of white splotching.** Feet dusky on outer edge, white on inner side, or entirely white, or dusky in black animals. Young like adult.

Variation. In some parts of Panama blackish (melanistic) individuals are common, with great individual variation in the degree of darkening of the body, underparts, and feet. Similar to and probably should be placed in the same genus as spiny rats (*Proechimys*).

Similar species. Spiny rats (*Proechimys* spp.) are similar but do not have conspicuous, upstanding spines except in SE Brazil; see spiny rat.

Sounds. Makes whining and squealing calls when disturbed.

Natural history. Nocturnal; terrestrial. Feeds on fruits. These rats are the most numerous mammal in the pluvial rainforests

Map 176

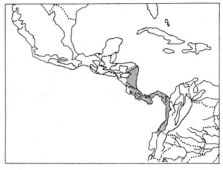

Armored rat, *Hoplomys gymnurus*

of western Colombia, the wettest forest habitat in the New World. They frequent fallen logs, brush piles, rocks, and streamsides in forest and seem to have an affinity for water. In mature and disturbed rainforest.

Geographic range. Central and South America: Honduras south along the Atlantic slope to Costa Rica and Panama and west of the Andes on the Pacific slope of Colombia and Ecuador. To about 800 m elevation in Panama, lower in Colombia.

Status. Locally common.

Local names. Rata éspinosa (Span).

References. Tesh, R. B. 1970. Notes on the reproduction, growth, and development of echimyid rodents in Panama. *J. Mammal.* 51:199–202.

White-faced Tree Rat
Echimys chrysurus
Plate 26, map 177
Identification. Measurements: HB = 232–300; T = 293–370; HF = 42–54; E = 11–20; WT = 415–890 g.

Upperparts **pale gray-brown; densely covered with wide flat spines from neck to rump,** all other pelage stiff bristles. **Head with white blaze down center of face** from crown to nose, either completely covering face between eyes or a narrow stripe, or small white tuft on crown; crown beside blaze usually dark brown; cheeks and sides of neck sometimes reddish. Ears small, thinly haired, brown; eyes dark brown, eyeshine weak, dull red. **Tail fully haired, basal half to two-thirds dark brown, distal half to one-third white,** yellow or chestnut. Chin, throat, inguinal region, and

Map 177

□ White-faced tree rat, *Echimys chrysurus*
▤ Dark tree rat, *E. saturnus*
▥ Speckled tree rat, *E. semivillosus*
■ Golden Atlantic tree rat, *Nelomys blainvillei*
▦ Pallid Atlantic tree rat, *N. lamarum*
▨ Black-spined Atlantic tree rat, *N. nigrispinis*
▦ Rusty-sided Atlantic tree rat, *Nelomys* sp.
▨ Drab Atlantic tree-rat, *N. dasythrix*
⬤ Giant Atlantic tree rat, *N. thomasi*

band across chest behind elbows white; rest
of underparts gray-brown not sharply differentiated from sides. Feet large, broad,
with strong claws, gray-brown above.
Young like adults.
Variation. Some animals from the Guianas
are tinged reddish. There is much individual variation within populations in color of
pale parts of the body (white, yellow, or
on tail chestnut), length of pale tip on tail,
and width of white facial stripe; underparts
may lack white.
Similar species. These are the only large,
heavily spined gray-brown rats in their
range with a white blaze and a fully furred,
white-tipped tail.
Sounds. Silent when seen; sometimes
drops pieces of fruit.
Natural history. Nocturnal; arboreal; usually solitary. Feeds on fruit and probably
leaves. White-faced rats use the middle and
upper levels of the forest, especially in
viny areas, but also in open forest. These
rats can run quickly through the branches,
but they usually move slowly and are difficult to spot. When disturbed they creep
quietly into thick vegetation and sit motionless for many minutes, then sneak away

silently. They make nests of leaves in can-
opy tree holes. Mature rainforest.
Geographic range. South America: the
Guianas and Amazon Basin of Brazil east
of the Rios Negro and Xingu.
Status. Unknown, usually uncommon;
widespread.
References. Miles, M. A., A. A. De
Souza, and M. M. Póvoa. 1981. Mammal
tracking and nest location in Brazilian for-
est with an improved spool-and-line de-
vice. *J. Zool., Lond.* 195:331–47.
Local names. Sauiá, conoco, bandeira
(Br); maka-alata (Su).

Dark Tree Rat
Echimys saturnus
Plate 26, map 177
Identification. Measurements: HB = 240–
335; T = 280–295; HF = 48–51; E =
17–18.
**Upperparts glossy brown; head, mid-
back, and tail base deep glossy black;
sides chestnut;** pelage stiff bristles, center
of back with wide, flat spines (not extending to rump). Ears short, buried in hair of
head. Underparts white, white-spotted, or
whitish with buff tinge; chin dark. **Tail
densely haired, black at base** with variable amount of **white on or near tip.** Feet
broad and strong, claws strong.
Variation. Some animals are entirely dark
brown and black.
Similar species. The only large blackish
rat in its range with fully furred tail. Black
(melanistic) squirrels have prominent ears
and bushy tails; Peruvian tree rats (*Echimys
rhipidurus*) are dull brown, with an almost
naked tail (the furry-tailed variety does not
occur in same area).
Natural history. Found in premontane and
lowland rainforests.
Geographic range. South America: Ecuador and N Peru, in the eastern Andean
foothills and adjacent lowlands. To at least
1,000 m elevation.
Status. Rare, known from fewer than 10
individuals, from a small geographic
region.

Map 178

▦ Giant tree rat, *Echimys grandis*
▤ Peruvian tree rat, *E. rhipidurus*
⦿ Painted tree rat, *E. pictus*

Giant Tree Rat
Echimys grandis
Plate 26, map 178
Identification. Measurements: HB = 260–330; T = 260–320; HF = 54–62; E = 18–22; WT = 582–586 g (and probably much heavier).

Upperparts golden black, a coarse mixture of black, yellow, and orange-tipped hairs, sometimes rusty; midback blackest, paling to brownish yellow sides; pelage of dry, stiff, straight, coarse bristles; shines with gold iridescence at some angles to light. **Head with blackish face, crown, and sides;** ear thinly haired, dark gray; whiskers long, coarse, black. **Tail fully haired with smooth, straight, flat hairs; same color as back for first one-sixth, abruptly changing to pitch black for distal five-sixths.** Underparts dirty whitish yellow, sometimes tinged with buff, often gray or pale brown on throat; inguinal region and ventral tail base usually rusty. Feet grizzled black and yellow, large, broad and strong, toes long, claws thick, sharp and curved; hindlegs thick at ankles. **A large, squirrellike rodent.** Young like adults.
Variation. Rare individuals have a white or rusty patch on the nape, a 1–2 cm white tip on the tail, or more blackish or rusty backs.
Similar species. Brush-tailed tree rats (*Isothrix* spp.) are olivaceous, with soft fur;

black (melanistic) squirrels have longer ears, bushy tails, and are diurnal.
Natural history. Nocturnal; arboreal. Nests in tree holes, often near water. Found only in lowland rainforest, probably in Várzea.
Geographic range. South America: Brazil along the Amazon on both banks from the Rio Negro to the Ilha Caviana at the mouth, and the lower Rio Negro; known only from near the main river.
Status. May be locally common in a small geographic range.
Local names. Toró preto.

Peruvian Tree Rat
Echimys rhipidurus
Plate 26, map 178
Identification. Measurements: HB = 210–247; T = 180–215; HF = 31–45; E = 11–17; WT = 315 g.

Upperparts reddish or yellowish brown, finely or heavily streaked with black hairs; head and forequarters more grayish or yellowish, rump and tail base more rusty or reddish; **fur slightly glossy, stiff, narrow bristles, and flexible spines with tips drawn out into hairlike processes; without prominent sharp, flat spines on forequarters or sides of body;** rump sometimes lightly sprinkled with a few white speckles from white-tipped spines. **Base of whiskers dull rust-red,** but not top of muzzle between eyes; whiskers coarse, numerous, black, long, reaching shoulder; ear small, almost naked. **Tail robust, 70–95% of head and body length, furred like back for 3.5 cm, rest either covered to tip with short, glossy, dark brown or grayish bristly hairs that do not quite hide scales,** or with a short line of such hairs on top surface of base, these diminishing in size for about half the length of tail, rest of tail thinly covered with inconspicuous pale brown hairs that form a tiny tuft beyond tip, and **from a distance tail appears naked.** Underparts entirely clear pale whitish, yellow or orange, or mixed with gray-brown especially around midbody; tail base below and rear of thighs dull to bright rust-red. Hindfeet short and broad, reddish above; forefeet small, gray above. Young less red than adults, underparts and hair on tail gray.
Variation. Large, hairy-tailed animals are from the Río Amazonas near Iquitos;

small, "naked"-tailed animals are from the Río Ucayali Basin near Pucallpa; they may represent two species.

Similar species. Red-nosed (*E. armatus*) and bare-tailed (*E. occasius*) tree rats are probably indistinguishable in the field from the naked-tailed form; the hairy-tailed forms are the only large red-brown rats with a hairy tail.

Natural history. Known only from lowland rainforest.

Geographic range. South America: Peru, central and northern Amazon Basin.

Status. Apparently uncommon.

Red-nosed Tree Rat
Echimys armatus
Plate 26, map 179

Identification. Measurements: HB = 200–260; T = 160–250; HF = 37–45; E = 16–21; WT = 340–405 g.

Upperparts drab **reddish brown** to yellowish brown heavily streaked with black, becoming more rusty on rump and tail base; midback and rump finely speckled by paler, reddish brown tips on blackish bristles; **dorsal fur a mixture of thin bristles and flat, flexible sharp spines;** sometimes with thin rusty hairs between spines. **Muzzle faint to bright rust-red,** including top to between eyes; whiskers coarse, reaching shoulder, brown; ears small, thinly haired; eyes dark brown, eyeshine faint dull red. **Tail usually shorter than head and body, thickly haired at base for 3–4 cm, rest almost naked, scaly,** thinly haired with inconspicuous brown hairs that do not extend beyond tip in a tiny tuft. Underparts pale orange, usually mixed with gray or brown at midbody; or entirely gray-brown, sometimes with white areas; rear of thighs and below haired tail base faint to bright rust. Feet short and broad, pale reddish or yellowish. Young like adults but without spines.

Variation. Animals from eastern Amazonia are small and more reddish, with brighter rusty parts of body; those from central Amazonia are more yellowish gray, with indistinct reddish muzzle and thighs; animals from Colombia, Venezuela, Guyana, and the central Amazon Basin are large, with heavily black-streaked backs and entirely gray-brown underparts. This species is sometimes placed in the genus *Makalata*.

Map 179

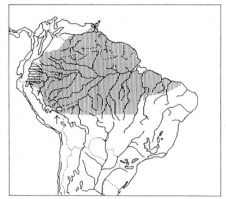

▨ Red nosed tree rat, *Echimys armatus*
▤ Bare-tailed tree rat, *E. occasius*

Similar species. See Peruvian tree rat (*E. rhipidurus*); bare-tailed tree rats (*E. occasius*) have fine whiskers, dense spines, conspicuously bare tail equal to or longer than head and body; speckled tree rats (*E. semivillosus*) are yellow-gray with white speckles; brush-tailed rats (*Isothrix* spp.) are olivaceous, with soft fur and a bushier tail; spiny tree rats (*Mesomys* spp.) are smaller, more heavily spined, with hairy tails.

Sounds. Silent when encountered; sometimes drops fruit fragments.

Natural history. Nocturnal; arboreal; solitary. Feeds on fruits and unripe seeds. Red-nosed tree rats use the canopy and middle layers of the forest but sometimes descend close to the ground. They favor dense, viny vegetation and have a habit of sitting motionless among thick foliage. When alarmed they will either sit motionless where they are or creep stealthily into a thicket and hide for many minutes before sneaking away upward. They travel quietly, without jumping or disturbing the vegetation. By day they nest in tree holes. Common in some areas, but always difficult to see. Found in mature and secondary rainforest and gardens, usually near water.

Geographic range. South America: east of the Andes from Colombia and Venezuela south to Peru and Bolivia, the Guianas, Amazon Basin of Brazil, Trinidad and Tobago. To about 700 m elevation.

Status. Widespread and locally common.

References. Charles-Dominique, P., et al. 1981. Les mamifères frugivores arboricoles nocturnes d'une forêt guyanaise: Inter-relations plantes-animaux. *Rev Ecol.* 35:341–435.

Bare-tailed Tree Rat
Echimys occasius
Plate 26, map 179
Identification. Measurements: HB = 218; T = 225; HF = 37; E = 40.

Upperparts dull buff-brown streaked with dark brown; body from shoulder to rump on both back and sides thickly covered with wide, flat, sharp, flexible spines, these lie flat and may not be apparent from a distance; spines on back brown or black, many with rusty tips, especially on rump; tips of spines not drawn out into hairlike processes, **spines on sides gray-brown; fine, soft, sparse, orange hair between spines on body.** Head clothed with stiff, thin bristles; ear short, naked; tip of muzzle and cheeks reddish; **whiskers fine,** thin, but long, reaching shoulder. **Tail robust, long,** equal to or longer than head and body length; **body hair ending abruptly 1–2 cm from base, rest of tail conspicuously naked, scaly. Underparts clear orange, often with white chest and inguinal region;** undersurface of tail base and rear of thighs not rust-red. Feet brownish with whitish toes.
Similar species. Red-nosed (*E. armatus*) and Peruvian (*E. rhipidurus*) tree rats have coarse, thick whiskers, tail always hairier than the present species and usually shorter than head and body, base of tail thickly haired for about 3 cm, and underparts usually gray-brown, but these species may not be distinguishable in the field; spiny tree rats (*Mesomys* spp.) are smaller, with long reddish brown hairs on tail.
Natural history. Unrecorded. A rat apparently of this species was photographed by Emmons as it sat motionless on a branch at 5 m height in whitewater floodplain forest at night. From lowland and perhaps montane rainforest.
Geographic range. South America: Ecuador and Peru east of the Andes. The type specimen is said to have come from the western slope of the Andes in N Ecuador at 1,300 m elevation, but this seems erroneous since all others known are from the lowlands of the eastern side.
Status. Rare; known from fewer than 10 individuals.

Speckled Tree Rat
Echimys semivillosus
Plate 26, map 177
Identification. Measurements: HB = 200–268; T = 210–261; HF = 36–43; E = 17–22; WT = 194–407 g.

Head, and often neck and shoulders, gray streaked with black; midback gray washed with fulvous; rump or rump and midback well speckled with white-tipped spines; sides like back or gray; **shoulder to rump heavily spined with flexible, flat spines.** Ear small, often with pale patch behind and below; whiskers numerous, long, to behind ear, black. **Tail robust, 80–120% of head and body length, appears naked,** scaly, thinly covered with stiff, pale brown hairs. Feet broad, gray or gray-yellow above. Underparts white or pale orange, sometimes with gray across abdomen. Young gray with gray-yellow-brown rump, soft fur and no spines.
Variation. Animals from Colombia have a white muzzle and an almost black crown.
Similar species. See red-nosed tree rat (*E. armatus*).
Natural history. Probably nocturnal; arboreal. Speckled tree rats den in tree holes. Found mostly in gallery forest, thorn forest, dry forest, and trees in llanos, but rarely also in rainforest.
Geographic range. South America: NE Colombia and NW Venezuela. To at least 600 m elevation.
Status. Locally common.

Painted Tree Rat
Echimys pictus
Plate 27, map 178
Identification. Measurements: HB = 250–295; T = 273–325; HF = 43–47; E = 16.

Upperparts with striking black and white markings; entirely dirty white except sharply defined glossy black saddle on top of back from shoulders to tail, across upper arms; and a wide black stripe on neck and crown forming points on nape and between eyes; fur coarse but not bristly, dense, of uneven lengths, brown at base with white or black tips, underfur long and wavy. Ears short, naked, not protruding above crown; whiskers very fine, longest to shoulder. **Tail much longer than head and body, thickly furred, black with tip shiny, silky white above,** golden yellow below. Feet extremely broad, toes long, whitish or gold above, soles black to base of toes.

Variation. This species was originally described as brown-and-white; this may have been due to fading of the fur, but possibly there are brown-and-white animals. This rat does not seem closely related to any other species; it has variously been placed in *Echimys,* to which it seems closest, or *Nelomys,* or *Isothrix.*

Similar species. Hairy dwarf porcupines (*Coendou* spp.) have spines, prehensile tails, and no striking black markings. They look like giant hair balls with huge pink noses.

Natural history. Found in Atlantic forest and cocoa plantations.

Geographic range. South America: E Brazil, known only from S Bahía around Ilhéus.

Status. One of the rarest rodents; known only from a few individuals from a small region.

Black-spined Atlantic Tree Rat
Nelomys nigrispinis
Plate 26, map 177
Identification. Measurements: HB = 215; T = 240; HF = 40; E = 18.
Upperparts warm reddish brown evenly and lightly streaked with black; sides brighter fulvous; pelage a mixture of stiff bristles and flexible spines with tips drawn out in hairlike processes, the **spines not visible, even at close range looks like normal coarse fur.** Ears short, naked; eye sometimes surrounded by dusky ring; whiskers medium to long, reaching ear or shoulder. **Tail longer than head and body, moderately robust, covered with smooth, shiny, pale brown hair that forms a smooth tuft at tip; tail looks paler than back.** Feet and hands dusky brown, toes and outside of hindfoot silvery white. **Underparts pale orange, hairs white at base,** chin, chest, and inguinal region may be white. Young like adults.
Variation. Another species or subspecies (*N. medius*), from Santa Catarina, Brazil,

is larger, but externally identical.
Similar species. Drab Atlantic tree rats (*N. dasythrix*) are mousy gray-brown with dirty yellow or gray underparts; pallid Atlantic tree rats (*N. lamarum*) are dirty yellow not streaked with black, with a thinly haired tail; rusty-sided Atlantic tree rats (*Nelomys* sp.) have heavily spined backs, bright rusty sides, gray belly, and tail shorter than head and body; southern bamboo rats (*Kannabateomys amblyonyx*) are larger, with square muzzles; other plain large rats in region have large ears and naked tails.
Natural history. Found in mature and secondary Atlantic coastal rainforest.
Geographic range. South America: SE Brazil, São Paulo and Rio de Janeiro. To at least 850 m elevation.

Giant Atlantic Tree Rat
Nelomys thomasi
Map 177
Identification. Measurements: HB = 270–287; T = 270–340; HF = 42–48; E = 17–20.
Upperparts rich reddish brown streaked with black; sides cinnamon; fur a mixture of bristles and flexible spines, spines not apparent. Ear short, with a few black hairs growing from inner rim. **Tail robust, thickly covered with dark brown hair, tail sometimes broken off or missing.** Underparts yellow-orange grading gradually from sides. **Very large.**
Variation. Sometimes considered a subspecies of the black-spined tree rat.
Similar species. These are the only giant tree rats in their range. Norway rats (*Rattus norvegicus*) have naked tails and large ears.
Natural history. Said to frequent coconut palms and to be considered an agricultural pest that eats mandioca.
Geographic range. South America: Brazil, São Paulo, Ihaa São Sebastião only.
Status. Unknown; geographic range miniscule, but there is currently considerable forest remaining on the island and the species apparently adapts to disturbed or secondary vegetation.
Local names. Cururuá, cururuá-sem-rabo.

Pallid Atlantic Tree Rat
Nelomys lamarum
Plate 26, map 177
Identification. Measurements: HB = 195–230; T = 210–233; HF = 32–35; E = 13–16.
Upperparts uniform drab, pale, grayish-yellow-brown (like clay or ocher); sides dirty yellowish; fur conspicuously spiny, wide, flat, sharp, flexible spines mixed with hairs, spines visible on rump. Ear small, whiskers fine, long. Tail thinly haired with short pale brown hairs that do not hide scales, slight tuft at tip darker brown; tail of large adults usually 90–96% of head and body length, longer in young. Hands and feet yellowish white. Underparts pure white, the white sometimes restricted to a narrow line down midline, or in splotches. Medial glandular stripe on midline of chest naked, surrounded by stained fur.
Variation. In other literature and in museums this species is generally confused with the golden tree rat (*N. blainvillei*).
Similar species. See black-spined Atlantic tree rat (*N. nigrispinis*); golden Atlantic tree rats (*N. blainvillei*) are less spiny, brighter gold, and have a thickly haired tail.
Natural history. From lowland Atlantic coastal forest.
Geographic range. South America: E Brazil, coastal Bahía at 300 m elevation.

Rusty-sided Atlantic Tree Rat
Nelomys sp.
Map 177
Identification. Measurements: HB = 215; T = 205; E = 14; WT = 225 g. Crown and neck dull orange brown heavily but finely streaked with black; back and top of rump black coarsely streaked and speckled with rust; thickly covered with wide, sharp, flexible spines, black tipped with rust, intermixed with pure rusty bristles; sides of body and rump and thighs bright rust; spines on entire body including underparts are pale gray at base. Muzzle and sides of head somewhat grayish; whiskers long, reaching shoulder; ear short, blackish. Tail shorter than head and body, moderately covered with brown hair, scales show through near base, hairs denser in slight tuft at tip.

Feet brown washed with gray-yellow, toes and side of hindfoot white. Underparts gray washed with whitish buff.
Variation. This rat has been called *Phyllomys brasiliensis* in the literature, but its correct name is not clear.
Similar species. See black-spined Atlantic tree rat (*N. nigrispinis*).
Natural history. Found in Atlantic forest.
Geographic range. South America: SE Brazil, Minas Gerais and perhaps coastal forests from S. Bahía to Rio de Janeiro.

Golden Atlantic Tree Rat
Nelomys blainvillei
Plate 26, map 177
Identification. Measurements: HB = 192–250; T = 155–220; HF = 33–40; E = 14–20.
Upperparts entirely pale yellow-orange to darker red-gold, head and back darkest, sides paler; pelage stiff bristles and soft, flat flexible spines, spines not apparent. Ears conspicuous, stand out from, but not above, crown; whiskers dense and long, to shoulder. Tail robust, 80–100% of head and body length, thickly haired, with longer, slightly wavy hairs forming bushy tuft at tip; either dark brown or dusky, darker than back, or pale brown close to color of back, but tip browner, below silver. Toes white. Underparts pure yellow-white.
Variation. This species has generally been confused with the pallid tree rat.
(*N. lamarum*).
Similar species. These are the only gold rats with thickly haired tail.
Natural history. Arboreal; probably nocturnal. Feeds on fruits. Golden tree rats live in small groups in holes in trees or in palms. Primarily found in caatinga vegetation.
Geographic range. South America: E Brazil, Ceará and Bahía.
Local names. Rabudo vermelho, rato-cuandu, rato-de-espinho.

Drab Atlantic Tree Rat
Nelomys dasythrix
Map 177
Identification. Measurements: HB = 180; T = 210; HF = 35; E = 16.
Upperparts entirely dull mousy gray-brown, slightly paler on thighs; with some ocher-yellow spine tips on rump; spines

inconspicuous, soft. **Tail longer than head and body, well covered with brown hairs that nearly cover scales and form a distinct pencil at tip.** Underparts dirty yellow-white to buff, hairs gray at base except throat, armpits and inguinal region may be white.
Similar species. Black-spined Atlantic tree rats (*N. nigrispinis*) are reddish; pallid Atlantic tree rats (*N. lamarum*) have conspicuous, wide spines.
Natural history. Found in Atlantic forest.
Geographic range. South America: Brazil, Rio Grande do Sul, perhaps to São Paulo.

Yellow-crowned Brush-tailed Rat
Isothrix bistriata
Plate 27, map 180
Identification. Measurements: HB = 240–275; T = 270–300; HF = 47–50; E = 16–22; WT = 320–570 g.
Upperparts olivaceous gray-brown, rump warmer brown with rusty tints; fur dense and soft, with no bristles. **Head with pale yellow medial spot on crown, bordered by a wide black stripe above each eye,** stripes broadening behind to nearly coalesce on nape; muzzle and cheeks pale gray; whiskers long, black, reaching shoulder; ear very short, sometimes outlined behind and below by yellowish; eye large, dark brown; eyeshine bright yellow. **Tail slightly longer than head and body, thickly and evenly covered with short hair that curls outward with a bottle-brush effect, bright rust-orange or yellowish for basal 10–50%, black for distal 50–70%,** or entirely orange or dusky. Feet grayish. Underparts pale yellow or orange with slate gray base of hair. Young like adults, but duller. Large, squirrellike rats.
Variation. There is much individual variation in the amounts of black and orange on the tail; some animals have completely blackish or orange tails. Crown patch is whitish in some populations or individuals. There seems to be more variation between individuals than between populations.
Similar species. Squirrels (*Sciurus* spp., *Microsciurus flaviventer*) have no facial stripes and bushier, olivaceous tails.
Natural history. Nocturnal; arboreal. Three individuals seen together sat on a branch 6 m high, two of them side by side in contact. They sat motionless for many min-

Map 180

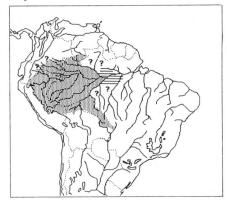

▒ Yellow-crowned brush-tailed rat, *Isothrix bistriata*
☰ Plain brush-tailed rat, *I. pagurus*

utes, behaving like other tree rats. They were in luxuriant floodplain vegetation. By day brush-tailed rats den in tree holes, often in hollow palms, on the borders of rivers. In late afternoon they are said to sit at the entrance of their dens with their heads poking out. Lowland evergreen rainforest.
Geographic range. South America: the Amazon basins of Venezuela, Colombia, Ecuador, Peru, Bolivia, and Brazil west of the Rios Negro and Upper Madeira.
Status. Unknown; appears uncommon everywhere.
Local names. Conocono pequeño (Pe).
References. Patton, J. L., and L. H. Emmons. 1985. A review of the genus *Isothrix*. Amer. Mus. Novitates, no. 2817.

Plain Brush-tailed Rat
Isothrix pagurus
Plate 27, map 180
Identification. Measurements: HB = 180–241; T = 170–271; HF = 37–45; E = 16.
Forequarters grizzled yellowish or brownish gray; hindquarters dull rusty brown; fur dense and soft. Head without prominent markings, face reddish; ear short; eye large; eyeshine bright yellow; whiskers long, reaching shoulder. **Tail much longer than head and body, grades from gray-brown at base to yellow-brown toward tip, completely covered with hairs that curl outward, on undersurface hairs**

grow laterally so naked tail bone is visible in field; sometimes curls tail in a loose coil. Underparts yellowish or pale orange with hair gray at base. Young like adults except tail gray-brown to dark brown.

Similar species. Local squirrels (*Sciurus aestuans*) have ears that protrude above crown and bushier tails that are flipped upward against the back; giant tree rats (*Echimys grandis*) have black head and tail; white-faced tree rats (*E. chrysurus*) have white facial blaze and white tail tip; no other large rodents in range have hairy tails.

Natural history. Nocturnal; arboreal. Plain brush-tailed rats will sit motionless for many minutes when disturbed. They den in tree holes and can be found in terra firme forest far from water. Lowland evergreen rainforest.

Geographic range. South America: Brazil, Amazon Basin east of the Rios Negro and Madeira; limits poorly known.

Status. Uncommon.

References. See yellow-crowned brush-tailed rat.

Rufous Tree Rats
Diplomys caniceps, D. labilis
Plate 27, map 181

Identification. Measurements: HB = 212–320; T = 178–267; HF = 40–48; E = 12–20; WT = 359–430 g.

Upperparts dull rusty red-brown faintly streaked with dark brown hairs, reddish brightest on rear half of body; forequarters sometimes fulvous; fur soft and dense. Head forward of ears gray streaked with black; face marked with small whitish spot over eye, white streak at base of whiskers, pale yellow spot behind and below ear; eyes large, eyeshine weak dull red; ear short, not protruding above crown; whiskers coarse, long, reaching shoulder, nose brown. Tail 70–105% of head and body length, robust, completely, moderately covered with dark brown hairs that curl outward to give a bristly look, scales of tail just visible beneath hair. Hindfeet broad, all feet yellow-brown, gray, or brown above. Chin often white; throat whitish or orange; rest of underparts pale to rich orange, midline of lower abdomen, underside and sides of tail

Map 181

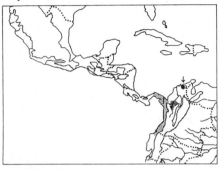

▨ Central American rufous tree rat, *Diplomys labilis*
▤ Colombian rufous tree rat, *D. caniceps*
⬤ Red crested tree rat, *D. rufodorsalis*

base often dark rust-orange. Young fulvous-brown, usually without red tones.

Variation. The two species are externally similar; their skulls appear to differ. There is much individual variation in the amount and brightness of the rusty colors of the back; the entire back may be rusty or only the rear half; or the entire back may be yellow-brown; facial markings may be prominent or indistinct. Some variation may be due to age.

Similar species. These are the only large reddish rats in their range with short ears and hairy tail.

Sounds. When disturbed makes a high-pitched whining sound.

Natural history. Nocturnal; arboreal; solitary. Feed on fruits and leaves. By day rufous tree rats den in tree holes, usually in trees near water. If disturbed they will stick their heads out the entrance (when they can be identified by their facial markings). They are agile climbers and aggressive when captured. Found in mature and secondary rainforest and deciduous forest, lowland and montane forest, mangroves, and plantations with trees. The information above refers to *D. labilis;* the natural history of *D. caniceps* is unrecorded.

Geographic range. Central and South America: Panama (*D. labilis*) and the Pacific coast of Colombia and N Ecuador (probably *D. labilis*); Colombia, near Medellín, the Cauca Valley, Cordillera Occidental of the Andes (*D. caniceps*). To at least 1,500 m elevation.

Status. Locally common; much of probable geographic range has been deforested.
Local names. Ratón maranero (Pa).
References. Tesh, R. B. 1970. Observations on the natural history of *Diplomys darlingi. J. Mammal.* 51:197–99.

Red Crested Tree Rat
Diplomys rufodorsalis
Plate 27, map 181
Identification. Measurements: HB = 190; T = 267; HF = 40; E = 12.
Upperparts bright rust-red; sides and legs paling to yellowish orange with gray underfur showing through parts; fur long and soft. Head with crest of long fur on crown between ears, bright red like back; muzzle and sides of head tawny yellow-gray streaked with black; whiskers fine, relatively short, to behind ear; ear short, almost naked, thin tufts of long brown hair sprouting from inner rim. Tail much longer than head and body, robust, thickly haired with rusty fur for 2 cm at base, abruptly changing to fine brown or black hair that covers scales for three-fifths of length, terminal two-fifths pure white. Feet brown washed with silver. Throat and chest pale orange, tip of chin white, belly gray washed with orange.
Variation. Probably belongs in the genus *Echimys.*
Similar species. Speckled tree rats (*Echimys semivillosus*) are spiny, grayish, with white-tipped spines and nearly naked tail; red-tailed squirrels (*Sciurus granatensis*) have much bushier tails and ears that protrude above crown.
Natural history. This beautiful rodent may be one of the rarest Neotropical mammals. It was reported from localities at about 700 m and 2,000 m elevation. It is presumably largely a montane species of humid forests.
Geographic range. South America: Colombia, known only from the Sierra Nevada de Santa Marta; could occur on the Venezuelan side of this massif. To at least 2,000 m elevation.
Status. Rare, apparently known from fewer than five specimens; geographic range is tiny, but much of it is protected within Tayrona National Park and Sierra Nevada Biosphere Reserve.

Map 182

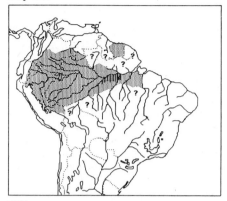

▓ Spiny tree rat, *Mesomys hispidus*
▥ Tuft-tailed spiny tree rat, *Lonchothrix emiliae*

Spiny Tree Rat
Mesomys hispidus
Plate 25, map 182
Identification. Measurements: HB = 155–210; T = 140–220; HF = 27–37; E = 12–16; WT = 130–220 g.
Upperparts pale to medium brown; midback often heavily streaked with black; pelage of conspicuous, short, wide, flat, flexible spines from shoulder to rump, each spine brown with a pale tip, spines on head and neck are softer and narrower than spines on back. Muzzle blunt; ears short and rounded, thinly haired, brown; a small tuft of longer hairs at base; eyes large, eyeshine moderately bright yellow; whiskers fine, long, reaching shoulder. Tail robust, brown, thinly covered with red-brown hairs that do not hide spines and form a sparse tuft at tip; tail often partly or completely missing. Underparts uniform pinkish orange, sharply demarcated from sides. Feet short and broad, claws sharp, strongly curved; soles pink. Juveniles gray-brown. A small, robust rat with a long back and short legs.
Variation. Upperparts vary from pale tan to dark brown with black and reddish tips on spines (some animals from the western Amazon Basin). Underparts of an animal from Colombia are creamy white. A small race or species (*M. stimulax*) is found south of the Amazon and east of the Rio Tapajós in Brazil, and one with narrow spines (*M. leniceps*) occurs at higher eleva-

tions in Peru. It is not yet clear how many species are in this taxon.

Similar species. Bare-tailed tree rats (*Echimys occasius*) are larger, with a smooth, naked tail; red-nosed (*E. armatus*) and Peruvian (*E. rhipidurus*) tree rats are larger, with inconspicuous spines; tuft-tailed spiny tree rats (*Lonchothrix emiliae*) are gray-brown, with a large tuft on tail; other small rats have soft fur.

Natural history. Nocturnal; arboreal; solitary. Feeds on fruit, insects, and browse. These small rats are often in the low understory near the ground on logs, vines, shrubs, or treefalls, but they also use the high canopy. They favor areas of lush, dense undergrowth and vines, such as old treefall gaps. They nest in tree holes and can sometimes be found in the thatched roofs of abandoned houses in the forest. Mature, disturbed, and secondary lowland rainforest.

Geographic range. South America: east of the Andes in the Guianas and most of the Amazon Basin of Colombia, Venezuela, Ecuador, Peru, and Brazil.

Status. Generally uncommon or rare, common in a few localities; widespread.

Tuft-tailed Spiny Tree Rat
Lonchothrix emiliae
Plate 25, map 182
Identification. Measurements: HB = 155–220; T = 150–230; HF = 30–37; E = 13.

Upperparts gray-brown, sides and rump speckled yellowish white; **body covered with dense wide, flat, flexible spines** mixed with fine buffy hairs; **wide spines on sides and rump with pale tips.** Ears short, naked, brown, buried in spines of head. **Tail robust, scaly, speckled with short, dark, evenly spaced, flat, scalelike hairs, tip with a large tuft of long, coarse, red-brown hairs** up to 4.5 cm long, but tail tip often missing. Underparts white, grayish, or buff, often tending to buff on midline; chest often pale gray-brown, throat with white-tipped hairs. Feet dirty white; hindfoot broad and short, with pink soles, forefoot small.

Similar species. Spiny tree rats (*Mesomys hispidus*) are brown with sharply contrasting orange belly and tail entirely hairy; tip with only a small tuft of fine hairs.

Natural history. Collectors' notes suggest that this is a nocturnal, arboreal, species that lives in mature forest and nests in tree holes. Found in lowland rainforest.

Geographic range. South America: Brazil, known only from south of the Amazon on the lower reaches of the Rios Tapajós and Madeira.

Status. Probably locally common in a small geographic range.

Amazon Bamboo Rat
Dactylomys dactylinus
Plate 27, map 183
Identification. Measurements: HB = 280–330; T = 345–437; HF = 50–65; E = 15–22; WT = 600–700 g.

Upperparts grizzled yellow-olivaceous streaked with black, midback with blackish hairs; sides paler, grayish; **rear of thighs and sides and bottom of tail base rust-orange;** head and back of neck sometimes with brown "hood" or nape patch; **fur long and soft, without bristles.** Head gray to beige, often with an indistinct dark stripe between eye and ear, and a paler stripe above it; ears small; eyes brown; eyeshine faint, dull red; whiskers long; **upper lips enlarged to give blunt, square profile. Tail robust, longer than head and body, furred at base, the rest naked, scaly. Hands and feet monkeylike; hands with four long fingers, middle fingers separated by a gap, digits with nails,** not claws. Underparts white. **Strong, musky odor** usually permeates their living areas. Young like adults. **Very large, stocky rats with square muzzles.**

Variation. Museum specimens bleach severely with age and show great color variation, with hood or nape patch from almost white to brown or tan and color of back sometimes with rusty central stripe; the true variation is therefore difficult to determine; but live animals in Ecuador, SE Peru, and E of the Xingu in Brazil appeared similar (to Emmons), and there may be relatively little color variation among living animals.

Similar species. Red-nosed tree rats (*Echimys armatus*) are smaller, with stiff or spiny fur, usually a reddish muzzle, fingers with claws, upper lips not enlarged, and warm brown upperparts; larger tree rats in region all have hairy tails. Montane bam-

Map 183

░░ Amazon bamboo rat, *Dactylomys dactylinus*
≡ Southern bamboo rat, *Kannabateomys amblyonyx*

boo rats (*Dactylomys peruanus*) occur at 1,000–3,000 m elevation on the eastern Andean slopes of SE Peru and Bolivia; they are olivaceous, with completely furred tails.

Sounds. Calls from the trees and thickets at night with loud, staccato bursts of about 3–46 pulses, each burst about 5–20 sec in length, given once and not repeated. The most calls can be heard between 7:00 and 7:30 P.M. This call varies geographically in number and timing of pulses given; it is often mistaken for a frog by foreigners, but always known to be a rat by local people. When disturbed, also calls with a series of faint, single grunts.

Natural history. Nocturnal; arboreal; solitary and in pairs. Feeds on leaves and shoots of a variety of plants and on bamboo stems. Usually quite slow-moving, bamboo rats often "freeze" for many minutes when disturbed. Their grasping hands and feet make them adept at climbing thin vertical vines and slick bamboos, over which they move extremely quietly. By day they rest hidden in thick vine tangles or clumps of vegetation, but they will run out readily if disturbed. They can be active in late afternoon and are thus sometimes seen by day. They are difficult to spot at night and are much more often heard calling than seen; they can sometimes be found by careful search of an area where their odor is powerful. Where they feed in bamboo, the stems are broken with large,

jagged gnawed holes. In most regions they are found only in patches of bamboo or in dense riverside or lakeside vegetation, especially canebrakes; but on rich soils in Ecuador they can be found throughout the high forest canopy or in disturbed vegetation. Found in rainforest only.

Geographic range. South America: the Amazon Basin of Colombia, Ecuador, Peru, Bolivia, and Brazil.

Status. Locally common where habitat is suitable, but distribution highly patchy, absent over large areas of unfavorable forest.

Local names. Toro (Br, Pe); cocopitzu (Ec); cono cono (Pe).

References. Emmons, L. H. 1981. Morphological, ecological, and behavioral adaptations for arboreal browsing in *Dactylomys dactylinus* (Rodentia, Echimyidae). *J. Mammal.* 62:183–89.

Southern Bamboo Rat
Kannabateomys amblyonyx
Plate 27, map 183

Identification. Measurements: HB = 230–347; T = 300–420; HF = 45–57; E = 16–27; WT = 350–570 g.

Upperparts uniform glossy olivaceous to fulvous, finely streaked with black; sides of body and tail sometimes fulvous; **fur long, soft,** hairs blackish with buff tips. Head large, muzzle square; whiskers long, dense, and coarse, reaching shoulder; cheeks grayish. **Tail much longer than head and body, robust, furred to tip, tapering from thick body hair on base to sparse hair distally, bicolored paler below, above dark brown at base paling through dusky to white or buff near tip, tip sometimes with tuft of dark brown hairs. Hands and feet monkeylike; hands with four long fingers, center two separated by a wide gap; nails,** not claws, **on digits.** Underparts orange with variable amount of white on throat and chest; lips around mouth white.

Variation. Animals from Argentina and Paraguay are dull, buff yellowish, with belly and tail almost white.

Similar species. These are the only large arboreal rats in their range with square muzzle and four long fingers with nails; Atlantic tree rats (*Nelomys* spp.) are smaller, with tails shorter or only marginally longer than head and body.

Sounds. Calls in duets at night, with repeated short, two-syllable, rasping sounds.
Natural history. Nocturnal; arboreal; solitary and in pairs. Feeds on young shoots and leaves of bamboo and other plants. Inhabits thickets of bamboo, especially at watersides, and dense thickets without bamboo in swamps. Said to make large nests. Atlantic coastal forests, inland rainforests, wet gallery forests, and bamboo patches such as planted bamboo hedgerows between fields.
Geographic range. South America: SE Brazil, SE Paraguay, and Misiones, Argentina.
Status. Apparently common in local patches, but such patches may be uncommon.
Local names. Rato-do-taquara (Br); rata de las taquaras, guaiquica (Ar).
References. Hensel, R. 1872. Beiträge zur Kentniss der Thierwelt Brasiliens. *Zool. Gart.* 13:76–87.

Crespo, J. A. 1982. Ecología de la comunidad de mamíferos del Parque Nacional Iguazu, Misiones. *Rev. Mus. Argent. Cien. Nat. "Bernardino Rivadavia"* 3:48–162.

Rabbits (Lagomorpha)

Leporidae

Dental formula: I2/1, C0/0, P3/2, M3/3 = 28. Forefeet with five toes, hindfeet with four toes. Members of this order are superficially rodentlike. They can be distinguished from all other mammals by the pair of miniature upper incisor teeth hidden behind the larger front incisors. Rabbits and hares (Leporidae) have long oblong ears, flexible necks, slit-like nostrils that can be opened and closed, large eyes, soles of feet thickly furred, long hindlegs, small button tails, and soft fur and fragile skin. They are small- to medium-sized and run with a jumping gait, propelled by simultaneous strong kicks of the hind legs. They are grazers and browsers especially fond of tender, protein-rich young leaves. They give birth to altricial or slightly precocial young. Members of the genus *Sylvilagus* are generally solitary and make nests on the ground surface. They shelter in thickets of dense vegetation, and their feeding areas are marked with many groups of round pellets. There are about 10 genera and 43 species worldwide, with only 1 species in Neotropical rainforest.

Brazilian Rabbit or Tapiti
Silvilagus brasiliensis
Plate 29, map 184
Identification. Measurements: HB = 268–395; T = 10–35; HF = 64–85; E = 40–61; WT = 450–1,200 g.
Upperparts dark, variegated black and tawny, sometimes with a reddish tinge; nape of **neck behind ears pure russet. Ears long** (though relatively short for a rabbit), **oblong, set close together on the top of the head,** brown. **Tail inconspicuous, buttonlike, very short to miniscule,** brown above, somewhat paler below. Lower legs and feet russet or red-brown. Underparts white. Small young dark brown with russet nape, back not variegated. Eyeshine bright red to yellow; animal often seen from side with only one red eye visible.
Variation. Body and foot color more or less strongly reddish.
Similar species. These are the only rabbits in most of their range, and the only species in rainforest. In savannas, grasslands, and fields of Venezuela and Colombia and Central America there are cottontails (*S. floridanus*). Cottontails are larger and paler, with conspicuous tails with white tuft beneath, longer ears, and whitish feet. Acouchys (*Myoprocta* spp.) have long, thin legs, small ears, and pencilthin tail and are diurnal.
Sounds. None usually heard in field.
Natural history. Nocturnal; terrestrial; solitary. Feeds on grass and browse. These little rabbits are most commonly seen just

Map 184

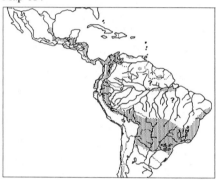

Brazilian rabbit or tapiti, *Sylvilagus brasiliensis*

after nightfall in grassy areas around houses or in gardens and plantations. When disturbed they hop into a thicket and crouch under thick vegetation. In deep rainforest away from human disturbance they occur in swamps and along river edges. They are not generally found among the trees of mature terra firme rainforest except near a habitat edge. They are attracted to salt and human urine (which contains salt). Forested habitats from the lowlands to cloud forest, dry inter-Andean valleys, paramo, and chaco.
Geographic range. Central and South America: Atlantic coast of Mexico from Taumalipas south through wetter parts of Central America; in South America to Tucuman, Argentina, and Rio Grande do Sul, Brasil. To 4,500 m in the northern Andes.
Status. Widespread and often common;

distribution in the Amazon Basin rainforest is highly patchy.

Local names. Conejo (Span); coelho (Br); tapetı́ (Su, Pa).

References. Hershkovitz, P. 1950. Mammals of northern Colombia. Preliminary report no. 6: Rabbits (Leporidae), with notes on the classification and distribution of the South American forms. *Proc. U.S. National Museum* 100:327–75.

Appendix A
Glossary

Adult. A fully developed individual capable of breeding; it usually has all its permanent teeth erupted.

Agouti. A grizzled color produced by bands of alternating black and yellow or orange on individual hairs; also the name of an animal.

Allopatric. Populations that are geographically separated such that they cannot potentially interbreed.

Altricial. Young born in an early stage of development, usually with eyes and ears sealed, body naked, and unable to walk (but can usually crawl and squeal).

Arboreal. Lives and finds its food above the level of the ground in the trees or shrubs.

Banded. Lines or streaks running around the body perpendicular to the axis of the backbone.

Basal. The part of a structure (e.g., a hair or tail) near its base; see also **proximal.**

Biogeography. The geographic distribution of species.

Blackwater. Streams and rivers with clear dark brown, or tea-colored water (stained with organic acids, e.g., the Rio Negro). These originate in areas of very poor, leached-out soils, often white sands. Forests seasonally flooded with blackwaters are called *igapó* in Brazil.

Broadleaf forest. In Central America used to distinguish an evergreen forest of other trees from conifer forests, which are also evergreen.

Browse, browser. Leaves, shoots, and twigs of plants other than grass; an animal that eats these.

Caatinga. An arid region of E Brazil with highly unpredictable rainfall and thorny woodlands and grasslands.

Campo. Literally, field. Small savannas usually surrounded by forest, often seasonally wet, but may be dry.

Cerrado. Literally, closed. An arid region of Brazil, adjacent to the caatinga, with savannas; dense, closed, thickety shrub-woodlands (the true cerrado); dry forests; and evergreen gallery forests along rivers, where there are rain-forest mammal species.

Chaco. A large flat region of Brazil, Paraguay, and part of Bolivia with dry shrub woodlands, marshes, evergreen gallery forest, and palm savannas.

Character. Any discrete feature of an organism, especially one that might be used to distinguish it from other similar organisms.

Clearwater. Rivers and streams with clear water with no sediment load or dark color from organic acids (e.g., the Rio Tapajós); see also **whitewater** and **blackwater.**

Cloud forest. High-elevation forest that is almost perpetually wet owing to the rain and dense mists caused by cool temperatures. These forests have short trees, deep mats of moss, and heavy growth of epiphytes on branches and trunks.

Conspecific. Belonging to the same species.

Crepuscular. Most active in reduced daylight at dusk and/or dawn.

Cursorial. Adapted for fast running on the ground. These animals usually have long, thin legs and walk on their toes.

Deciduous forest. Forest in which most of the trees lose their leaves simultaneously in the dry or cold season.

Digitigrade. The body weight when standing is borne only on the toes, and the heel does not touch the ground. Typical of long-legged, cursorial species.

Dimorphic (usually sexually dimorphic.) When individuals in the same population, such as males and females, fall into two classes that are distinctly different in any feature such as size, color, or shape (e.g., Guianan saki monkey males are black, females are gray; male deer are antlered, females are not).

Distal. The part of a structure (e.g., hair, tail) farthest from the center of the body.

Disturbed forest. Mature forest that has been artificially disturbed by people cutting some of the natural vegetation (such as selective logging or small garden patches); or naturally disturbed such as by floods, hurricanes, or fires.

Diurnal. Active in the daytime.

Dorsal. On the back, usually top, side of the body.

Edaphic. Relating to soil type.

Endemic. Found only in a given region.

Evergreen forest. Forest in which the canopy of leaves is intact and green year-round and only a small proportion of the trees lose their leaves at any one time.

Eyeshine. The reflection from an animal's eyes when light is shined on them at night. It is not visible at extremely close range. Its color varies depending on the angle of the eye to the light: paler, more whitish when the animal is viewed from directly in front, more reddish when seen from the side or rear. Eyes do not shine by themselves.

Fauna. The collection of animal species that together inhabits a particular place.

Floodplain. The land on the sides of a river or the river valley bottom that is susceptible to being flooded during high water and across which the river meanders through time.

Form. One or a population of variant individual(s); often used as a way of avoiding a statement of its rank as a species or subspecies (e.g., a small form of raccoon lives on Cozumel Island).

Fossorial. Adapted for digging and living in burrows and tunnels underground. Fossorial mammals have short legs and tail, long claws on their forefeet, and small eyes and ears.

Frugivore. An animal that eats mostly fruit.

Fulvous. Reddish yellow.

Gallery forest. Forest that grows alongside a river or stream and on its floodplain. Regions that are too dry to support a forested habitat overall will often have gallery forests along watercourses between savannas. These are corridors where forest animals can live in otherwise inhospitable habitat.

Grazer. An animal that eats mainly grasses, sedges, and forbs.

Grizzled. Colors that on close inspection are formed by mixtures of separate colors (usually black, white, yellow, or red) either because each hair has bands of different color (most common) or because hairs of different colors are intermixed.

Harem. A (polygynous) social organization in which a single male is attached to a group of several females; he usually defends them by driving away other males.

Hispid. Bristly or spiny, or looks that way because of coarse streaks of black in fur.

Home range. The area of land (or water or treetops) regularly used by an animal, in which it lives and finds all the necessities. It may or may not also be a territory.

Igapó. Forest that is flooded by blackwaters for a part of the year.

Juvenile. A young mammal that is still nursing or has just been weaned.

Llanos. Large grassland regions in Colombia, Venezuela, and Brazil that are partly flooded with standing water in the rainy season; forest grows on hummocks of high ground, and gallery forest along rivers.

Mammae. Nipples or teats.

Mature forest. Forest old enough to be more or less at its climax; its species composition will not change much over time. See also **primary forest;** much primary forest is also mature forest.

Meatus. An opening; auditory meatus: the hole in the head that leads to the inner ear.

Melanistic. Darker than normal, or black. Some species have occasional dark or black individuals (e.g., jaguar, squirrels).

Mesic. Moist or humid; opposed to xeric, or dry (often incorrectly used as median).

Monogamous. A social system in which a male/female pair forms a long-lasting exclusive relationship.

Morphology. Structure; an animal's structure or any portion of it (e.g., ear morphology).

Nocturnal. Active in the nighttime.

Olivaceous. A greenish brown color usually produced by a mixture of black and yellowish or fulvous hairs, or hairs individually banded with those colors.

Pantanal. A large region of seasonally flooded grasslands interspersed with forested high ground in SW Brazil and Bolivia.

Parous female. A female that is reproducing or has reproduced.

Plantigrade. Stands with the heels and palms touching the ground and bearing weight.

Platanillos. Large-leaved undergrowth plants, usually *Heliconia* spp. (like bananas), that may dominate the understory, especially in swampy or river-edge areas.

Polymorphism. The coexistence in the same population of individuals with characters of different discrete types (e.g., blue-eyed and brown-eyed).

Population. A group of individuals of the same species that live in the same area and interbreed.

Precocial. Young born at an advanced stage of development with eyes and ears open, body covered with hair, and able to walk or run shortly after birth.

Prehensile tail. A muscular tail which can wrap around objects and grasp them tightly, supporting the weight of the body. Such a tail usually has a naked gripping surface near its tip. An adaptation of arboreal animals.

Primary forest. Forest that has not been cut over or disturbed by man. This may also be mature forest.

Proximal. The part of a structure (e.g., hair, tail) closest to the center of the body.

Savanna. Natural grassland, often with scattered trees, especially palms in the Neotropics.

Scansorial. Lives both on the ground and in the trees, like many opossums and rodents.

Scat. Feces, often used for carnivore droppings.

Secondary forest. Young forest that has grown back over areas formerly cut down by man or stripped by natural causes such as floods. The plant species composition of such forest will change as it matures. It is often dense and scrubby.

Striped. Lines or streaks running lengthwise parallel to the axis of the backbone.

Subadult. An adolescent between juvenile and adult, with permanent teeth not completely erupted and not yet of breeding age.

Sympatric species. Species whose geographic and habitat ranges overlap such that individuals of each can come into contact during the breeding season.

Syntopic. Animals that live in the same habitat on a fine scale such that their individual home ranges overlap.

Systematics. The study of the evolutionary relationships of organisms, usually entailing a process of organizing taxa to show their relationships.

Taxon. Any standard unit of classification (e.g., species, genus, family, order); the plural is taxa.

Taxonomy. The description and classification of organisms.

Terra firme. High ground that is never flooded during the rainy season. Many species of plants and some animals are restricted to this ground.

Terrestrial. Lives and finds its food on the ground.

Territory. A space (often the whole home range) that is actively defended by its owner to exclude other individuals from using it.

Ticked. Fine mixture of more than one color; see also **grizzled**.

Type. The individual animal, usually a museum specimen, from which a species was described for the first time, and to which the species' scientific name forever afterward refers. The type locality is the place the type specimen came from.

Variety. An imprecise term for a distinctive population, often used when it is not clear to what rank it should belong (species, subspecies, etc.).

Várzea. Forest that is flooded by whitewater for a large part of the year.

Ventral. On the belly or undersurface.

Whitewater. Streams or rivers that carry a heavy load of sediment and deposit rich alluvial mud along their beds (e.g., the Rio Amazonas). They drain more fertile soils and are completely opaque and usually pale brown. Forest seasonally flooded with such waters is called *várzea* in Brazil.

Appendix B

Keys to the Families and Genera of Rainforest Mammals

The following keys are based on external features that can be seen on a live animal, but the animal must be in the hand for close inspection and measurement. In some cases it is necessary to open its mouth and look at the teeth. A metric ruler, pocket hand lens, or small balance is needed in some cases. The characters often necessary to identify small mammals accurately include details of the skull and teeth not usually visible on a whole animal in the hand. Because we often cannot see the truly diagnostic features externally, in some cases the keys below may be ambiguous; they are intended not for rigorous scientific identification, which requires a prepared study specimen, but as a general guide. The keys are designed for use with Neotropical rainforest mammals only; they may or may not work at any level outside the region.

Orders of Mammals

1   Forelimbs in the form of wings with membranes .... Bats (Chiroptera)
1'   Forelimbs not in the form of wings 2
2   Entirely aquatic, with no hindlimbs but with a horizonal tail paddle or flukes 3
2'   At least partly terrestrial, with hindlimbs and no tail paddle or flukes 4
3   With dorsal fin, breathes through blowhole on back of head ..... Dolphins (Cetacea)
3'   Without dorsal fin, breathes through nostrils on muzzle .... Manatees (Sirenia)
4   Hindfeet with two large, approximately equal-sized, weight-bearing hooved toes .....
    .... Peccaries and deer (Artiodactyla)
4'   Hindfeet with three or more weight-bearing toes or claws 5
5   Upper lip elongated into a proboscis, adult weight > 100 kg .... Tapirs
    (Perissodactyla)
5'   Upper lip not a proboscis, weight < 100 kg 6
6   Incisor teeth absent .... Anteaters, sloths, armadillos (Xenarthra)
6'   Incisor teeth present 7
7   Canine teeth absent 8
7'   Canine teeth present 9
8   One large pair of chisellike upper incisors with a tiny second pair hidden behind them;
    incisors separated by a large gap from row of cheekteeth .. Rabbits (Lagomorpha)
8'   Same as above but only one pair of large upper incisors, no second pair hidden behind
    them .... Rodents (Rodentia)
9   Muzzle short, mouth wide and short, wider than long, nose set on flat face ....
    .... Monkeys (Primates)
9'   Muzzle long, mouth longer than wide, nose on pointed muzzle 10
10   First toe of hindfoot opposable, set widely apart from four other toes ....
    .... Opossums (Marsupialia)
10'   First toe of hindfoot not opposable .... Carnivores (Carnivora)

Genera of Opossums (Marsupialia, Didelphidae)

1   Tail much shorter than head and body length ....
    .... Short-tailed opossums, *Monodelphis* (p. 27)
1'   Tail much longer than head and body 2
2   Hindfeet with complete webs between all toes, back with four wide black bands from
    side to side across top connected by a single stripe down midback ....
    .... Water opossum, *Chironectes* (p. 18)

2' Hindfeet not webbed, back without four black bands from side to side 3
3 Top of tail thickly haired for at least nine-tenths its length 4
3' Tail tip naked for at least terminal third 5
4 Two wide black stripes running from forearm over front of shoulder and down each side of spine, underside of distal third of tail naked, no black stripes through eye .... Black-shouldered opossum, *Caluromysiops* (p. 13)
4' No black stripes over shoulder and down back, underside of tail hairy to near tip, face with a prominent black stripe through each eye onto crown ....
.... Bushy-tailed opossum, *Glironia* (p. 14)
5 Size large (HB usually > 300 mm), fur with two distinct layers visible from a distance: short, dense, woolly yellow or white underfur thinly and incompletely covered by long, coarse, stiff, black or white overhairs ....
.... Common opossums, *Didelphis* (p. 14)
5' Size usually smaller (HB < 300 mm), fur uniform, soft and dense, without two prominent layers visible from a distance 6
6 Face with a prominent pale yellow or cream spot over each eye 7
6' Face without pale spots over eyes 8
7 Color gray or black, rarely brownish, fur finely grizzled with white, tail thickly furred for first 5–8 cm, tail tip usually pure white separated by a sharp line from dark gray base, females have a pouch ....
.... Gray four-eyed opossums, *Philander* (p. 17)
7' Color brown, fur finely grizzled with buff, not white, tail thickly furred at base for less than 4 cm; if tail tip pale, it becomes so gradually, without a sharp line; females have no pouch .... Brown four-eyed opossum, *Metachirus* (p. 19)
8 Face with a dark line extending up middle between eyes to crown, with or without dark eye rings, tail may be thickly furred on top for half its length ....
.... Woolly opossums, *Caluromys* (p. 11)
8' Face without a dark line extending between eyes to crown, dark eye rings always present, tail never thickly furred for more than one-quarter of its length 9
9 Fur long, dense and woolly, tail tip may be white or spotted, tail base thickly furred for 2–4 cm, size larger (HB > 150 mm) ....
.... Woolly mouse opossums, *Micoureus* (p. 20)
9' Fur short and velvety or size much smaller than HB = 150 mm, tail thickly haired at base for less than 2 cm, not white or usually heavily spotted at tip, almost always HB < 150 mm 10
10 Second tooth behind upper canine much larger than third tooth, lower canine short and rectangular, hardly differentiated from adjacent teeth, color usually gray-brown or gray with white or gray belly .... Slender mouse opossums, *Marmosops* (p. 21)
10' Second and third teeth behind upper canine about the same size, third tooth may be slightly larger, lower canine taller than neighboring teeth, pointed or squarish, color usually warm brown to reddish with pink, orange, yellow, or white belly 11
11 Lower canine pointed and well differentiated from neighboring teeth and larger than second tooth behind it, fur usually short, HB = 97–183, usually > 120 (common species) .... Mouse opossums, *Marmosa* (p. 23)
11' Lower canine rectangular and barely larger than second tooth behind it, fur long and lax, size always tiny (HB = 72–116), South America only ....
.... Gracile mouse opossums, *Gracilinanus* (p. 25)

Families and Genera of Anteaters, Sloths, and Armadillos (Xenarthra)

1 Head and body covered with armor plates, usually little or no hair ....
.... Armadillos (Dasypodidae) ... 6
1' No armor plates on body, head and body covered with hair 2

2   Muzzle elongated, head conical, limbs short and thick, no teeth ....
.... Anteaters (Myrmecophagidae) ... 3
2' Muzzle short, head round, limbs long and thin, teeth present ....
.... Sloths (Bradypodidae, Megalonychidae) ... 5
3   Size very large (HB > 1,000 mm), tail bushy with hanging plume ....
.... Giant anteater, *Myrmecophaga* (p. 31)
3' Size smaller (HB < 700 mm), tail slender, not bushy 4
4   Size large (HB > 500 mm), fur stiff, glossy bristles ....
.... Collared anteaters, *Tamandua* (p. 32)
4' Size small (HB < 300 mm), fur soft and woolly ....
.... Pygmy anteater, *Cyclopes* (p. 34)
5   Forefeet with three claws, short stumpy tail, face flat ....
.... Three-toed sloths (Bradypodidae), *Bradypus* (p. 36)
5' Forefeet with two claws, no visible tail, muzzle protruding ....
.... Two-toed sloths (Megalonychidae), *Choloepus* (p. 37)
6   Ears arising close together on top of head, bases nearly touch ....
.... Long-nosed armadillos, *Dasypus* (p. 42)
6' Ears set far apart on sides of head, wide armored crown between 7
7   Upperparts sprinkled with long, conspicuous, stiff, whitish hairs, foreclaws not greatly
enlarged .... Yellow armadillo, *Euphractus* (p. 39)
7' Upperparts without conspicuous hairs, foreclaws greatly enlarged 8
8   Size enormous (HB > 700 mm), hindlegs disproportionately large and thick ....
.... Giant armadillo, *Priodontes* (p. 41)
8' Size smaller (HB < 500 mm), hindlegs not enlarged ....
.... Naked-tailed armadillos, *Cabassous* (p. 40)

Families of Bats (Chiroptera)

1   With a freestanding, spear-shaped noseleaf behind nostrils, or if without, with naked,
grotesquely wrinkled face and ladderlike pattern in wing ....
.... Leaf-nosed bats (Phyllostomidae)
1' Without spear-shaped noseleaf or wrinkled face and ladder pattern in wing 2
2   With a pair of low, horseshoe- or M-shaped folds behind nostrils, one behind the other,
and central incisor teeth larger than canine teeth ....
.... Vampire bats (Phyllostomidae)
2' Without any folds behind nostrils, incisors smaller than canines 3
3   With long, robust tail projecting free beyond edge of tail membrane for at least one-
third its length .... Free-tailed or mastiff bats (Molossidae)
3' With tail either completely or almost completely (more than three-fourths of its length)
within tail membrane, or shorter than tail membrane and with any free tip protruding
above the middle of the membrane 4
4   With sucker disks on base of thumb and on ankle ....
.... Sucker-footed bats (Thyropteridae)
4' Without sucker disks on thumb or ankle 5
5   Chin with a large concave plate on each side, or lips meaty and flared out forward and
wings sometimes meeting at midline of back so that back looks naked ....
.... Leaf-chinned and mustached bats (Mormoopidae)
5' Lips and chin not as above, back not naked 6
6   Upper lip split near nose, forming drooping, bulldoglike folds, size large (FA > 60),
fur extremely short, one pale stripe down midback ....
.... Bulldog bats (Noctilionidae)
6' Upper lip not split and drooping, size smaller (FA < 58), fur not extremely short, back
not striped or with two stripes 7
7   Thumb a rudimentary stub encased in membrane, without claw, tail half as long as tail
membrane, entirely enclosed within it .... Thumbless bats (Furipteridae)

7'   Thumb well developed, with claw outside membrane, tail either as long as tail membrane or, if shorter, with tip protruding free above it 8

8   Tail much shorter than legs or tail membrane, tip sticks up free above from middle of membrane, glandular sac sometimes present in propatagium ....
.... Sheath-tailed bats (Emballonuridae)

8'   Tail longer than legs, reaches edge of tail membrane, wing sac never present in propatagium 9

9   Two phalanges on third finger, legs very long, ears funnel-shaped, wings and body always pale .... Funnel-eared bats (Natalidae)

9'   Three phalanges on third finger, legs not greatly elongated, ears not funnel-shaped, wings and body black to pale .... Vespertilionid bats (Vespertilionidae)

Genera of Sheath-tailed Bats (Emballonuridae)

1   Color white, grayish white, or whitish buff .... Ghost bats, *Diclidurus* (p. 49)
1'   Not whitish 2
2   Nose elongated in a proboscis, forearm with spaced tufts of white hair ....
.... Long-nosed bat, *Rhynchonycteris* (p. 45)
2'   Nose not in a proboscis, forearm without white tufts 3
3   Wing membrane attaches high on foot near base of toe 4
3'   Wing membrane attaches at ankle 5
4   Fur long and shaggy, covers face and chin, ear tips pointed, no wing sac, color yellowish to orangish .... Shaggy bat, *Centronycteris* (p. 46)
4'   Fur not shaggy, face not hairy, ear tips rounded, large wing sac present, blackish to reddish brown .... Chestnut sac-winged bat, *Cormura* (p. 48)
5   Wing sac absent, ears short and rounded .... Smoky bat, *Cyttarops* (p. 50)
5'   Wing sac present or indicated by a fold, ears normal, triangular and slightly pointed
.... 6
6   Wing sac close to bone of forearm beyond elbow, back with a pair of wavy pale lines often present, fur on forehead gradually decreases down face and muzzle ....
.... White-lined sac-winged bats, *Saccopteryx* (p. 45)
6'   Wing sac in middle or forward edge of membrane, back without white lines, tuft of fur on crown ends abruptly on line between ears 7
7   Wing sac extends from forward edge of propatagium, muzzle short and broad, calcar twice as long as foot .... Doglike sac-winged bats, *Peropteryx* (p. 47)
7'   Wing sac in center of propatagium, muzzle narrow and pointy, calcar only slightly longer than foot .... Least sac-winged bats, *Balantiopteryx* (p. 48)

Genera of Leaf-chinned, Mustached, and Naked-backed Bats (Mormoopidae)

1   Chin with large, platelike folds, ears short, rounded, and joined by a band across forehead
.... Leaf-chinned bats, *Mormoops* (p. 53)
1'   Chin without plates, ears pointed and widely separated, not joined by band across brow
.... Mustached and naked-backed bats, *Pteronotus* (p. 52)

Subfamilies of Leaf-nosed Bats (Phyllostomidae)

1   With a pair of low, horseshoe- or M-shaped noseleaves one behind the other behind nostrils and central incisor teeth larger than canine teeth ....
.... Vampires (Desmodontinae)

1′ With a freestanding, spear-shaped noseleaf behind nostrils, or if without, with naked, grotesquely wrinkled face and ladderlike pattern in wing, incisor teeth smaller than canine teeth 2

2 Muzzle elongated, narrow, lower jaw often longer than upper, tongue very long, noseleaf short, spear-shaped, with no free horseshoe below and beside nostrils, size small (FA ≤ 45) .... Long-tongued bats (Glossophaginae)

2′ Muzzle and tongue not greatly elongated; if spear-shaped noseleaf present, nostrils at least partly flanked by free fold or horseshoe; size small to large 3

3 Tail membrane nearly as long to longer than legs, tail usually present, ears usually large, muzzle narrow .... Spear-nosed bats (Phyllostominae)

3′ Tail membrane much shorter than legs or absent, tail usually absent, ears medium, muzzle usually broad 4

4 Tail always absent, muzzle short and broad, pale facial stripes and/or midback stripes often present .... Neotropical fruit bats (Stenodermatinae)

4′ Tail present or absent, muzzle narrow, stripes never present, size always small (FA ≤ 45) .... Little spear-nosed and short-tailed fruit bats (Carolliinae)

Genera of Spear-nosed Bats (Phyllostominae)

1 Tail absent or a short stub hidden in base of membrane, size large (FA > 77; HB > 94), noseleaf and horseshoe raised to form a deep hollow cup with unbroken rim around nostrils 2

1′ Tail always present and obvious, size large to small, noseleaf not forming an unbroken cup around nostrils 3

2 Giant size (FA > 98; HB > 135; WT = 125 g), fur medium length, dark brown to orange, pale stripe down back .... False vampire bat, *Vampyrum* (p. 61)

2′ Large size (FA = 77–87), fur long and woolly, gray-brown, no stripe on back .... Woolly false vampire bat, *Chrotopterus* (p. 60)

3 Tail longer than legs, reaches edge of tail membrane 4

3′ Tail much shorter than legs and tail membrane 5

4 Noseleaf extremely long, narrow, sword-shaped, as long as ears, ears very large, pointed, tail membrane comes to a point at tail tip .... Sword-nosed bats, *Lonchorhina* (p. 55)

4′ Noseleaf broad, spear-shaped, much shorter than ears, ears large and rounded, tail membrane square at end, does not come to a point at tail tip .... Long-legged bat, *Macrophyllum* (p. 56)

5 Chin, lips, and muzzle around nose studded with long projecting tubercules .... Fringe-lipped bat, *Trachops* (p. 60)

5′ Lips and muzzle around nose not studded with projecting tubercules 6

6 Chin tip with single, long, smooth pads angled to form a V, size usually small .... Little big-eared bats, *Micronycteris* (p. 54)

6′ Chin tip with rows of flat, roundish warts, size usually medium to large 7

7 Noseleaf very long, lance-shaped, much longer than broad, ears very large, pointed .... Hairy-nosed bats, *Mimon* (p. 57)

7′ Noseleaf spear-shaped, not much longer than broad; ears, if large, rounded; if medium, pointed 8

8 Ears large, rounded, hair long and woolly, tail membrane longer than legs .... Round-eared bats, *Tonatia* (p. 56)

8′ Tail membrane equal to or shorter than legs, fur medium to short and velvety, ears broadly triangular 9

9 Horseshoe around nostrils with free flange completely around base, lips and noseleaf dark .... Spear-nosed bats, *Phyllostomus* (p. 58)

9′ Horseshoe around nostrils with free flange only on sides, continuous with upper lip in center, lips and noseleaf pinkish mottled brown .... Pale-faced spear-nosed bat, *Phylloderma* (p. 59)

Genera of Long-tongued Bats (Glossophaginae)

1 Tail absent or greatly reduced, tail membrane virtually absent, reduced to a narrow, hairy band much shorter than knees ....
.... Hairy-legged long-tongued bats, *Anoura* (p. 64)
1' Tail always present (though sometimes short), tail membrane longer than knees ... 2
2 Lower jaw about the same length as upper, not prominently overshot forward 3
2' Lower jaw longer than upper, distinctly overshot forward 4
3 Fur bicolored, pale at base and dark at tip, wing attached at ankle opposite calcar, color pale brown or gray .... Common long-tongued bats, *Glossophaga* (p. 62)
3' Fur unicolored chocolate-brown or chestnut to base, wing attached halfway down foot near base of toe beyond origin of calcar ....
.... Chestnut long-tongued bat, *Lionycteris* (p. 63)
4 Dorsal fur bicolored, pale at base and dark at tip 5
4' Dorsal fur tricolored, dark at base (sometimes faint, like a shadow) and tip, pale in middle 7
5 Muzzle only moderately elongated, noseleaf large, reaches eye when flattened ....
.... Spear-nosed long-tongued bats, *Lonchophylla* (p. 63)
5' Muzzle greatly elongated, noseleaf does not reach eye 6
6 Distal thumb joint (outside membrane) longer than proximal (inside membrane), long whisker in front of ear, color blackish, South America only ....
.... Ega long-tongued bat, *Scleronycteris* (p. 65)
6' Thumb joints about equal in length, no long whisker in front of ear, color dark brown, Central and South America ....
.... Long-nosed long-tongued bats, *Choeroniscus* (p. 66)
7 Muzzle robust, elbows thinly furred above, underparts paler than back, Central and South America .... Dark long-tongued bats, *Lichonycteris* (p. 65)
7' Muzzle narrow and tubelike, elbows thickly furred above and below, underparts about the same as back, Central America only ....
.... Underwood's long-tongued bat, *Hylonycteris* (p. 67)

Genera of Short-tailed and Little Fruit Bats (Carolliinae)

1 Tail present, tip sticks up freely from middle of tail membrane, Central and South America .... Short-tailed fruit bats, *Carollia* (p. 67)
1' Tail absent, South America only .... Little fruit bats, *Rhinophylla* (p. 68)

Genera of Neotropical Fruit Bats (Stenodermatinae)

1 Tail membrane virtually absent, a narrow hairy band down leg, no facial or midback stripes, no white spots where wings join shoulder, fur prominently tricolored dark at base and tip, pale in middle, central upper incisors much longer than outer incisors, broad and buck-toothed .... Yellow-shouldered bats, *Sturnira* (p. 70)
1' With tail membrane or, if almost without, with facial stripes and/or short central incisors
.... 2
2 Pure white spot where wing meets shoulder 3
2' No white spot on shoulder 6
3 Noseleaf well developed, spear-shaped, no folds on brow 4
3' Noseleaf not spear-shaped, or noseleaf absent, folds present on brow 5
4 Upper lip with a fold from base of noseleaf to corner of mouth forming a "double" lip
.... Double-lipped bat, *Pygoderma* (p. 77)
4' Upper lip without fold .... Little white-shouldered bat, *Ametrida* (p. 77)
5 Forehead above eyes with large fleshy horizontal fold that presses down on noseleaf, ears without horizontal lobe extending over brow, no ladderlike pattern in wing ...
.... Visored bat, *Sphaeronycteris* (p. 78)

5' Whole face covered with complex folds, no united noseleaf, horizonal lobes from ears cover brow, ladderlike pattern in wing .... Wrinkle-faced bat, *Centurio* (p. 78)

6   Tongue blackish if bat medium or large, central upper incisors bilobed, only slightly longer than outer incisors 7

6' Tongue pink, central upper incisors bilobed or single-pointed, much longer than outer incisors 8

7   Midback and facial stripes always present ..... Tent-making bats, *Uroderma* (p. 70)

7' No midback stripe, facial stripes present or absent, upper facial stripes usually end at forward edge of ear, rarely to middle of ear .... Fruit-eating bats, *Artibeus* (p. 75)

8   Fur pale at base 9

8' Fur with dark band at base, sometimes faint, like shadow 10

9   No facial or midback stripes, tail membrane naked, without hairy fringe on rear edge, forequarters white, hindparts grayish, Central America only .... Honduran white bat, *Ectophylla* (p. 75)

9' Pale facial stripes present, tail membrane with hairy fringe on edge 12

10 Large and brawny (FA ≥ 38), with or without stripes, central upper incisors long, narrow, with single points, outer incisors tiny, calcar about three-fourths length of foot .... Big-eyed bats, *Chiroderma* (p. 72)

10' Small (FA ≤ 39), with or without midback stripe, central upper incisors much longer than outer, bilobed, calcar about half length of foot 11

11 Facial stripes present, midback stripe present only if bat large, if tiny, then without back stripe and tail membrane edge with slight hairy fringe .... Yellow-eared bats, *Vampyressa* (p. 73)

11' Size tiny, no facial or midback stripes, tail membrane naked, without hairy fringe .... Macconnell's bat, *Mesophylla* (p. 74)

12 Size large (FA > 45; WT > 25 g), color warm cinnamon, back and face stripes bright and sharp, posterior edge of facial stripes reach rear of ear, sides of noseleaf and base of ears yellow .... Great stripe-faced bat, *Vampyrodes* (p. 72)

12' If large (FA > 45), color blackish brown, warm brown if smaller only, sides of noseleaf and base of ears cream or gray .... White-lined bats, *Vampyrops* (p. 71)

Genera of Vampires (Desmodontinae)

1   Tail membrane almost absent, reduced to a narrow, very hairy band down side of leg, fur soft and dense .... Hairy-legged vampire, *Diphylla* (p. 80)

1' Tail membrane well developed, forms a complete band joining legs, fur short, thin, and straight .... Common and white-winged vampire bats, *Desmodus* (p. 79)

Genera of Vespertilionid Bats (Vespertilionidae)

1   Ears enormous (> 26 mm), short stub of tail tip extends free beyond membrane .... Big-eared brown bats, *Histiotus* (p. 86)

1' Not as above 2

2   Tail membrane densely furred for at least half its length, ears short and rounded .... Hairy-tailed bats, *Lasiurus* (p. 87)

2' Not as above 3

3   Pale yellow frosted brown above, pale yellow below, fur pale at base, two upper incisors, size tiny .... Little yellow bats, *Rhogeessa* (p. 85)

3' Black, brown, or red, fur dark at base, four upper incisors 4

4   Fur deep blackish at base with tips slightly or strongly paler, first tooth behind upper canine large, no gap between canine and the first large tooth, FA = 37–54 .... Big brown bats, *Eptesicus* (p. 85)

4′   Fur usually uniformly dark to base, if frosted FA ≤ 37, first tooth behind upper canine
       tiny, leaving an apparent gap in front of first large tooth, FA = 29–44 . . . .
       . . . . Little brown bats, *Myotis* (p. 84)

Genera of Free-tailed Bats (Molossidae)

1   Upper lip with deep vertical wrinkles 2
1′   Upper lip smooth 3
2   Ears joined broadly at midline of brow, four lower incisors . . . .
       . . . . Broad-eared free-tailed bats, *Nyctinomops* (p. 91)
2′   Ears almost meet but are not joined on midline of brow, six lower incisors . . . .
       . . . . Free-tailed bats, *Tadarida* (p. 90)
3   Midline of muzzle between eye and nose raised in a ∧ ridge 4
3′   Midline of muzzle flat, not raised in a ridge 5
4   Hair on crown longer than on neck, forming a slight crest, four lower incisors, tail
       usually > 60% of head and body length, palate between upper tooth rows deeply
       concave . . . . Crested mastiff bats, *Promops* (p. 92)
4′   Hair on crown not longer than neck, two lower incisors, tail usually 50–60% of head
       and body length, palate between upper tooth rows flat . . . .
       . . . . Mastiff bats, *Molossus* (p. 93)
5   Ears joined broadly at midline of brow, reach to nose when flattened forward . . . .
       . . . . Bonneted bats, *Eumops* (p. 91)
5′   Ears may meet or be separate but are not joined, reach midway between nose and eye
       when flattened 6
6   Forearm skin above sprinkled with bumps, head and body extremely flattened, size tiny
       . . . . Flat-headed bat, *Neoplatymops* (p. 90)
6′   ˋNot as above 7
7   Ears with pointed tips, widely separated, slight accordianlike fold where ear meets crown,
       two lower incisors, hair of crown diminishes gradually down face, tail close to 50% of
       head and body length . . . . Dog-faced bats, *Molossops* (p. 88)
7′   Ears rounded, no fold at junction with crown, hair of crown ends abruptly in a line, four
       lower incisors, tail usually 40–50% of head and body length . . . .
       . . . . Doglike bats, *Cynomops* (p. 89)

Families and Genera of Monkeys (Primates)

1   Size tiny to small, claws, not nails, on hands and feet . . . .
       . . . . Callitrichidae, Callimiconidae 2
1′   Size small to large, nails, not claws, on hands and feet . . . . Cebidae 6
2   Size tiny, WT < 150 g, west and central Amazon Basin only . . . .
       . . . . Pygmy marmoset, *Cebuella* (p. 96)
2′   Size larger, WT > 200 g 3
3   Lower incisors narrow at tips, tightly clustered to form a scoop, lower canine little dif-
       ferentiated from incisors and clustered with them, ears often with tufts, south of the
       Amazon and east of the Madeira including SE Brazil . . . .
       . . . . Marmosets, *Callithrix* (p. 97)
3′   Lower incisors broad at tips, not clustered into a scoop, lower canine distinctly larger than
       incisors, may have mane but not ear tufts, whole region 4
4   Bright gold or gold and black with prominent, centrally parted mane on head, SE Brazil
       only . . . . Lion tamarins, *Leontopithecus* (p. 107)
4′   Mane, if present, not centrally parted, color not gold or gold and black, or found outside
       SE Brazil only 5

5   Color black, face entirely black, hair on head forms two tiers: a shorter even cap bordered on nape and sides by longer ruff .... Goeldi's monkey, *Callimico* (p. 108)

5' Black or colored, if black, then muzzle usually grizzled or patterned with white, hair on head not two-tiered .... Tamarins, *Saguinus* (p. 100)

6   Tail nonprehensile, not carried coiled at tip, size small to medium (WT = 0.6–4 kg) 7

6' Tail prehensile, often coiled at tip, size medium to large (WT = 2–15 kg) 12

7   Eyes very large, crown with three black stripes, nocturnal ....
    .... Night monkeys, *Aotus* (p. 110)

7' Eyes normal size, crown not striped, diurnal 8

8   Fur very short, tail slender, white mask around eyes, small, slender, and agile
    .... Squirrel monkeys, *Saimiri* (p. 113)

8' Fur medium to long, tail not slender, no white mask around eyes, if small (~ 1 kg), then body not slender 9

9   Size small (WT = 0.9–1.7 kg), tail thickly furred but not bushy ....
    .... Titi monkeys, *Callicebus* (p. 111)

9' Size larger (WT = 1.6–4 kg), tail distinctly bushy 10

10 Tail much shorter than head and body .... Uakari monkeys, *Cacajao* (p. 123)

10' Tail longer than head and body 11

11 Fur on tail, head, and body long and shaggy, tail carried low ....
    .... Saki monkeys, *Pithecia* (p. 119)

11' Fur on body not long and shaggy, bushy tail often carried in high arc over back ....
    .... Bearded saki monkeys, *Chiropotes* (p. 121)

12 Size medium (WT = 1.2–4.5 kg), top of head with black or brown cap or wedge, underside of tail tip hairy .... Capuchin monkeys, *Cebus* (p. 116)

12' Size large (WT = 3.5–15 kg), crown without contrasting dark cap, underside of tail tip naked 13

13 Head large, chin with prominent beard, throat swollen ....
    .... Howler monkeys, *Alouatta* (p. 124)

13' Head and throat not large, chin without beard 14

14 Limbs and tail robust, not thin and greatly elongated, thumbs well developed, does not usually travel by swinging by arms or tail below the branches ....
    .... Woolly monkeys, *Lagothrix* (p. 128)

14' Limbs and tail long and thin, thumb almost absent, often swings below branches...
    15

15 Ears almost naked, inconspicuous, does not occur in SE Brazil ....
    .... Spider monkeys, *Ateles* (p. 130)

15' Ears thickly furred, stand out from side of head, SE Brazil only ....
    .... Woolly spider monkey, *Brachyteles* (p. 132)

Families and Genera of Carnivores (Carnivora)

1   All feet with four weight-bearing toes, fifth toe if present a dewclaw that does not touch the ground 2

1' Feet with five weight-bearing toes 5

2   Claws of forefeet sharp and retractable, fifth toe present, tail not bushy ....
    .... Cats (Felidae) ... 3

2' Claws of forefeet blunt and nonretractable, fifth toe absent in wild species, tail bushy
    .... Dogs (Canidae) ... 4

3   Size large (HB > 1,000 mm), body spotted including neck or black with shadows of spots .... Jaguar, *Panthera* (p. 153)

3' Size large or medium, plain or spotted, if spotted, size medium and neck striped
    .... Other cats, *Felis* (p. 149)

4   Tail long, brush touches the ground, legs and neck long ....
    .... Small-eared dog, *Atelocynus* (p. 134)
4' Tail short and stumpy, does not reach ground, legs and neck short ....
    .... Bush dog, *Speothos* (p. 135)
5   Large webs between all toes, tail very thick at base, tapering to tip ....
    .... Otters (Mustelidae) ... 6
5' No webs or at most partial webs between toes, tail not much thicker at base than near
    tip 7
6   Size very large (HB > 1,000 mm), tail strongly flattened dorsoventrally, throat usually
    spotted, belly dark .... Giant otter, *Pteronura* (p. 147)
6' Size smaller (HB < 900 mm), tail not strongly flattened, throat not spotted, belly pale
    .... River otter, *Lutra* (p. 147)
7   Tail short, 60% of head and body length or less 8
7' Tail long, 80% of head and body length or more .... Procyonidae ... 12
8   Large (HB > 400 mm), face with black mask around eyes, tail with sharp black and
    pale rings .... Raccoons (Procyonidae), *Procyon* (p. 136)
8' Large or small, tail never with dark and pale rings .... Mustelidae ... 9
9   Sharply bicolored black or dark brown with pure white hood over top of head and neck
    and down back as stripe, tail pure white except at base ....
    .... Hog-nosed skunk, *Conepatus* (p. 145)
9' Not sharply bicolored, tail not pure white 10
10 Head sharply patterned tricolor, crown gray, white stripe over eye down side of neck,
    black muzzle, cheek and throat ....Grisons, *Galictis* (p. 143)
10' Head not tricolored, if white stripe or spot present, crown not gray 11
11 Large (HB > 400 mm), doglike, legs quite long, throat usually with pale patch, un-
    derparts blackish .... Tayra, *Eira* (p. 144)
11' Small, (HB < 400 mm), not doglike, legs extremely short, underparts pale ....
    .... Weasels, *Mustela* (p. 142)
12 Tail prehensile .... Kinkajou, *Potos* (p. 140)
12' Tail nonprehensile 13
13 Snout and front claws greatly elongated .... Coatis, *Nasua* (p. 138)
13' Snout and front claws not elongated 14
14 Rings on tail indistinct or absent, no black and white eye rings, Central and South
    America .... Olingos, *Bassaricyon* (p. 139)
14' Rings on tail sharp black and pale, prominent black-and-white eye rings, Central America
    only ....Cacomistle, *Bassariscus* (p. 141)

Suborders of Rodents

1   Tail always long, bushy, completely covered above with straight, multibanded hairs
    > 10 mm long that stand out perpendicular to backbone, four to five cheekteeth on
    each side of each jaw .... Squirrels (Sciuromorpha: Sciuridae)
1' Tail long or short, if completely covered with long hair above, then hair not multibanded,
    never more than four cheekteeth 2

2	Caviomorpha	Myomorpha
Four cheekteeth on each side of each jaw	+	− (except Heteromyidae)
Three or fewer cheekteeth	−	+
May have < 5 hindtoes	+	−
May have lateral mammae	+	−
May have HB < 150 mm	−	+
May have HB > 300 mm	+	−
May be HB > 150 mm and have pelage of spines or bristles	+	−

May have HB < 150 mm and pelage of spines or bristles	−	+
May have externally opening cheek pouches	−	+
May have fringe of downcurling hair on side of foot	−	+
May have HB > 200 mm and a hairy tail	+	−

Genera of Squirrels (Sciuromorpha: Sciuridae)

1   Size tiny (HB < 115 mm; WT < 50 g) ....
.... Neotropical pygmy squirrel, *Sciurillus* (p. 176)
1' Size larger (HB > 115; WT > 50 g) 2
2   Ears long, standing well up above crown ...... Common squirrels, *Sciurus* (p. 167)
2' Ears short, not standing high above crown 3
3   Tail hairs dense and bushy, Talamanca highlands Panama and Costa Rica only ....
.... Montane squirrel, *Syntheosciurus* (p. 176)
3' Tail relatively slender, Central and South America ....
.... Neotropical dwarf squirrels, *Microsciurus* (p. 174)

Families and Genera of Myomorph Rodents

Many of the small rodents are externally similar, and although those familiar with them can usually tell the genera and species apart in the field by differences in size, shape, fur texture, and behavior, these features are difficult to express in a key that has to work for a huge array of species over an enormous geographic range. Some characters are difficult to interpret, and in one case below the key is designed to bring out a genus from different possible interpretations of the same character. An individual species is given by name where one, but not other, species in a genus terminates in the key. This key is the most difficult to use in this book and may not always result in success.

1   Large externally opening cheek pouches present, fur spiny ....
.... Spiny pocket mice (Heteromyidae), *Heteromys* (p. 177)
1' No cheek pouches present, spiny or not spiny .... Muridae ... 2
2   Dorsal fur includes stiff bristles or spines, not apparent visually, detectable when fur is
rubbed backward... 3
2' Dorsal fur without spines 4
3   Tail at least as long as head and body, color brown with orange sides, underparts white
and/or orange .... Spiny mice, *Neacomys* (p. 181)
3' Tail much shorter than head and body, color gray, underparts gray ....
.... Gray spiny mouse, *Scolomys* (p. 182)
4   Tail conspicuously hairy, with or without tuft at tip 5
4' Tail looks naked, may have fine hairs that protrude very slightly beyond tip 12
5   Tail entirely covered with flat-lying hairs that completely cover scales, tip without thicker
tuft of hairs, muzzle blunt, eyes and ears small 6
5' Tail hairs not adpressed (flat-lying), or hairs not completely hiding scales, and/or tail
tip with conspicuously larger tuft, muzzle pointed, ears and eyes normal 8
6   Underparts dark gray or brown not sharply contrasting with sides, hindfeet not strongly
paddle-shaped, hindtoes not partially webbed ....
.... South American water mice, *Neusticomys* (p. 187)
6' Underparts white or silver sharply contrasting with sides, hindfeet paddle-shaped, hindtoes partially webbed at base 7
7   Forefeet with four pads on palm, Central America only ....
.... Central American water mice, *Rheomys* (p. 186)

7′ Forefeet with five pads on palm, Central and South America ....
.... Crab-eating rats, *Ichthyomys* (p. 185)

8 Underparts (and upperparts) entirely rich orange-rufous, first toe of hindfoot with rounded nail, SE Brazil only .... Brazilian arboreal mouse, *Rhagomys* (p. 184)

8′ Underparts white, gray, yellow, or faintly orange, first toe of hindfoot with sharp claw
.... 9

9 Color of back and/or sides bright orange or red, belly white 10

9′ Color dull red, brown, or tawny yellowish 11

10 Tail much longer than head and body, foot with broad white patch across top, SE Brazil only .... Rio de Janeiro rat, *Phaenomys* (p. 183)

10′ Tail slightly longer than head and body, foot dusky with white toes, eye very large and ringed with black, Central America only .... Vesper rat, *Nyctomys* (p. 189)

11 Tail thickly or thinly haired, with distinct slight or prominent tuft of denser hair at tip, whiskers coarse and long, several reach shoulder, females have six mammae
.... Climbing rats, *Rhipidomys* (p. 184)

11′ Tail finely haired, hairs protruding slightly beyond tail tip, but not forming a distinctly thicker tuft, whiskers dense but fine, reach ear tip but few or none reach shoulder, females with eight mammae .... Arboreal rice rats, *Oecomys* (p. 181)

12 Hindfeet broad and paddle-shaped, with fringe of downcurling silver hairs along side, hindtoes partly webbed at base, whiskers fine and short, ears small, size large, fur with gloss or sheen, tail hairs often longer and denser below than above 13

12′ Hindfoot not paddle-shaped, without fringe or webs, whiskers long or short, size large or small, fur glossy or dull, tail hairs similar above and below 14

13 Tail slightly longer than head and body, ears naked at tips, sole of hindfoot to heel covered with roundish scales .... Water rats, *Nectomys* (p. 182)

13′ Tail usually shorter than head and body, ears hairy to tips, sole of heel without scales
.... Marsh rats, *Holochilus* (p. 191)

14 Tail much shorter than head and body, 66% or less 15

14′ Tail 66% to longer than head and body 18

15 Muzzle blunt, eyes and ears tiny, lost in fur, tail < 50% of head and body, SE Brazil only .... Brazilian shrew mouse, *Blarinomys* (p. 187)

15′ Muzzle pointed, eyes and ears normal, tail 50–80% of head and body 16

16 Muzzle greatly elongated, claws greatly elongated ....
.... Long-nosed mice, *Oxymycterus* (p. 191)

16′ Nose pointed but not elongated, claws not elongated 17

17 Color dark brown to olivaceous, South America only ....
.... Grass mice, *Akodon* (p. 190)

17′ Color blackish with reddish tints, Central America, west coast of Colombia and Ecuador, N Venezuela only .... Rice rat, *Oryzomys caliginosus* (p. 178)

18 Tail tip white, sharply divided from dark base, size large (HB > 180; WT > 150 g)
.... Naked-tailed climbing rats, *Tylomys* (p. 188)

18′ Tail tip not white, or size smaller 19

19 Size tiny (HB < 110; adult WT < 40 g, usually < 30 g) 20

19′ Size larger (adult WT > 40 g) 23

20 Hindfeet broad and short, with pink soles, color uniform reddish, with pure white underparts, females with eight mammae ....
.... Bicolored arboreal rice rat, *Oecomys bicolor* (p. 181)

20′ Hindfeet long and narrow, soles dark, color tawny, gray or brown 21

21 Tail about the same length as head and body, underparts gray or grayish not sharply contrasting with sides, females with 10 mammae ....
.... House mouse, *Mus musculus* (p. 195)

21′ Tail much longer than head and body, underparts white or gray contrasting with sides, females with fewer than 10 mammae 22

22 Incisor teeth grooved, sides of head, neck or body often orange or cinnamon, females with six mammae .... Harvest mice, *Reithrodontomys* (p. 193)

22′ Incisors not grooved, sides of head and neck tawny gray, females with eight mammae
.... Pygmy rice rats, *Oligoryzomys* (p. 180)
23 Large and robust, fur sparse and coarse, with very long guard hairs, tail naked, slightly
bristly, with coarse, large scales, females with 10 or more mammae ....
.... Brown and black rats, *Rattus* (p. 194)
23′ Small, fur soft and dense, tail either slender, with fine, inconspicuous scales, or dark
and shiny, with rings of scales, females with fewer than 10 mammae 24
24 Ears greatly enlarged, tail completely naked, dark and shiny as if varnished, with
conspicuous rings of scales, Central America only ....
.... Big-eared climbing rat, *Ototylomys* (p. 188)
24′ Ears not greatly enlarged, tail not shiny, without broad rings of scales 25
25 Tail longer than head and body, whiskers long, reaching shoulder, females with six
mammae, Central America only 26
25′ Tail shorter than head and body, or if not, whiskers not reaching shoulder and females
with eight mammae, Central and South America 27
26 Brown to tawny, tail slightly longer than head and body ....
.... Mexican deer mouse, *Peromyscus mexicanus* (p. 192)
26′ Cinnamon to orange, tail much longer than head and body ....
.... Isthmus rats, *Isthmomys* (p. 193)
27 Hindfeet short and broad, soles entirely pink, tail finely haired, hairs protruding slightly
beyond tip in a tiny tuft, ears relatively short, muzzle blunt ....
.... Arboreal rice rats, *Oecomys* (p. 181)
27′ Hindfeet long and narrow, soles dark, ears large, muzzle pointed 28
28 Hind toepads pink, midback sometimes with black stripe, SE Brazil and NE Argentina
only, females with six or eight mammae ....
.... Atlantic forest mice, *Delomys* (p. 185)
28′ Hind toepads black, midback not striped, females always with eight mammae, entire
region .... Rice rats, *Oryzomys* (p. 178)
Note: The final genera in (28) may not be distinguishable in the hand.

Families and Genera of Caviomorph Rodents

1 Body size large (WT > 0.8 kg), body shape and tail not ratlike or squirrellike, sometimes
piglike, tail long or short 2
1′ Body size smaller (WT < 0.8 kg), body and tail ratlike or squirrellike, tail always
naturally long but may be broken off .... Spiny rats (Echimyidae) ... 8
2 Clothed with stout, sharp spines (these may be partially hidden by woolly hair), tail
over one-third of head and body length, thick, tapered toward tip, prehensile ... 3
2′ Not spiny, tail almost absent, or if present, nonprehensile 4
3 Spines on hindquarters not barbed, flexible and dry, like broom straws, SE Brazil only
.... Bristle-spined porcupine, *Chaetomys* (p. 202)
3′ Spines on hindquarters stiff and barbed, throughout rain forest region ....
.... Neotropical porcupines (Erethizontidae), *Coendou* (p. 197)
4 Spotted, always large (WT > 5 kg) 5
4′ Not spotted, large to small 6
5 Tail large, about 30–50% of head and body length, color black with rows of white spots
from shoulder to rump .... Pacarana (Dinomyidae), *Dinomys* (p. 205)
5′ Tail miniscule, 5% of head and body length, color brown or red with rows of white
spots from neck to rump .... Pacas (Agoutidae), *Agouti* (p. 204)
6 Size very large (> 20 kg), piglike, with stout legs, webs between toes ....
.... Capybara (Hydrochaeridae), *Hydrochaeris* (p. 204)
6′ Smaller (WT < 6 kg), with slender legs and high rump, toes not webbed ....
.... Dasyproctidae 7
7 Large (WT = 2–6 kg), tail a tiny, naked, inconspicuous blackish stump ....
.... Agoutis, *Dasyprocta* (p. 206)

7' Smaller (WT = 0.8–1.5 kg), tail short, hairy, but conspicuous, pencil-thin with a white tip .... Acouchys, *Myoprocta* (p. 210)

8   Fingers of hands elongated, with nails, not claws, middle fingers separated by a gap ... 9

8' Fingers not elongated or separated by a gap, with claws, not nails 10

9   Tail naked, scaly, from Amazon Basin (or tail hairy and from Andean montane habitat in S Peru or N Bolivia) .... Amazon bamboo rats, *Dactylomys* (p. 224)

9' Tail hairy, from SE Brazil and N Argentina ....
    .... Southern bamboo rat, *Kannabateomys* (p. 225)

10 Muzzle long, ears distinctly longer than broad, protruding above crown, hindfeet long and narrow, tail slender, usually naked, and ratlike (may be missing) 11

10' Muzzle short and blunt, ears short, not much if at all longer than wide, do not protrude above crown, hindfeet short and broad, tail robust, naked, tufted, or hairy 12

11 Lower back with stiff spines standing upward in a geometrically spaced pattern, Central America and Pacific coastal Ecuador and Colombia only ....
    .... Armored rat, *Hoplomys* (p. 214)

11' If spiny, spines lie flat and are inconspicuous, except in SE Brazil; Central and South America .... Spiny rats, *Proechimys* (p. 212)

12 Fur soft, not bristly when brushed backward, tail fully furred and bushy 13

12' Fur with stiff bristles or flexible spines 14

13 Color olivaceous, tail often orange and/or dusky; Amazon Basin only ....
    .... Brush-tailed rats, *Isothrix* (p. 221)

13' Color rusty, or rusty especially on rump, Central America, Pacific Coast, and mountains of N Colombia and Venezuela only .... Rufous tree rats, *Diplomys* (p. 222)

13''Color dirty white with striking black markings, SE Brazil only ....
    .... Painted tree rat, *Echimys pictus* (p. 218)

14 Densely covered with flat, flexible spines, size medium to small (HB = 155–220), Amazon Basin and the Guianas, tail hairy 15

14' If densely spined, then either large (HB > 230), or tail naked-looking and scaly, or from Atlantic coastal forests of Brazil from Ceará south 16

15 Size small, HB < 210, brown, tail entirely sparsely covered with red-brown hairs that form a small tuft at tip (but tail sometimes missing) ....
    .... Spiny tree rats, *Mesomys* (p. 223)

15' Slightly larger (HB ≥ 220), gray-brown, tail scaly except for long, coarse hairs in prominent tuft at tip, Brazil, S of Amazon and E of Madeira only ....
    .... Tuft-tailed spiny tree rat, *Lonchothrix* (p. 224)

16 From the Amazon Basin, Colombia, Venezuela, and the Guianas ....
    .... Tree rats, *Echimys* (p. 214)

16' From Brazil southeast of the Amazon Basin .... Atlantic tree rats, *Nelomys* (p. 219)

Appendix C

Classification, Study, Biogeography, and Conservation of Neotropical Rainforest Mammals

The System of Classification of Organisms

You will see in this book that a Venezuelan cannot talk to a Peruvian about a paca and be understood: each uses an entirely different name for the same animal. Scientific names are a necessity for unambiguous communication, even on the same continent among people who speak the same language. Only because the scientific names are given for each species can anyone in the world read this book and know exactly what animal is meant—now or a century hence.

Organisms are scientifically classified in a universal system. It is a practical system devised before the theory of evolution was developed, originally based on how people perceive the world, and very similar to the way peoples worldwide classify living things in their own native languages. Humans naturally tend to group and name everything they see according to visible relationships or similarities, such as trees and oak trees; birds and sparrows. Science has refined the natural way we classify things into a precise system. The system is hierarchical: organisms are grouped into a series of nested, increasingly inclusive categories. Recent classifications often have the underlying principle, or goal, that the pattern of the classification should reflect true evolu-

tionary relationships: all the animals grouped under one name at a given level of classification are more closely related to each other than to any animals under another name at the same level. For example, all members of the cat family (Felidae) are more closely related to each other (share a more recent common ancestor) than to any member of the dog family (Canidae); within the Canidae, all members of the dog genus *Canis* are more closely related to each other than they are to any members of the fox genus *Dusicyon*.

The level of classification that a name refers to is sometimes shown by the ending of the name. The classification of a rice rat (*Oryzomys capito*) is shown below, including the categories of classification used in this book. There are other levels, such as superfamily or subtribe, not used here. Subcategories (subfamily, subgenus, subspecies, etc.) have been defined for only some cases; many taxa do not have any. A taxon (plural: taxa) is the group of organisms under a single name at any of the levels below (e.g., family, genus, species, or subspecies).

The binomial or two-part name (genus and species, e.g., *Oryzomys capito*) is always used to designate a species. According to the rules, every generic (genus) name has to be different, but the species names do not (as long as they apply to different genera), therefore the species name

Kingdom: Animalia	Animals
Class: Mammalia	Mammals
Order: Rodentia	Rodents
Suborder: Myomorpha	Ratlike rodents
Family: Mur**idae**	Rats and mice
Subfamily: Sigmodont**inae**	New World rats and mice
Tribe: Oryzomy**ini**	Rice rat group
Genus: *Oryzomys*	True rice rats
Subgenus: *Oryzomys*	Typical rice rats
Species: *capito*	Common Amazon rice rat
Subspecies: *velutinus*	The populations in the Guiana region

alone does not unambiguously designate a species. The generic name is always capitalized; the species (and subspecies) name is always in lowercase; both names are always italicized. The subgenus is placed in parentheses following the genus. The names are usually abbreviated as follows, from the full name, *Oryzomys (Oryzomys) capito velutinus; O. capito* to designate the species, or *O. c. velutinus* to designate the subspecies; *O. (Oryzomys)* to designate the subgenus. The singular abbreviation sp., meaning a species, is used after the generic name when it is not clear what the specific name is, or when it is a new, unnamed species (usually written sp. nov.); the plural, spp., is used after the generic name to designate all or a number of the species in that genus.

On Species, Speciation, Why We Need Museums, and Why Scientists Cannot Agree about Names

Species. If you opened this book, you are probably curious about what kinds of mammals live in American rainforests. You perhaps couldn't care less about that scientific name attached to each mammal, especially since even the experts often cannot agree on what it should be. However, this and every other guidebook to living organisms in the world could not have been written without the sciences that concern themselves with what those names should be: systematics, taxonomy, and evolutionary biology. This book describes the kinds of animals. The kinds of animals are species. To write a book like this, we need to know what species there are, where they are, and how they differ from each other so that we can tell them apart.

A species is a real thing. It can be defined as "groups of actually or potentially interbreeding organisms separated from other such groups," in other words, a bunch of related animals that can breed among themselves so that genes are exchanged among them and not among other such groups. There are other definitions of species, but this is not the place to discuss them. Species are composed of populations of interbreeding individuals; if geographically separated populations or groups of populations of a species have evolved sets of different characteristics (such as different color patterns) so that by looking at an individual it is always possible to tell which group of populations it came from, then the different populations are often called subspecies. Subspecies are arbitrary units that are defined for the sake of convenience; two neighboring subspecies may blend into each other with no clear dividing line between them.

Speciation. The great discovery of Charles Darwin and Alfred Russel Wallace was that individual animals and plants vary, that their variations can be inherited by their offspring, and that by the process of natural selection, some variations make the individuals that carry them more successful than other individuals in producing offspring. Through time, the perpetual process of variation of individuals—the weeding out of some individuals and selective success in producing successful offspring of others—causes populations to change or evolve into sets of individuals with characteristics quite different from those of their ancestors. At some point in the process, a population can become so different from its ancestors that it becomes (has evolved into) a new species. We now know (which Darwin did not) that heritable characteristics are encoded in genes on molecules in reproductive cells, and that changes in the genes (mutations) create the genetic (heritable) variations among individuals.

If a population of a species becomes physically separated from others of its kind, such as by living on a separate island or on different sides of a mountain range too high to cross, then the evolutionary changes that occur within it can no longer be freely shared by interbreeding with members of other populations (there is no "gene flow" between them). An isolated population can gradually become more and more different from its nearest relatives, perhaps adapting to a different environment, until it is so changed that they have become different species: they look different, act differently, have different sets of genes, and if they come into contact with their former relatives again, either they cannot interbreed, or if they do, the "hy-

brid'' offspring are less successful in reproducing than offspring of members of
either parent population. Speciation has occurred, and if the two species differ
enough in their ecological roles, they could
now share the same geographic range as
noninterbreeding, separately evolving populations. It is not surprising that both Darwin and Wallace discovered natural selection during voyages among islands, where
steps in the processes of the evolution of
species (speciation) are most clearly laid
out.

**Why Scientists Disagree about Species
Names.** If we could stand in one place in
the Amazon rainforest and look at all the
mammals that live together in that patch of
forest—perhaps as many as 150 species—
it would be relatively easy to sort them all
out, count the number of species, describe
them, and give each one a name. The individuals of each species in that forest interbreed constantly, mixing their genes at
each generation, so that all members of
that species in the same population look
much alike. If we could then jump across
the River Amazon—so wide in places that
you cannot see across it—and again gather
up all the mammals from one forest on the
other side, we could again easily sort
them, separating the individual species.
But then, if we could take the animals
from both sides, line them up together, and
again try to sort them, we would immediately have a problem. Some kinds of animals would occur on only one side of the
river, with none at all like them on the other side; but for most of the remaining animals we would find that each side of the
river has animals of a very similar kind
(clearly members of the same genus), but
not exactly alike. Similar animals from different sides of the river may have slightly
or very different colors; they might be different sizes or have tails of different
lengths. We can tell by looking at them
which side of the river they came from.
The differences between the populations
are maintained because the physical barrier
between them (a river) prevents interbreeding. Are they different species or not?

What we are seeing is geographic variation, often an indication of slight genetic
isolation among populations. Geographic
variation across a species' range may represent stages in the process of the separation of lineages that can lead to formation
of new species. South America is a continent with exceptionally large numbers of
barriers that animals and plants cannot easily cross. The rainforest region is split into
many pieces: the Andes divide the west
coast and Central America from the interior
of the continent; the Amazon splits the region across the middle from east to west;
on each side a series of giant tributaries
cuts the forest into long strips often 300
km wide and over 1,000 km long; and on
the north and southeast of the forest, wide
areas of grasslands or dry scrub divide the
Amazon Basin rainforest from rainforests
in wet coastal regions of Venezuela and
southeastern Brazil (maps 1, 2, 3). It is
quite natural then, and even to be expect-
ed, that the populations of animals we find
on different sides of these barriers do not
look the same. How do we decide which
of these slightly or distinctly different-
looking populations of animals are species
and which are simple variations
(subspecies)?

The best way to go about it is to gather
together, side by side, representatives of
every different population of that kind of
animal, making sure we include representatives of all the kinds of individuals that
occur within the same unbroken population, as well as from all the ''separated''
populations. For example, for the animals
found between the Amazon, the Madeira,
and the Purus rivers, we might want to
look at individuals from the south bank of
the Amazon, from 200 km upstream on
each side of each tributary, from 400 km
upstream, and from 1,000 km up in the
headwaters. We then have before us what
we need to make a study of the variation
that occurs in those populations. We might
see immediately that although animals from
one side of the Madeira at its mouth near
the Amazon are small and pale brown and
those from the other side are large and
black, that as we go farther and farther
southwest toward the river source, the animals become more and more similar until
they merge into a single medium-sized,
dark brown population 1,200 km away in
Peru and Bolivia around the headwaters. In
this case, we could infer that we were looking at two populations of the same variable
species. On the other hand, if we find that
everywhere we look there are either small,

pale brown animals or large black ones, with no intermediate-looking individuals, and that at the headwaters there is even a small zone where both kinds of animals are found together in the same forest but do not interbreed, then we can be confident that we are dealing with two species.

To find out what species there are, and also to study the evolutionary relationships among them, we therefore need to compare individuals from different places in as great detail as possible: not only their size or color, but every feature of their anatomy that we find to vary—each bone of the skull, the bumps on the teeth, the soles of the feet, the anatomy of the reproductive tract, and increasingly, the very structure of their molecules. This is done with museum specimens. You will soon notice if you leaf through the "variation" headings in the accounts above that, for a very large number of genera, scientists do not agree on how many species there are: whether different-looking populations represent true species or just geographically separated varieties of the same species. One of the main reasons for such disagreements is that we simply do not have enough specimens in museums, from enough places in the animals' ranges, to be able to do the simple study outlined in the previous paragraph. We may have individuals from a few locations along the Amazon but none at all from any headwaters; we often do not even know where the geographic limits of a species lie. Often the museum specimens do exist, scattered throughout many museums in Europe, the United States, and South America, but no one has yet undertaken the job of comparing them all. In such cases we simply have to make educated guesses about how to divide the populations into species. In other cases we know that the geographic ranges of the different varieties do not overlap, and thus we cannot ever observe whether they might interbreed in nature. In this case also, whether we call them a single variable species or several species is our best educated opinion. When there is not enough information available to truly resolve the question, scientists' opinions on species vary depending on their views about different processes of evolution and slightly differing concepts of what degrees of difference species are likely to show. Some prefer to "lump" vari-

eties under one species name until they have been proved separate, while others tend to "split" varieties into as many individual units as are recognizable and call them all species. Despite these philosophical differences, the study of the nature and boundaries of species in natural populations presents major scientific problems of great theoretical and practical importance, as well as describing the world we live in. In summary, it is the natural variations among organisms, and lack of sufficient knowledge about them, that causes the most problems when we try to classify them into species.

If two species are so similar looking that we can barely tell them apart, who cares whether we give them the same or different scientific names? Why worry about it? Species that look morphologically alike can have strikingly different ecologies and behaviors (sometimes it is these differences that have led to the discovery that more than one species was lumped under the same name). Closely related species can have drastically different effects in agro-economics or in the epidemiological cycles of human diseases. The most similar species are also the best ones for studies of the finest levels of the workings of ecology and evolution.

Most field guides sweep the messy unresolved problems of taxonomy under the rug of authority and present to the reader a static world where all species are sharply defined and ordered. But this is not true for any set of organisms anywhere in the world. We have decided to present the *taxonomie réalité* not to exasperate and confuse readers and convince them that scientists are nitwits, but to impress upon them how little we know, how much still needs to be done before we can write that perfect field guide, and most of all how tropical America is a living example of the changeable nature of organisms and mechanisms of evolution, where mammals are as diverse and diversifying as anywhere on earth.

Why We Need Museums. Museums of scientific specimens are libraries where organisms are carefully preserved so that they can be studied in the future. Many of the specimens collected by great scientific explorers such as Darwin, Aubudon, von Humboldt, and Wallace are still preserved today. We can directly compare them now, side by side with species discovered decades or centuries later. As we saw in the previous section, we cannot study species, or evolution, or the distribution of organisms without museum collections. The differences between species of mammals can be subtle and easily overlooked: the lobes on a tooth, the number of nipples, or the position of a whisker can be a distinguishing feature. We may have to look again and again over the years before we discover important differences between two taxa. We must be able to reexamine an animal when a new method is discovered by which we can study the relationships between organisms, or when a new machine allows us to look at what we have not been able to see before. Variation is characteristic of all forms of life on earth and part of their essential nature, but a single specimen can tell us little or nothing about it.

If we want to understand much about species, we need to preserve enough individuals from different populations and geographic locations to encompass their ranges of variation. This means collecting animals—not one, but a whole series. Many people object strongly to killing wild animals, for sentimental reasons. This is certainly an understandable viewpoint. But collections of animals are also vital for the preservation of living ones.

Most of us, including those who object to all killing of animals, wish to conserve the other species of life on earth. Man has already overrun the earth, and it seems likely that by the end of our own lifetimes nothing will remain of the richest natural ecosystems on earth, such as the rainforests, except those places deliberately chosen for preservation. We now face the irrevocable choices: to condemn this forest and all its inhabitants to death and replace it with crops to feed more people; or to save that one for keeping a few species. Could you choose which group of forest preserves would save the most species?

To make a wise choice, you would need to know what kinds of species there were in each forest and whether those same species were found in many other forests, or in none. For the majority of species in the world, that information is derived solely and directly from museum collections. If the information cannot be found for the forests one wants to save or exploit, then new collections of specimens need to be made there. These constitute the only credible and lasting record and proof of where organisms are or, increasingly, used to be. The collecting of mammals for museums is on far too small a scale to harm any of the species or individual populations concerned. Unlike hunters for meat or commerce, museum collectors usually visit an area for only a short period, take a limited number of specimens, and then do not return to the same area again for years, if at all. Unlike the situation with butterflies or birds' eggs, mammal specimens have no commercial value, and specimens are not taken for private collection.

This book was written based on museum specimens. The range maps are drawn by compiling where they came from. The illustrations were painted by examining series of specimens and choosing typical individuals as models of the color patterns. Museum collections are the origin of the lists of species that are our only way of actually measuring the species richness or diversity of a habitat. Their vital importance for both science and conservation should be understood by everyone interested in the inhabitants of our planet.

The Biogeography of Neotropical Mammals

The geographical distribution of species, or biogeography, is the combined result of the geological history of the earth and the biological history of the evolution of organisms. The collection of animals and plants that are present at any one place and time always tells a fascinating story of events that we cannot see but that can be unlocked like a mystery. The study of biogeography asks simple questions: (1) What group of species (fauna or flora) is found in a region (such as the Neotropics)? (2) How does the fauna change from place to place within the region? (3) Where did the different species come from, that is, what is the geographic history of the elements of the fauna? and (4) What can we infer about the events and mechanisms that led to the distribution of species on the landscape at a particular time? Answering these questions is not simple.

There are only two ways for a species to occur at a given geographic place: either it came there by immigration (or transport) from somewhere else, or it evolved in that place by speciation from ancestors that were already there. In the second case, the ancestors could themselves have reached that place by immigration at any time; between a remote era, say 100 million years ago when a primitive marsupial may have traveled, to just yesterday, say about 10,000 years ago, when the rainforest grew back over former arid areas and a rainforest mammal could move with it.

Whether a species can spread from one region to another essentially depends on the kinds of barriers there are to prevent such spread. Barriers can be of many kinds: physical, such as wide oceans, rivers, mountain ranges, or inhospitable habitat types; or biological, such as a disease or a predator that might prevent a species from surviving or another species that might already be established and that uses the same resources.

Speciation itself, and also the assembly of lineages and species into regional faunas, occurs on geologic time scales: the fauna of the Neotropics today is the accumulated product of millions of years of events, some very old, some recent, some happening now. Several major events are responsible for the main features of the Neotropical fauna.

The most momentous factor that has determined what mammals we have in South America has been the history of its connections with other continents. The earliest ancestral mammals evolved and spread on the landmass of Pangaea before continental drift broke the continents apart and separated them by oceans, beginning in the early Jurassic period about 190 million years before the present (MYBP). By about 135 MYBP in the earliest Cretaceous, Pangaea had separated into the southern continent of Gondwanaland and the northern continent of Laurasia; by the end of the Cretaceous to the early Paleocene, about 65–55 MYBP, as the dinosaurs died out, Gondwanaland itself had split, and South America had become an island. It remained so for about the next 50 million years.

This period was the dawn of the time when, finally free of the dominance of the dinosaurs, mammals exploded into an exuberance of evolution and filled the earth's ecosystems with creatures as strange and wonderful-seeming to us now as were the dinosaurs themselves. In South America, several endemic orders of mammals radiated into diverse assemblages with many bizarre-looking animals entirely unlike those that were evolving on other continents, much as Australia today is full of beasts unlike those anywhere else. Most of these orders are now extinct, but the last remnants of Xenarthra (anteaters, sloths, armadillos) are still with us. The splitting up of the continents was a slow process, and all the while mammals were evolving. Some early members of orders also found on other continents somehow managed to immigrate to South America before the continents got very far apart, but once there they were isolated and evolved into endemic families of species. These included the marsupials, primates, and caviomorph rodents, which still dominate the rainforest mammal fauna today.

About 7 MYBP an event took place that dramatically changed the history of South America: the Isthmus of Panama rose above sea level, and for the first time in about 130 million years animals could walk on land between North and South America. The "great faunal interchange" then began. Xenarthra, marsupials, primates, and caviomorph rodents moved north, and carnivores, deer, peccaries, squirrels, and murid rodents went south. This completed the major groups of the fauna as they are today. Other major geologic events, such as the uplifting of the Andes, had profound effects on the temperate-climate and high-elevation species, but relatively little effect on lowland rainforest genera, apart from their roles in influencing the weather and presenting a barrier to migration.

The present fauna of the Neotropics is thus a mixture of families that evolved in the Northern and Southern hemispheres. We still do not know much about the history of some important families of mammals in South America, such as the leaf-nosed bats and murid rodents.

If we look in general at the faunas of the New World, some interesting facts about the rainforest appear. It is common knowledge that there are more species of mammals in rainforest than in other environments, but what kinds of species are these? Of the approximately 500 rainforest species, 60% belong to endemic South American families. Among these are all of the living sloths and platyrrhine monkeys and most of the species of leaf-nosed bats, echimyid rodents, anteaters, and marsupials, so that the rainforest is now a refuge for the last or only representatives of some of the old, endemic radiations. If the rainforest is lost, a large and perhaps final chapter in the evolution of mammals in South America will be closed forever.

More recent events shaped the fauna at the species level. The extent of the Neotropical rainforest was not always as we see it today. During the turbulent Pleistocene, when glaciers advanced and retreated many times in the Northern Hemisphere, the Neotropics (and many other tropical areas) experienced a series of alternating wetter and drier periods. There is evidence that during drier periods the rainforest shrank and became fragmented, perhaps like a series of forest islands (called refugia) with drier savannas or scrub between. The rainforest animals and plants were then isolated in these refugia. With the exception of a few, mostly large species, most rainforest mammals can live only in rainforest climates and habitats. A savanna or desert is as great a barrier as a wide river. As we have seen, this scenario of isolation of populations of animals (and plants) on islands is a situation likely to allow animals to evolve differences that can result in the formation of new species.

As described in the earlier section on speciation, the Neotropical rainforest region is also today divided into many pieces by barriers of mountains, rivers, grasslands, and arid zones. Certain barriers are large and almost complete (the Andes, the arid regions between Amazonia and SE Brazil); others allow animals to cross from time to time (rivers). Many scientists think that this dynamic history of past and present fragmentation explains the high species richness of the Neotropical rainforest today, which has more species in one place than do the world's other rainforests.

When we look at the present distribution of species in the Neotropics, we see that we can divide the rainforest region into large sections that have series of endemic species within them. They have their own faunas of birds, mammals, lizards, butterflies, and so forth. The faunas of adjacent sections are often very similar, with adjacent sections having what one might call "sister" species: species that are clearly nearest relatives and that may have the same ecological roles (also known as parapatric species).

There are four main faunal areas for rainforest mammals (maps 1, 2): (1) Central America and the Pacific lowlands of Colombia, Ecuador, and northern Peru; (2) the great block of the Amazon Basin; (3) the Atlantic rainforests of southeastern Brazil; and (4) the Caribbean coastal rainforests of Colombia and Venezuela. Each rainforest block is separated from the others by a combination of barriers, including mountains, rivers, and dry habitats. As one might expect, the vast forests of the Amazon harbor by far the most species. The Central American rainforest fauna is relatively small, with a number of endemic species that are close relatives of those in Amazonia, but few endemic rainforest genera. The northern Caribbean coastal forests have even fewer species, a number of them also endemic. The barriers between the latter two regions are leaky, and a number of exclusive species and even some endemic genera (e.g., *Diplomys*) are shared between them. In contrast, the rainforest fauna of the Atlantic forest of southeastern Brazil is now, and has for a long time been, totally isolated from the other regions, and is extraordinarily different from that of the Amazon Basin, with many endemic species and genera and unique animals of unclear relationship, such as the bristle-spined porcupine.

These four large regions by no means have homogeneous faunas: they can once again be divided into areas that each have a number of their own endemic species. For example, within the Amazon forest, the "Guiana" region, which includes all the rainforest east of the Rio Negro and north of the Amazon, and south of the Amazon east of the Xingu, has a series of species of its own, such as the Guianan saki monkey, brown bearded saki, and Brazilian bare-face and midas tamarins; but north of the river it also lacks many species or genera found elsewhere in Amazonia, such as red squirrels, night monkeys, titi monkeys, and olingos.

In numbers of mammal species present, the richest part of the Neotropical rainforest is the area along the base of the Andes in Peru and Ecuador, and the poorest is the Guiana region. But it is essential to reemphasize that even this relatively "poor" region is very rich and includes its own endemic species that are not found elsewhere.

Conservation of Neotropical Rainforest Mammals

Conservation is the protection of species against the actions of man. Many factors are involved in what makes a given species more or less vulnerable to extinction or serious damage by human action. The main ones are outlined briefly below. As stated in the epigraph of this book, we have now become responsible for the fate of the other species on earth. If we are to succeed in preserving many of them and allowing evolution to proceed on many fronts rather than stopping it in its tracks for most vertebrates and plants, we must understand how human actions may affect the organisms we wish to preserve.

Man exterminates organisms in three basic ways: (1) by singling them out and destroying them individually for human consumption (in the broad sense), as he has done for Steller's sea cows, passenger pigeons, and dodos (among many others); (2) by destroying resources they need in order to survive, such as habitat or food supply, as is happening to giant pandas and golden lion tamarins; and (3) by introducing exotic species that kill or outcompete the native flora or fauna. Two or all three factors can be involved in specific cases, as in the extinction of California condors and decline of woolly spider monkeys. The first factor was the primary one historically, when human populations were low; the second factor is becoming and will in the future be by far the most important one as human populations reach catastrophic levels. The third is a special case that will not

be discussed below because it is not at the moment of great importance in tropical rainforests, except on islands.

For the rainforest mammals described here, we can define what characteristics make a species more or less vulnerable to extinction from each cause. As can be seen repeatedly throughout the book, hunting of individual species for food, skins, or commerce has affected populations of many species throughout the Neotropics. Several features of a species biology determine how intensely it is hunted, how well it can withstand hunting pressure, and for how long. The first is size; on the whole animals are shot for meat only if there is enough meat on them to be worth the price of a shotgun shell (a relatively large sum for subsistence hunters). The cutoff point is 0.5 to 1 kg. In addition, a hunter will tend not to risk scaring away larger game by shooting at smaller animals until he is at the end of a day's fruitless hunt. The amount of meat on an animal determines its value, which in turn determines the amount of special effort a hunter will give to finding it. Because of their weight, the largest species, such as manatees, tapirs, spider monkeys, and woolly monkeys, are relentlessly pursued and have become locally extinct in many inhabited areas.

A second important characteristic determining vulnerability is conspicuousness; this determines whether hunters will be able to find and kill every individual. Animals that live in noisy groups are more conspicuous than solitary ones, and large animals that live in the canopy are much easier to see and follow than terrestrial ones. Secretive animals that favor dense thickets, such as deer or agoutis, are often able to escape detection or flee out of range and sight before a hunter can shoot, and some individuals always escape.

A third factor is whether an animal's habitat is easily reached by man. There are few roads in the Amazon Basin, and rivers are the main access routes into the forest; any place that a motor canoe can reach is accessible to hunters. The only species that appear genuinely endangered in the Amazon Basin today are those restricted to riverine habitats: giant otters, manatees, black caiman, crocodiles, giant river turtles, giant osteoglossine fish, and so forth.

Giant otters have every characteristic to make them vulnerable: valuable skins, large size, large, noisy groups, and riverine habits; they are now extinct over much of their former range. The Amazon Basin is so large and thinly inhabited that most terrestrial species can still find remote refuges regardless of how much they are worth.

The fourth major characteristic that affects a species' ability to withstand hunting pressure is its reproductive rate. This determines how quickly animals that are shot can be replaced by new individuals in the population and how well the population can withstand steady removal of individuals. Deer, where females can first breed as yearlings and can have one or twin young every year thereafter, have a strong ability to maintain their populations under some hunting pressure, especially when this is coupled to solitary, cryptic habits. Spider monkeys, where females may first breed only at 5 or 6 years old and have one young only every three years thereafter, have almost no ability to replace individuals shot under even light hunting pressure, especially since their low reproductive rate is coupled with large body size and conspicuous arboreal behavior which allows hunters to find almost every individual.

Susceptibility to the other cause of extinction—loss of resources—mainly involves two characteristics of species: (1) the size of their geographic range and (2) their ecological resilience or flexibility. The first of these is fairly self-evident: the smaller an area a species inhabits, the more likely that a single catastrophic event or human project can wipe it out. The second determines how well a species can adapt (as a living animal, not in the sense of adapt as evolutionary change) to human-caused changes in its habitat. For example, a species that can live in secondary forest and hedgerows may not be threatened by cutting the primary forest, but a species that can live only in primary forest will be. An example is the near extinction of lion tamarins, while in the same region many populations of tufted-ear marmosets (which can live in secondary scrub) are succeeding well.

A final characteristic of species affects their susceptibility to extinction by any means, that is, their natural population size and density of individuals. Population density is a result of many factors; while there is not space to discuss them here, suffice it to say that the fewer individuals there are, both in total and per square kilometer, the more easily the species seems to be eliminated by habitat destruction and fragmentation or by hunting.

The intact rainforest could produce an indefinitely sustainable yield of its myriad resources of plants and animals useful to man. For it to do so, each exploited species needs to be regulated with respect to its individual ability to support cropping. If rainforests can be managed for production of products, then many more of them may be saved.

When regional conservation plans are made the biogeography of a region must be considered, because, as we saw in the last section and can see from the accounts in this book, one piece of the Amazon rainforest may have a quite different set of species in it than a superficially similar piece somewhere else: to save all the species, areas need to be saved throughout the region. Certain regions of Neotropical rainforest are currently far more threatened than others. Because of high human populations, deforestation, and relatively small geographic size, southeastern Brazil, northern Colombia, the west coasts of Peru, Ecuador, and Colombia, much of Central America, and the mid- and high-elevation Andean slopes are now in a critical stage where only immediate action seems likely to save many endemic species from extinction. Because so many of its species are endemic and so little of its forests are left, the Atlantic forests of Brazil are probably the most severely endangered of all Neotropical rainforests. The Amazon Basin and the Guianas are not at present in much danger relative to the peripheral regions mentioned above; except for their riverine faunas. However, the Amazon Basin rainforest is not a limitless frontier; its resources are being rapidly wasted by uncontrolled local exploitation.

Although the rainforest is of economic importance to man, both for its global effects on climate and water and for its many individually valuable species, the chief reason to save it must remain a philosophical one: for the species themselves.

Appendix D
Tracks of Large Mammals

The tracks of large animals in rainforest habitats are far more often seen than are the animals themselves. The presence of tracks can be the only sign that a species is present in an area. Tapirs, for instance, are shy, silent, and largely nocturnal and like to lurk in inaccessible swamps, but they are great walkers, and their weight sinks their tracks deep in the mud. Such tracks can remain easily visible for days or weeks. A single tapir can thus leave a deceptively large number of tracks throughout its haunts.

The feet of most mammals are flexible, dynamic structures that both mold to the terrain and adjust to the speed of travel. The amount of a foot that touches the ground or is impressed in a track varies with both the gait and the slope and softness of the ground: peccary tracks on hard ground show only the tip of the toes; in softer ground the entire outline of the main hooves will show; and in deep mud the tiny side toes will appear. Tracks of the same animal are thus highly variable. Be-cause they are so variable, there is a lot of information in them, and sometimes the species, age, sex, activity, and time it was made can be read from a single footprint.

The best places to look for tracks are on stream and river banks and beaches near the vegetation line, the fringes of swamps, and trails and roads after rain. A cheap, easy, and portable method of making and keeping records of tracks is simply to lay a sheet of clear, stiff plastic (such as thin Mylar or overhead-projector transparency sheets) over the track and trace it with a fine-point permanent marking pen (nonpermanent markers will rub off). The date, place, and any notes can be written directly on the same sheet. The tracks shown on the following pages (figs. 8–11) were traced by this method in Manu National Park, Madre de Dios Department, Peru. An excellent reference is J. M. Aranda, *Rastros de los mamíferos silvestres de México* (Xalapa: Instituto Nacional de Investigaciones sobre Recursos Bioticos, 1981).

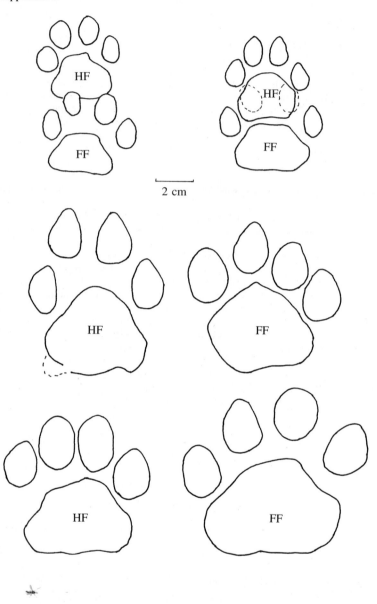

2 cm

Figure 8. (*top*) **Ocelot,** *Felis pardalis;* left, female; right, male. Track positions as found in nature, with hindfoot forward of forefoot during walk; left feet. Note forefoot wider than hindfoot; in **margay,** *F. wiedii,* hind and fore are about the same size. (*center*) **Puma,** *Felis concolor;* male, hind on left; note slightly pointed toes, length of foot slightly greater than width (both measured at maximum dimension), large gap between center hind toes and pad. (*bottom*) **Jaguar,** *Panthera onca;* male, hind on left; note round toes, foot wider than long.

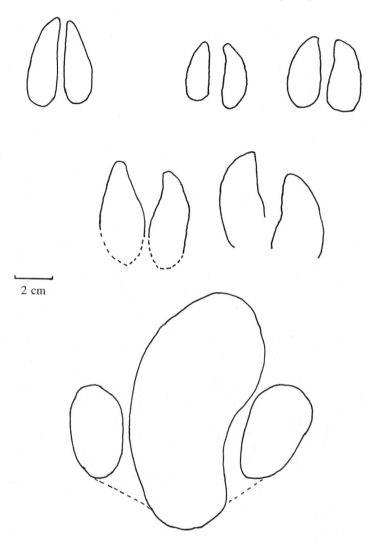

2 cm

Figure 9. (*top*) Far left, **red brocket deer,** *Mazama americana.* Center and right, collared **peccary,** *Tayassu tajacu;* hind foot on right; left feet. (*center*) **White-lipped peccary,** *Tayassu pecari;* hindfoot on left; left feet. (*bottom*) **Brazilian tapir,** *Tapirus terrestris;* hindfoot.

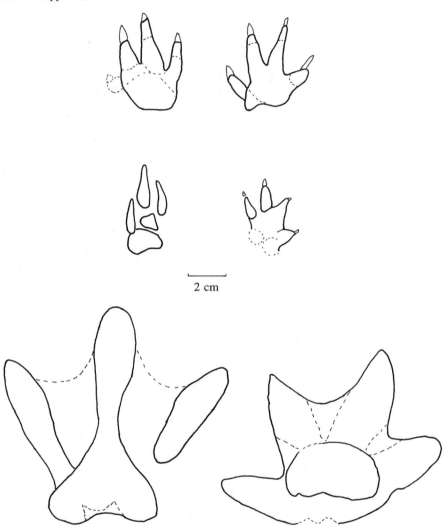

Figure 10. (*top*) **Paca,** *Agouti paca;* hindfoot on left; left feet. Note that the outer digit often does not appear in tracks (dotted on hindfoot). (*center*) **Agouti,** *Dasyprocta variegata;* hindfoot on left; left feet. Tracks of **acouchys,** *Myoprocta,* are similar but much smaller. (*bottom*) **Capybara,** *Hydrochaeris hydrochaeris;* hind on left, left feet.

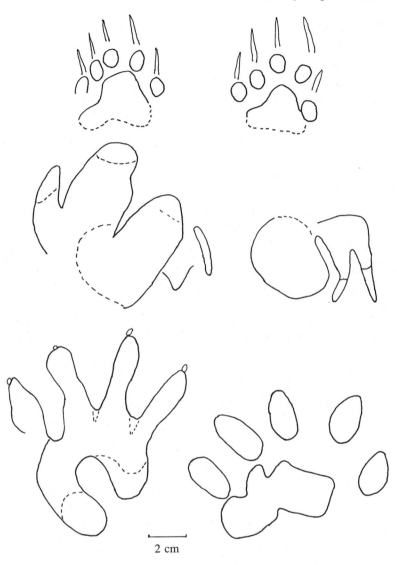

Figure 11. (*top*) **Coati,** *Nasua nasua;* hindfoot on left; left feet. (*center*) **Giant anteater,** *Myrmeco-phaga tridactyla;* two left forefoot tracks from same series on different ground (and perhaps different gait). (*bottom*) **Giant otter,** *Pteronura brasiliensis;* hindfoot on left lacks the impression of one toe (the foot has five toes). The large webs between the toes to their tips did not show in either of two sets of tracks traced but may show on other substrates. The tracks are associated with drag marks of the tail.

Appendix E
General References

Following is a short list of general reference works that cover large parts of the fauna.

Aranda, J. M. 1981. *Rastros de los mamíferos silvestres de México*. Xalapa: Instituto Nacional de Investigaciones sobre Recursos Bioticos.

Carleton, M. D., and G. G. Musser. 1984. Muroid rodents. In S. Anderson and J. K. Jones, Jr., eds., *Recent mammals of the world*, 289–379. New York: John Wiley.

Coimbra-Filho, A. F., and R. A. Mittermeier, eds. 1981. *Ecology and behavior of Neotropical primates*. Vol. 1. Rio de Janeiro: Academia Brasiliera de Ciências.

Hall, E. R. 1979. *The mammals of North America*. Vols. 1 and 2. New York: John Wiley.

Hershkovitz, P. 1977. *Living New World monkeys (Platyrrhini)*. Vol. 1. Chicago: University of Chicago Press.

Janzen, D. H., ed. 1983. *Costa Rican natural history*. Chicago: University of Chicago Press.

Linares, O. J. 1987. *Murciélagos de Venezuela*. Caracas: Cuadernos Lagoven.

Mammalian species. Published in series by, and available from, the American Society of Mammalogists.

Mendez, E. 1977. *Los principales mamíferos silvestres de Panamá*. Panama: E. Mendez.

Montgomery, G. G., ed. 1985. *The evolution and ecology of armadillos, sloths, and vermilinguas*. Washington, D.C.: Smithsonian Institution Press.

Moojen, J. 1952. *Os roedores do Brasil*. Biblioteca Cientifica Brasileira ser. A. II.

Mora, J. M., and I. Moreira. 1984. *Mamíferos de Costa Rica*. San José: Editorial Universidad Estatal Distancia.

Olrog, C. C., and M. M. Lucero. 1981. *Guía de los mamíferos Argentinos*. Ministerio de Cultura y Educación, Fundación Miguel Lillo, Tucuman.

Appendix F

Checklist and Index of Scientific Names

Species in the lowland rainforest fauna are shown in **boldface.** Species or genera from other habitats that are treated or mentioned in the text are in regular type. Plate numbers are shown in **boldface** following page numbers.

Index of Genera and Common Names

Boldface numbers and letters refer to plates.